T0198602

SECOND EDITION

# Computer
# Architecture

## Fundamentals and Principles
## of Computer Design

SECOND EDITION

# Computer Architecture

## Fundamentals and Principles of Computer Design

## Joseph D. Dumas II

University of Tennessee at Chattanooga
Chattanooga, TN, USA

CRC Press
Taylor & Francis Group
Boca Raton   London   New York

CRC Press is an imprint of the
Taylor & Francis Group, an **informa** business

*To Chereé, my best friend and life partner, who "always and forever" makes this "a wonderful world." I love you as big as the sky.*

# Contents

# Preface

Digital electronic computer systems have gone through several generations and many changes since they were first built just before and during World War II. Machines that were originally implemented with electromechanical relays and vacuum tubes gave way to those constructed with solid-state devices and, eventually, integrated circuits containing thousands, millions, or billions of transistors. Systems that cost millions of dollars and took up large rooms (or even whole floors of buildings) decreased in price by orders of magnitude and shrank, in some cases, to single chips less than the size of a postage stamp. CPU clock speeds increased from kilohertz to megahertz to gigahertz, and computer storage capacity grew from kilobytes to megabytes to gigabytes to terabytes and beyond.

Although most people have noticed the obvious changes in modern computer system implementation, not everyone realizes how much has remained the same, architecturally speaking. Many of the basic design concepts and even the advanced techniques used to enhance performance have not changed appreciably in 30, 40, even 50 years or longer. Most modern computers still use the sequential, von Neumann programming paradigm that dates to the 1940s; they accept hardware interrupts that have been a standard system design feature since the 1950s; and they store programs and data in hierarchical memories that are, at least conceptually, very similar to storage systems built in the 1960s. Although computing professionals obviously need to stay abreast of today's cutting-edge architectural breakthroughs and the latest technical wizardry, it is just as important that they study historical computer architectures—not only because doing so gives a valuable appreciation for how things were done in the past, but also because, in many cases, the same or similar techniques are still being used in the present and may persist into the future.

Over many years of teaching the computer architecture course to an audience of mostly undergraduate computer science (and a few computer engineering) students, I have observed that few—if any—of the students in a typical CS (or even electrical engineering) program ever go to work designing microprocessors, memory devices, or other ICs, let alone complete computer systems. Many, probably most, of them instead become system administrators, programmer/analysts, technical managers, etc. In these positions, one is not generally called upon to design hardware, and so there is no need to know where each individual transistor goes on a particular chip. However, it is quite likely that at some point almost every computing professional will have to specify or purchase a computer system to run a particular application(s). To do so, he or she must know enough about computer architectures and implementation technologies to be able to understand the characteristics of the machines under consideration, see through manufacturer hype, intelligently compare system performance, and ultimately select the best and most cost-effective system for the job. A course designed around this textbook should prepare students to do exactly that, without getting them lost in the myriad technical details characteristic of other excellent, but lengthy and involved, texts.

My philosophy in developing—and now revising—this book was to concentrate on the fundamental principles of computer design and performance enhancement that have proven effective over time, and to show how current trends in architecture and implementation rely on these principles while in many cases expanding them or applying them in new ways. Because specific computer designs tend to come and go quickly, this text does not focus on one particular machine or family of machines. Instead, important concepts and techniques are explained using examples drawn from a number of different computer architectures and implementations, both state-of-the-art and historical. In cases in which explanations based on real machines would include too many "trees" for the reader to see the "forest," simpler examples have been created for pedagogical purposes. The focus is not on understanding one particular architecture, but on understanding architectural and implementation features that are used across a variety of computing platforms. The author's underlying assumption is that if students have a thorough grounding in what constitutes high performance in computers, a good concept of how to measure it, and a thorough familiarity with the fundamental principles involved in making systems perform better, they will be able to understand and evaluate the many new systems they will encounter in their (hopefully long) professional careers.

This book is primarily designed to be used in a one-semester upper (usually senior)-level undergraduate course in computer architecture as taught in most traditional computer science programs, including the one at the University of Tennessee at Chattanooga where I have been on the faculty since 1993. It is also suitable for use in an undergraduate electrical engineering or computer engineering curriculum at the junior or first semester senior level, with the idea that it would likely be followed by an additional hardware-oriented course(s) covering topics such as advanced microprocessor systems, embedded systems, VLSI design, etc. (Refer to the chapter breakdown for suggestions on how to incorporate the text into courses based on the quarter system.) This book can also be effectively used (perhaps with some supplementary materials provided by the instructor and/or an additional text chosen to meet specific needs) in an introductory, master's level graduate course. I believe it would be particularly useful for what appears to me to be the increasing number of students seeking a master's degree in computer science after obtaining an undergraduate degree in another field.

To get the maximum benefit from this text, students should have had a previous course(s) covering introductory topics in digital logic and computer organization. Although this is not a text for a programming course, it is assumed that the reader is quite familiar with computer programming concepts. Ideally, students should not only be well versed in at least one high-level programming language, such as C, C++, or Java, but they should also have had some exposure to machine-level programming in assembly language (the specific architecture covered is not all that important, but the concepts are highly relevant). Previous courses in operating systems and/or systems programming would be helpful, but are not essential in following the material presented here.

To use computer architecture terminology, this textbook is a RISC design. The idea is to help students achieve success in understanding the essential concepts not by including a lot of extra "bells and whistles," but by providing an easily understood, almost conversational text illustrated with simple, clear, and informative figures. Each chapter begins with an introduction that briefly lays out the ideas to be covered and the reasons why those topics are important to the study of modern computer systems. The introduction is followed by several sections explaining the material in detail and then a "chapter wrap-up." This section briefly reviews the key concepts that the student should have gleaned from reading the material and participating in class, placing them in context to help reinforce the student's learning by helping him or her to see the "big picture" or broader context into which the detailed material that has just been presented fits.

Finally, the end-of-chapter review questions include not only problems with concrete, numerical, right and wrong answers, but also "fill in the blank" items that reinforce key concepts and terminology, as well as open-ended short answer and essay-type questions. In years past, when using other texts, my students have often complained about textbook exercises that were either too trivial or too involved to be of much use in preparing for exams (which, of course, is a major concern of students). One of my goals in developing this book was to provide review questions that would be sufficiently thought provoking to challenge students, but manageable within a reasonable time frame (as they would have to be to serve as viable testing items). To this end, many of the review questions are drawn from actual exam questions used over the years I have taught the computer architecture course. By working out answers to a variety of problems and questions comparable to items that their instructor would be likely to include on a test, students should not only master the material thoroughly, but also (of equal if not greater importance to them) be prepared to demonstrate their mastery when called upon to do so.

Chapter 1 provides a background for the topics in the rest of the book by discussing the difference between architecture and implementation and the ways in which they influence each other. It also includes a brief history of computing machines from the earliest, primitive computers up to the present day. (To understand where the field is now and where it may go in the future, it is important to know where it has been.) The various types of single-processor and parallel, general- and special-purpose systems that will be covered in following chapters are introduced. Then the reader, who will likely at some point in the future be responsible for specifying and selecting computer systems, is introduced to some concepts of architectural quality and other factors that may cause particular systems to succeed or fail in the marketplace. Last, but certainly not least, methods for quantifying and measuring the performance of computers and their major subsystems are discussed.

Chapters 2 through 5 are the heart of the text; they cover, in appropriate detail, the architecture of traditional, single-processor computer systems. (Even though multicore chips and other parallel systems have become commonplace, understanding the operation of a uniprocessor machine is still fundamental.) Chapter 2 deals with the important topic of memory systems. It begins by explaining the characteristics of an ideal memory system (which of course does not exist) and then discusses how various memory technologies approximate ideal behavior in some ways but not others. This naturally leads to an explanation of how a hierarchical storage system may be used to maximize the benefits of each type of device while hiding their less desirable characteristics, thus enabling the overall memory system to keep up with the demands of the processor (and the needs of the programmer).

Because CPU architecture and implementation are so complex and so critical to the performance of a system, both Chapters 3 and 4 are devoted to this topic. Chapter 3 explains the basics, including the design of CISC and RISC machine instruction sets, the datapath hardware that is used to carry out those machine instructions by operating on integer and real number values, and the control unit, which develops the control signals that operate the datapath as well as the rest of the machine using either a microprogram or hardwired logic. Chapter 4 then discusses techniques that can be used to enhance the performance of the basic CPU design, with particular emphasis on pipelining. Both arithmetic and instruction-unit pipelines are covered; because so many modern microprocessors make extensive use of pipelined implementation, we pay particular attention to RISC, superpipelined, superscalar, VLIW, and multithreaded designs. Chapter 5 completes the coverage of single-processor system design considerations by discussing I/O related topics, including basic interfacing approaches, exceptions and interrupts, and the use of DMA and input/output processors to offload I/O-related tasks from the main system processor. These concepts are further illustrated via a discussion of how the ubiquitous Universal Serial Bus (USB) interface works.

The final two chapters deal with approaches that may be adopted when even the most advanced conventional, single-processor systems do not provide the desired performance or are not well suited to the intended application. Chapter 6 covers the most common types of high-performance systems, most of which are parallel to varying degrees. Flynn's venerable taxonomy of computer architectures is discussed in relation to vector and array processors, graphics processing unit (GPU) computing, shared-memory multiprocessors, and message-passing cluster (multicomputer) systems. Because communication between processors is critical to the performance of parallel systems, static and dynamic interconnection networks are discussed at some length. Finally, Chapter 7 goes beyond Flynn's classifications to explore the characteristics of unconventional architectures of the past, present, and future. From dataflow machines to artificial neural networks to fuzzy systems to quantum computers, students will see that there are other approaches, in addition to von Neumann's, that can be taken to solve particular types of computational problems.

As mentioned earlier, most instructors and their students should be able to cover the entire contents of this book in a one-semester course, given satisfactory completion of the prerequisites suggested above. If it is desired to spend more time covering conventional architectures at a leisurely pace, Chapter 7 may be omitted or left for outside reading. At institutions using the quarter system, it is suggested that Chapters 1 through 5 might be covered in one quarter while Chapters 6 and 7, plus perhaps some additional, advanced topics added by the instructor—or a substantial research or design project—could be reserved for a second, quarter-long course.

Writing (and updating) a textbook is a challenging task, and no matter how hard one tries, it is impossible to avoid error or to please every reader. I have, however, done my best, and I hope you enjoy the result. All known errors in the first edition have been corrected, and (I hope) not too many new ones have been introduced. I would like to hear from instructors and students alike about how you found this book useful and, conversely, of any mistakes or misconceptions you may have encountered, or any parts of the text that were less than clear. Please feel free to suggest any improvements that I might make in future revisions of the text. I welcome your comments via e-mail at Joe-Dumas@utc.edu.

**Joe Dumas**
*Signal Mountain, Tennessee*

# Acknowledgments

No book makes it to print without a lot of help from knowledgeable people in the publishing industry. Stan Wakefield deserves my thanks for encouraging me to write the first edition of this book and for helping me find the right people to publish it. Throughout the creation and revision of both editions, the folks at Taylor & Francis/CRC Press have been upbeat, personable, and most helpful to this author. I hope they didn't find my questions too annoying! I would particularly like to thank Nora Konopka for signing me up as an author in the first place and then encouraging me to update the text when the time was right. Nora's editorial assistant Kyra Lindholm and project coordinator Jessica Vakili were helpful and enthusiastic throughout the process of producing the second edition. Ladies, it has been a pleasure working with you.

Many outstanding professors challenged and inspired me as I traveled the long and winding road of higher learning on the way to becoming an educator myself. Although (over the course of many years and three degrees) they number far too many to mention here, I learned something from each and every one. To those who remember me (and even those who don't), I express my sincere appreciation for your dedication to your profession and to your students. I recall with particular gratitude Ted Bogart of the University of Southern Mississippi, Jim Harden of Mississippi State University, and Harry Klee of the University of Central Florida, who served as my advisors and mentors. I couldn't have made it this far without you.

My UTC colleagues and administrators at the department, college, and university level have been incredibly supportive of this project. I would especially like to thank the University of Chattanooga Foundation for supporting the sabbatical leaves during which the original text was written and the revisions for the second edition were made. But, most of all, it was my students who inspired me to undertake this project, so I'd like to thank all the young (and some not so young) people who have ever studied and worked hard to make it through a course with the infamous "Dr. Doom." I hope each of you found the time you spent in my classes worthwhile. It has been my privilege to help you expand your knowledge and understanding, and I have learned as much (or more) from you as you have from me. I especially acknowledge the computer architecture students at UTC who used the original draft and then the first edition of this book. By reading it thoroughly, pointing out errors, and offering suggestions for improvement, you helped me create a better text than would otherwise have been possible. On behalf of my future readers, I thank you for your valuable input.

Last, but not least, one can never undertake a labor of love like this one without having been given much love in the first place. For starting me out on the right path in life, I thank my parents, who stressed the importance of a good education and made many sacrifices to make sure I got one. They always encouraged me to pursue excellence and to follow my dreams wherever they might lead me; their love and support will never be forgotten. I also thank my dear wife for enduring, with love and as much patience as she could muster, my spending many long hours in front of a computer writing this book when I might otherwise have shared them with her. Honey, you're the best!

# Author

Joe Dumas earned a PhD in computer engineering from the University of Central Florida, where he received the first Link Foundation Fellowship in Advanced Simulation and Training, in 1993. Previously, he earned an MS degree in electrical engineering from Mississippi State University in 1989 and a BS degree in electronics engineering technology, with a minor in computer science, from the University of Southern Mississippi in 1984.

Dr. Dumas is a faculty member at the University of Tennessee in the Chattanooga's College of Engineering and Computer Science, where he holds the rank of UC Foundation Professor and has served as a Faculty Senator and Chair of the Graduate Council, among a number of campus leadership positions. He was chosen Outstanding Computer Science Teacher in 1998, 2002, and 2009. Dr. Dumas' areas of interest include computer architecture, embedded systems, virtual reality, and real-time, human-in-the-loop simulation.

Dr. Dumas is a member of several academic honor societies, including Upsilon Pi Epsilon (computer science), Eta Kappa Nu (electrical engineering), Tau Beta Pi (engineering), and Tau Alpha Pi (engineering technology). He was a founding member of the Chattanooga chapter of the IEEE Computer Society and served for several years as faculty advisor for the UTC student chapter of IEEE-CS. An avid downhill skier, tennis player, and distance runner with more than 30 completed marathons, Joe Dumas lives in Signal Mountain, Tennessee, with his wife Chereé.

# chapter one

# Introduction to computer architecture

"Computer architecture" is not the use of computers to design buildings (although that is one of many useful applications of computers). Rather, computer architecture is the design of computer systems, including all of their major subsystems: the central processing unit (CPU), the memory system, and the input/output (I/O) system. In this introductory chapter, we take a brief look at the history of computers and consider some general topics applicable to the study of computer architectures. In subsequent chapters, we examine in more detail the function and design of specific parts of a typical modern computer system. If your goal is to be a designer of computer systems, this book provides an essential introduction to general design principles that can be expanded upon with more advanced study of particular topics. If (as is perhaps more likely) your career path involves programming, systems analysis or administration, technical management, or some other position in the computer or information technology field, this book provides you with the knowledge required to understand, compare, specify, select, and get the best performance out of computer systems for years to come. No one can be a true computer professional without at least a basic understanding of computer architecture concepts. So let's get underway!

## 1.1    What is computer architecture?

Computer architecture is the design of computer systems, including all major subsystems, including the CPU and the memory and I/O systems. All of these parts play a major role in the operation and performance of the overall system, so we will spend some time studying each. CPU design starts with the design of the instruction set that the processor will execute and includes the design of the arithmetic and logic hardware that performs computations; the register set that holds operands for computations; the control unit that carries out the execution of instructions (using the other components to do the work); and the internal buses, or connections, that allow these components to communicate with each other. Memory system design uses a variety of components with differing characteristics to form an overall system (including main, or primary, memory and secondary memory) that is affordable while having sufficient storage capacity for the intended application and being fast enough to keep up with the CPU's demand for instructions and data.

I/O system design is concerned with getting programs and data into the memory (and ultimately the CPU) and communicating the computational results to the user (or another computing system) as quickly and efficiently as possible. None of these subsystems of a modern computer is designed in a vacuum; each of them affects, and is affected by, the characteristics of the others. All subsystems must be well matched and well suited to the intended application in order for the overall system to perform well and be successful in the marketplace. A system that is well designed and well built is a powerful tool that multiplies the productivity of its users; a system that is poorly designed, or a poor implementation of a good design, makes an excellent paperweight, doorstop, or boat anchor.

### 1.1.1   Architecture versus implementation

It is important to distinguish between the design (or architecture) of a system and the implementation of that system. This distinction is easily understood through an analogy between computer systems and buildings.

Computer architecture, like building architecture, involves first of all a conceptual design and overall plan. Architects ask, "What is this building (or computer) going to be used for, and what is the general approach we will take to fulfill the requirements of that application?" Once these general decisions are made, the architect has to come up with a more specific design, often expressed in terms of drawings and other specifications showing the general composition and layout of all the parts of the building or computer. This tells how everything will fit together at the higher levels.

When the design gets down to the level of specification of the actual components, the building or computer architect needs to have engineering knowledge (or enlist the help of a construction or computer engineer) in order to make sure the paper design is feasible given available materials. A construction engineer needs to make sure the building foundation and beams will carry the required loads, that the heating and air conditioning units have sufficient capacity to maintain the temperature of the building, and so on. A computer engineer must make sure the electrical, mechanical, thermal, timing, control, and other characteristics of each component are sufficient to the job and compatible with the other components of the system. The result of all this architectural and engineering effort is a design specification for a building or for a computer system. However, this specification exists only on paper or, more likely, as computer-aided design (CAD) files containing the drawings and specifications.

If one wants to have an actual building to occupy or a computer system to use, it must be built in the physical world. This, of course, is what we mean by implementation. In the case of a building, the design documents prepared by the architect and engineer are given to a contractor who uses components made of various materials (steel, wood, plastic, concrete, glass, brick, etc.) to construct (implement) an actual building. Likewise, the design produced by a computer architect or engineer must be put into production and built using various electrical and mechanical components. The end result—the implementation—is a working computer system.

It should be obvious that, for the end product to be a properly functioning building or computer, both the architectural plan and the physical implementation must be done well. No amount of attention to quality in construction will turn an inadequate design into a building or computer that meets the requirements of the intended application. However, even the best and most well–thought out design can be ruined by using substandard components or shoddy workmanship. For a good end result, all aspects of design and implementation must come together.

We should also recognize that although a clear distinction should be made between architecture and implementation, they are intimately interrelated. Neither architecture nor implementation exists in a vacuum; they are like two sides of a coin. Architectural vision affects the type of technologies chosen for implementation, and new implementation technologies that are developed can broaden the scope of architectural design. Taking the example of building architecture again, in the distant past all buildings were made of materials such as stone and wood. When iron was discovered and first used as a building material, it allowed architects to design new types of buildings that had previously been impossible to build. The advent of steel and concrete enabled the design and construction of skyscrapers that could never have existed before the invention of those materials. Modern materials, such as polymers, continue to expand the possibilities available to building architects. Sometimes the desire to include certain architectural features in buildings has led to the development of new construction materials and techniques or at least to new applications for existing materials.

Building architecture and implementation have progressed hand in hand over the course of human civilization, and the vision of computer architects and the available implementation technologies have likewise (over a much shorter time but at a much faster pace) moved forward hand in hand to create the computers we used in the past, the ones we use today, and the ones we will use tomorrow.

## 1.2   Brief history of computer systems

Computing devices of one type or another have existed for hundreds of years. The ancient Greeks and Romans used counting boards to facilitate mathematical calculations; the abacus was introduced in China around A.D. 1200. In the 17th century, Schickard, Pascal, and Leibniz devised mechanical calculators. The first design for a programmable digital computer was the analytical engine proposed by Charles Babbage in 1837. Given the technology available at the time, the analytical engine was designed as a mechanical rather than electronic computer; it presaged many of the concepts and techniques used in modern machines but was never built due to a lack of funding. Ada Augusta, Countess of Lovelace, developed "cards" (programmed instructions) to demonstrate the operation of the proposed analytical engine and, to this day, is revered as the first computer programmer. Although Babbage's machine was never commercially successful, mechanical punched-card data processing machines were later developed by Herman Hollerith and used to tabulate the results of the 1890 U.S. census. Hollerith's company later merged with another to become International Business Machines (IBM).

While Babbage's design was based on the decimal (base 10) system of numbering used by most cultures, some of his contemporaries, such as George Boole and Augustus DeMorgan, were developing a system of logical algebra (today known as Boolean algebra) that provides the theoretical underpinnings for modern computers that use a two-valued, or binary, system for representing logical states as well as numbers. Boolean algebra was an intellectual curiosity without practical application until a young mathematician named Claude Shannon recognized (in his 1937 master's thesis at the Massachusetts Institute of Technology) that it could be used in designing telephone switching networks and, later, computing machines. Boolean algebra is the logical basis for virtually all modern digital computer design.

In addition to Babbage's engine and Hollerith's punched-card machine, another type of "computer" predated the existence of modern digital computers. From the early 20th century, mechanical and electrical "analog computers" were used to solve certain types of problems that could be expressed as systems of differential equations with time as the independent variable. The term *computer* is in quotes because analog computers are more properly called *analog simulators*. They do not actually perform discrete computations on numbers, but rather operations such as addition, subtraction, integration, and differentiation of analog signals, usually represented by electrical voltages. Analog simulators are continuous rather than discrete in their operation, and they operate on real values; numbers are measured rather than counted. Thus, there is no limit, other than the tolerances of their components and the resolution of the measuring apparatus, to the precision of the results obtained. During the 1930s, 1940s, 1950s, and even into the 1960s, analog simulators were widely used to simulate real-world systems, aiding research in such fields as power plant control, aircraft design, space flight, weather modeling, and so on. Eventually, however, analog simulators were rendered obsolete by digital computers as they became more powerful, more reliable, and easier to program. Thus, in the rest of our discussion of the history of computing devices we will restrict ourselves to the electronic digital computing devices of the 20th and now the 21st centuries.

## 1.2.1   The first generation

During the late 1930s and early 1940s, mostly as part of the Allied effort to win World War II, digital computers as we know them today got their start. The first generation (approximately late 1930s to early 1950s) of computer systems were one-of-a-kind machines, each custom built for a particular purpose. Computers of the early 1940s, such as the Mark-I (also known as the IBM Automatic Sequence Controlled Calculator or ASCC) and Mark-II machines built by Howard Aiken at Harvard University, were typically built using electromagnetic relays as the switching elements. This made them very slow. Later machines were built using vacuum tubes for switching; these were somewhat faster but not very reliable. (Tubes, like incandescent light bulbs, have a nasty habit of burning out after a few hundred or a few thousand hours of use.) Two of the first electronic computers built using vacuum tube technology were the Atanasoff-Berry Computer (ABC) developed at Iowa State University and the Electronic Numerical Integrator and Calculator (ENIAC) built by John Mauchly and J. Presper Eckert at the University of Pennsylvania for the U.S. Army Ordnance Department's Ballistic Research Laboratories. ENIAC, which was used to calculate bomb trajectories and later to help develop the hydrogen bomb, was more similar to today's pocket calculators than to our general-purpose computers as it was not a stored-program machine. Connections had to be rewired by hand in order to program different calculations.

The first modern computer designed to run software (program instructions stored in memory that can be modified to make the machine perform different tasks) was the Electronic Discrete Variable Computer (EDVAC) designed by Mauchly, Eckert, and John von Neumann of Princeton University. EDVAC was designed to perform sequential processing of instructions that were stored in memory along with the data, characteristics of what has become known as the von Neumann architecture (to be discussed further in Section 1.3.1). It is debatable whether von Neumann deserves primary credit for the idea—the report outlining EDVAC's design mysteriously bore only his name although several other researchers worked on the project—but what is certain is that the stored-program concept was a major step forward in computer design, making general-purpose machines feasible. To this day, most computers are still basically von Neumann machines with some enhancements.

Although the original EDVAC design was never built (it was eventually modified and built as Princeton's Institute for Advanced Studies [IAS] machine), its concepts were used in many other machines. Maurice Wilkes, who worked on EDVAC, built the Electronic Delay Storage Automatic Calculator (EDSAC), which became the first operational stored-program computer using the von Neumann architecture. In 1951, the first commercially available computer was produced by the Remington-Rand Corporation. Based on the stored-program designs of the EDVAC and EDSAC, this computer was known as the UNIVAC I (for Universal Automatic Computer). The first of 46 of these machines were delivered to the U.S. Census Bureau in 1951; the following year, another UNIVAC found the spotlight as it was used to predict the outcome of the Eisenhower–Stevenson presidential race on election night. With only 7% of the vote counted, the machine projected a win for Eisenhower with 438 electoral votes (he ultimately received 442).

## 1.2.2   The second generation

The second generation (approximately mid-1950s to early 1960s) of digital computer systems included the first machines to make use of the new solid-state transistor technology. The transistor, invented in 1947 by John Bardeen, Walter Brattain, and William Shockley of Bell Laboratories, was a major improvement over vacuum tubes in terms of size, power consumption, and reliability. This new implementation technology paved the way for many architectural enhancements, mainly by allowing the total number of switching elements in machines to increase.

Vacuum tube-based computers could never have more than a few thousand switching elements because they were constantly malfunctioning due to tube failures. The mean time between failures (MTBF), or average lifetime, of vacuum tubes was only about 5000 hours; a system containing 5000 tubes could thus be expected to have a hardware-related crash about once per hour on average, and a system with 10,000 of the same type of tubes would average one failure every half hour of operation. If the machine were to run long enough to do any meaningful calculations, it could only contain a limited number of tubes and (because all logic required switching elements) a very limited set of features.

Then, along came the transistor with its much longer life span. Even the earliest transistors had a typical MTBF of hundreds of thousands or millions of hours—one or two orders of magnitude better than vacuum tubes. They were also much smaller and generated less heat, allowing components to be more densely packaged. The result of these factors was that second-generation computers could be more complex and have more features than their predecessors without being as large, expensive, or power-hungry and without breaking down as often. Some architectural features that were not present in first-generation machines but were added to second-generation computers (and are still used in modern machines) include hardware representation of floating-point numbers (introduced on the IBM 704 in 1954), hardware interrupts (used in the Univac 1103 in 1954), general-purpose registers used for arithmetic or addressing (used in the Ferranti Pegasus in 1956), and virtual memory (introduced on the University of Manchester's Atlas machine in 1959). The IBM 709, released in 1958, featured asynchronous I/O controlled by independent, parallel processors as well as indirect addressing and hardware interrupts.

Another technology that came into use with the second generation of computer systems was *magnetic core memory*. Core memory stored binary information as the magnetized states of thousands of tiny, doughnut-shaped ferrite rings or "cores." This technology reduced the space and power required for large amounts of storage. Although typical first-generation computers had only 1 to 4 kilobytes (KB) of main memory in the form of delay lines, vacuum tubes, or other primitive storage technologies, second-generation computers could have comparatively huge main memories of, for example, 16 KB up to 1 megabyte (MB) of core. This increase in the size of main memory affected the types of instructions provided, the addressing modes used to access memory, and so on.

Despite technological advances, second-generation machines were still very bulky and expensive, often taking up an entire large room and costing millions of dollars. The relatively small IBM 650, for example, weighed about a ton (not counting its 3000-pound power supply) and cost $500,000 in 1954. Of course, that would be several millions of today's dollars! Core memory cost on the order of a dollar per byte, so the memory system alone for a large computer could have a cost in the seven-figure range.

With such large, expensive systems being built, no organization could afford to run just one program, get an answer, then manually load the next program and run it (as was done with the first-generation machines). Thus, during this time came the advent of the first batch-processing and multiprogramming operating systems. Batch processing meant that programs were loaded and executed automatically by the system instead of the operator loading them manually. Multiprogramming meant that more than one program was resident in memory at the same time (this, of course, was made possible by the larger main memory space). By keeping multiple programs ready to run and by allowing the system to switch between them automatically, the expensive CPU could be kept busy instead of being idle, awaiting human intervention. Programs were still executed one at a time, but when one program was completed or encountered an I/O operation that would take some time to complete, another program would be started to occupy the CPU's time and get useful work done.

The first attempts to make human programmers more productive also occurred during this time. Assembly language, as a shorthand or mnemonic form of machine language, was first developed in the early 1950s. The first high-level languages, Formula Translation (Fortran), Algorithmic Language (Algol), and Common Business-Oriented Language (COBOL), came along a few years later. It is interesting to note that computer pioneer John von Neumann opposed the development of assemblers and high-level language compilers. He preferred to employ legions of human programmers (mostly low-paid graduate students) to hand-assemble code into machine language. "It is a waste of a valuable scientific computing instrument," von Neumann reportedly said, "to use it to do clerical work." Fortunately, his point of view did not win out!

### 1.2.3   The third generation

The third generation (approximately mid-1960s to early 1970s) marked the first use of *integrated circuits* in computers, replacing discrete (individually packaged) transistors. Integrated circuits (ICs) are semiconductor chips containing multiple transistors, starting with one or two dozen transistors comprising a few gates or flip-flops (small scale integration [SSI]) and later increasing to a few hundred transistors (medium scale integration [MSI]), which allowed an entire register, adder, or even an arithmetic logic unit (ALU) or small memory to be fabricated in one package. Core memory was still common in third-generation computers, but by the 1970s, core was beginning to be replaced by cheaper, faster semiconductor IC memories. Even machines that used magnetic core for main storage often had a semiconductor *cache memory* (see Section 2.4) to help improve performance.

Integrated circuits continued the trend of computers becoming smaller, faster, and less expensive. All first- and second-generation machines were what would come to be called *mainframe* computers. The third generation saw the development of the first *minicomputers*, led by Digital Equipment Corporation's (DEC) PDP-8, which was introduced in 1964 and cost $16,000. DEC followed it with the more powerful PDP-11 in 1970. Data General brought the Nova minicomputer to market in 1969 and sold 50,000 machines at $8000 each. The availability of these small and relatively inexpensive machines meant that rather than an organization having a single central computer shared among a large number of users, small departments or even individual workers could have their own machines. Thus, the concept of a *workstation* computer was born.

Also during the third generation, the first of what came to be called *supercomputers* were designed. Control Data Corporation introduced the CDC 6600, widely considered to be the first supercomputer, in 1964. A key contributor to the machine's design was Seymour Cray, who later left CDC to form his own very famous supercomputer company. The 6600 was the fastest computer in the world at the time—roughly three times the speed of the IBM Stretch machine, which previously held that title. The main processor operated at a 10 MHz clock frequency (unheard of at the time) and was supported by 10 parallel execution units. It could execute a then-astonishing 3 million instructions per second and cost about $7 million in mid-1960s dollars.

The 6600 was followed in 1968 by the even more powerful CDC 7600, which topped out at 15 million instructions per second. The 7600 had a heavily pipelined architecture and is considered the first *vector processor* (Mr. Cray would go on to design many more). Other third-generation vector computers included the CDC Star-100 and Texas Instruments' Advanced Scientific Computer (TI-ASC), both announced in 1972. The third generation also saw the development of some of the first high-performance parallel processing machines, including Westinghouse Corporation's SOLOMON prototype and later the ILLIAC IV (a joint venture between the University of Illinois, the U.S. Department of Defense, and Burroughs Corporation).

The third generation also gave us the first example of a "family" of computers: the IBM System/360 machines. IBM offered multiple models of the 360, from low-end machines intended for business applications to high-performance models aimed at the scientific computing market. Although their implementation details, performance characteristics, and price tags varied widely, all of the IBM 360 models could run the same software. These machines proved very popular, and IBM sold thousands of them. Ever since then, it has become common practice for computer manufacturers to offer entire lines of compatible machines at various price and performance points.

Software-related developments during the third generation of computing included the advent of timesharing operating systems (including the first versions of UNIX). Virtual memory became commonly used, and new, more efficient computer languages were developed. Although third-generation hardware was becoming more complex, computer languages were being simplified. Combined Programming Language (CPL), developed circa 1964 at Cambridge University and the University of London, was an attempt to streamline the complicated Algol language by incorporating only its most important features. Ken Thompson of Bell Laboratories continued the trend of simplifying computer languages (and their names), introducing the B language in 1970. Move to the head of the class if you can guess which language he helped develop next!

## 1.2.4   The fourth generation

The fourth generation (approximately mid-1970s to 1990) saw the continuing development of large-scale integration (LSI) and very large-scale integration (VLSI) circuits containing tens or hundreds of thousands, and eventually millions, of transistors. For the first time, an entire CPU could be fabricated on one semiconductor microcircuit. The first *microprocessor*, or "CPU on a chip," was Intel's 4-bit 4004 processor, which debuted in 1971. It was too primitive to be of much use in general-purpose machines, but useful 8-, 16-, and even 32-bit microprocessors followed within a few years; soon essentially all computers had a single-chip microprocessor "brain." Semiconductor main memories made of VLSI RAM and ROM devices became standard, too. (Although core memory became extinct, its legacy lives on in the term *core dump*, which refers to the contents of main memory logged for diagnostic purposes when a crash occurs.) As VLSI components became widely used, computers continued to become smaller, faster, cheaper, and more reliable. (The more components that are fabricated onto a single chip, the fewer chips that must be used and the less wiring that is required. External wiring is more expensive and more easily broken than on-chip connections and also tends to reduce speeds.) The Intel 486 CPU, introduced in 1989, was the first million-transistor microprocessor. It featured an on-chip floating-point unit and cache memory and was in many ways the culmination of fourth-generation computer technology.

The invention of the microprocessor led to what was probably the most important development of the fourth generation: a new class of computer system known as the *microcomputer*. Continuing the trend toward smaller and less expensive machines begun by the minicomputers of the 1960s and early 1970s, the advent of microcomputers meant that almost anyone could have a computer in his or her home or small business. The first microcomputers were produced in the mid- to late 1970s and were based on 8-bit microprocessors. The Altair computer kit, introduced in 1975, was based on the Intel 8080 microprocessor. More than 10,000 Altairs were shipped to enthusiastic hobbyists, and the microcomputer revolution was underway. (Bill Gates and Paul Allen of Microsoft got their start in the microcomputer software business by developing a BASIC interpreter for the Altair.) Californians Steve Wozniak and Steve Jobs quickly joined the wave, developing the Apple I computer in Wozniak's garage using a Mostek 6502 as the CPU. They refined their design and created the Apple II in 1977; it outsold its competitors (the 6502-based Commodore Pet and the Radio Shack TRS-80, based on a Zilog Z80 processor) and became a huge success.

IBM saw the success of the Apple II and decided to enter the microcomputer market as well. The IBM Personal Computer (or PC) was an immediate hit, prompting other manufacturers to create compatible "PC clones" using Intel's 16-bit 8086 processor. (The IBM PC and PC/XT actually used the cheaper 8088 chip, which was architecturally identical to the 8086 but had an 8-bit external interface.) The availability of compatible machines from competing manufacturers helped bring down the price of hardware and make PCs a mass-market commodity.

While IBM was enjoying the success of the original PC and its successor the PC/AT (which was based on Intel's faster 80286 CPU), Wozniak and Jobs were developing new systems around Motorola's 16-/32-bit 68000 family of microprocessors. Their ambitious Apple Lisa (the first microcomputer with a graphical user interface) cost too much (about $10,000) to gain wide acceptance. However, its successor, the Apple Macintosh (launched in 1984), incorporated many of Lisa's features at about one-fourth the price and gained a large following that continues to the present day.

Acceptance of microcomputers was greatly increased by the development of office applications software. Electric Pencil, written by Michael Shrayer for the Altair, was the first microcomputer word processing program. Electric Pencil was not a big commercial success, but it was followed in 1979 by WordStar, which gained widespread acceptance. dBase, the database management package, and VisiCalc, the first microcomputer spreadsheet program (originally developed for the Apple II) also appeared in 1979. In particular, the spreadsheet program VisiCalc and its successor Lotus 1-2-3 (developed for the IBM PC) helped promote the use of microcomputers in business. Microsoft, which got its start with Altair BASIC, won the contract to develop the PC-DOS (later generically marketed as MS-DOS) operating system and a BASIC interpreter for the IBM PC. Soon Microsoft branched out into applications as well, introducing Microsoft Works (a combined word processor, spreadsheet, database, graphics, and communication program) in 1987. The Windows operating system and Microsoft Office (the successor to Works) gained massive popularity in the 1990s, and the rest, as the saying goes, is history.

While microcomputers were becoming smaller, less expensive, and more accessible to the masses, supercomputers were becoming more powerful and more widely used. The first new supercomputer of the fourth generation, the Cray-1, was introduced in 1976 by Seymour Cray, who had left CDC to form his own company, Cray Research, Inc. The Cray-1 cost about $8 million and could execute more than 80 million floating-point operations per second (MFLOPS). Within a few years, Cray followed this machine with the X-MP (Cray's first multiprocessor supercomputer) in 1982, the Cray-2 in 1985, and the Y-MP in 1988. The eight-processor Y-MP had a peak calculation rate of more than 2600 MFLOPS—about 30 times the performance of the Cray-1.

Cray Research was not the only company developing fourth-generation supercomputers. Most of Cray's competition in the area of high-end vector machines came from Japanese manufacturers. Some important Japanese supercomputers of the 1980s included the Nippon Electric Company (NEC) SX-1 and SX-2 systems, introduced in 1983; Fujitsu's VP-100 and VP-200 machines (1983), followed by the VP-400 in 1986; and Hitachi's S820/80 supercomputer released in 1987. Although the dominant supercomputers of this period were mostly pipelined vector machines, the fourth generation also saw the debut of highly parallel supercomputers. The Massively Parallel Processor (MPP), first proposed in 1977 and delivered in 1983 by Goodyear Aerospace Corporation to NASA Goddard Space Flight Center, was a one-of-a-kind machine constructed from 16,384 1-bit processors. A few years later (1986–1987), Thinking Machines Corporation entered the supercomputer market with its Connection Machine CM-1 and CM-2 systems. These machines contained 65,536 processors each; with more than 70 installations, they were the first commercially successful massively parallel supercomputers.

Although most of the new developments of the fourth generation of computers occurred at the high and low ends of the market, traditional mainframes and minicomputers were still in widespread use. IBM continued to dominate the mainframe computer market, introducing a series of upgrades (more evolutionary than revolutionary) to its System/370 line of machines, which had replaced the System/360 in 1970. Among IBM's workhorse fourth-generation mainframes were the 3030 series (1977–1981), the 3080 series (1980–1984), the 3090 series (1985–1989), and the 4300 series (1979–1989). All of these machines saw extensive use in a variety of medium-to-large business applications.

The dominant minicomputer of the fourth generation was Digital Equipment Corporation's VAX series, which was a successor to the popular PDP-11. DEC's VAX 11/780, released in 1977, was the first minicomputer with a 32-bit architecture, allowing large amounts of memory to be addressed at once by the programmer. It was also the first minicomputer to execute one million instructions per second. (The VAX acronym stood for Virtual Address eXtension, and the operating system, VMS, stood for Virtual Memory System.) To address compatibility concerns, early VAX models incorporated a PDP-11 emulation mode to ease migration to the newer system. The 11/780 was followed by models 11/750 and 11/730, which had close to the same performance but were smaller and cheaper. Higher performance needs were addressed by the dual-processor 11/782 and the 11/785. Largely due to the success of these computers, by 1982, DEC was the number two computer company in the world, behind only IBM. DEC remained a major force in the market as it continued to expand the VAX line through the 1980s, developing the higher performance 8x00 series machines and the microVAX line of microcomputers that were compatible with DEC's larger and more expensive minicomputers.

Characteristics of fourth-generation machines of all descriptions include direct support for high-level languages either in hardware (as in traditional Complex Instruction Set Computer [CISC] architectures) or by using optimizing compilers (characteristic of Reduced Instruction Set Computer [RISC] architectures, which were developed during the fourth generation). The MIPS and SPARC architectures gave rise to the first RISC microprocessors in the mid-1980s; they quickly rose to challenge and, in many cases, replace dominant CISC microprocessors such as the Motorola 680x0 and Intel x86 CPUs as well as IBM's mainframes and DEC's mini-computers. We study these competing schools of architectural thought and their implications for computer system design in Chapters 3 and 4.

With the advent of hardware memory management units, time-sharing operating systems and the use of virtual memory (previously available only on mainframes and minicomputers) became standard on microcomputers by the end of the fourth generation. By giving each program the illusion that it had exclusive use of the machine, these techniques made programming much simpler. Compilers were considerably improved during the 1970s and 1980s, and a number of new programming languages came into use. BASIC, a language simple enough to be interpreted rather than compiled, became popular for use with microcomputers. In 1984, Borland's inexpensive Turbo Pascal compiler brought high-level language programming to the personal computer mainstream. Meanwhile, C (developed in 1974 by Dennis Ritchie of Bell Laboratories as a refinement of Ken Thompson's B) became the dominant language for systems (especially UNIX) programming. Fortran 77 was widely used for scientific applications, and Ada (named for the Countess of Lovelace) was adopted by the U.S. military in 1983 for mission-critical applications. Finally, object-oriented programming got its start during the fourth generation. C++ was developed by Bjarne Stroustrup at Bell Laboratories during the early 1980s; the first version of the new OO language was released in 1985.

## 1.2.5    The fifth generation

The fifth generation (approximately 1990 to 2005) of computing systems can arguably be termed more evolutionary than revolutionary, at least in terms of architectural features. Machines of this era used VLSI and ultra large-scale integration (ULSI) chips with tens (eventually hundreds) of millions of transistors to perform many complex functions, such as graphics and multimedia operations, in hardware. Processors became more internally parallel (using techniques such as superscalar and superpipelined design to execute more instructions per clock cycle), and parallelism using multiple processors, once found only in mainframes and supercomputers, started becoming more common even in home and small business systems. Processor clock frequencies that reached the tens of megahertz in fourth-generation machines increased to hundreds and thousands of megahertz (1 GHz = 1000 MHz). Intel's Pentium microprocessor, introduced in 1993, had two pipelined execution units and was capable of executing 100 million instructions per second. Ten years later, its successor, the Pentium 4, reached clock speeds of 3.2 GHz on a much more highly concurrent internal microarchitecture, implying a microprocessor performance improvement of roughly two orders of magnitude (a hundredfold) over that time span.

With microprocessors becoming ever more powerful, microcomputers were the big story of computing in the fifth generation. Personal computers and scientific and engineering workstations powered by single or multiple microprocessors, often coupled with large amounts of memory and high-speed but low-cost graphics cards, took over many of the jobs once performed by minicomputers (which effectively became extinct) and even mainframes such as IBM's zSeries (which were mostly relegated to important but "behind the scenes" applications in business transaction processing). Even the supercomputers of this generation increasingly made use of standard microprocessor chips instead of custom-designed, special-purpose processors.

Supercomputing experienced a transformation during the fifth generation. The high-speed, pipelined vector processors of the late 1970s and 1980s only proved cost-effective in a limited number of applications and started to fall out of favor for all but extremely numerically intensive computations done by well-funded government agencies. By the 1990s, Fujitsu and other manufacturers quit making vector computers, leaving NEC and Cray as the only vendors of this type of system. Cray Computer Corporation was spun off from Cray Research, Inc., to develop the Cray-3, but only one machine was delivered before the company filed for Chapter 11 bankruptcy in 1995; plans for the Cray-4 were abandoned. Tragically, supercomputer pioneer Seymour Cray met the same fate as Cray Computer Corporation, perishing in an automobile accident in 1996.

Cray Research, Inc., began to move toward nonvector, massively parallel, microprocessor-based systems in the mid-1990s, producing its T3D and T3E machines before being bought by Silicon Graphics, Inc. (SGI), in 1996. SGI's Cray Research division produced the SV1 vector supercomputer in 1998 but was sold in March 2000 to Tera Computer Company and renamed Cray, Inc. Cray introduced the X1 (formerly codenamed the SV2) in 2002. Meanwhile, in 2001 NEC signed an agreement with Cray, Inc., to market its SX-5 and SX-6 supercomputers, effectively combining all remaining high-end vector machines into a single vendor's product line.

By June 2005, only 18 of the top 500 supercomputers were vector processors (including nine Crays and seven NEC machines). The other 482 were highly parallel scalar machines, mostly built by IBM (259),

Hewlett-Packard (131), and SGI (24). An increasing number of high-performance computer systems, including 304 of the June 2005 top 500 supercomputers, were classified as *cluster systems* composed of large numbers of inexpensive microcomputers connected by high-speed communication networks. Many of these clusters ran the open-source Linux operating system, which enjoyed a huge surge in popularity during the 1990s and early 2000s for all classes of machines from PCs (with which it began to make inroads into the popularity of Microsoft's Windows operating system) to supercomputers.

Perhaps the most important characteristic of fifth-generation computer systems, large or small, that distinguished them from systems of previous generations was the pervasiveness of networking. In previous generations, the vast majority of computers were designed as self-contained, standalone systems. The Internet was virtually unknown outside of government installations and academia. Networking capabilities, if they existed at all, were separately designed hardware and software additions tacked onto a system not designed with connectivity in mind. All this began to change in the fifth generation. By the mid-1990s, modems had become fast enough to allow reasonable dial-up connections between PCs (and their GUIs) and the Internet. E-mail clients, such as Eudora, allowed easy electronic communication, and GUI browsers, such as Mosaic (and later Netscape and Internet Explorer), let users conveniently "surf" for hyperlinked information all over the World Wide Web. Users quickly demanded more and better connectivity for even low-end systems; ISDN (integrated services digital network) and eventually DSL (digital subscriber line) and cable modem service were developed to meet this demand.

Fifth-generation computers and their operating systems were designed "from the ground up" to be connected to wired and/or wireless local area or wide area networks (LANs and WANs) and the Internet. UNIX, from which Linux was derived, had been network-friendly for many years; Apple and Microsoft followed suit by integrating networking capabilities into their operating systems as well. Even the programming languages of this era, such as Java (introduced in 1995) were designed with a networked environment in mind. This focus on connectivity was not confined to traditional computer systems; another characteristic of the fifth generation of computing was the blurring of distinctions between general-purpose computers and other communications devices, such as cellular telephones, pagers, personal digital assistants (PDAs), digital cameras and media players, and other devices with embedded microprocessors. Users increasingly expected all of their electronic devices to be able to conveniently exchange information, and especially during the latter part of the fifth generation, this started to become much easier than ever before.

## 1.2.6   Modern computing: the sixth generation

It is difficult to write about history when one is living it, and some computing professionals claim that we are still in the fifth (or even the fourth) generation of computing, but a good case can be made that we are currently (since approximately 2005) well into a sixth generation. What has changed since the systems of the 1990s and early 2000s? Quite a number of things—not only in terms of the computing hardware itself, but also in the ways systems are connected together and how they are used.

Let's start with the processors themselves. Although fifth-generation CPU chips could contain tens or (later) hundreds of millions of transistors, it has become common for sixth-generation processors to contain billions of these ever-tinier switching elements. Generally, these additional transistors have been used in two main ways: constructing ever-larger on-chip cache memories (to be discussed in Section 2.4) and adding additional processing cores (essentially creating single-chip multiprocessor systems of the type we will examine in Section 6.1). The results have been continued increases in processing power even as clock frequencies have leveled off in the low gigahertz range.

In 2006, Intel released a dual-core version of its Itanium 2 processor with 1.7 billion transistors, constructed with a feature size of 90 nanometers (nm). Four years later, its successor, the quad-core Itanium 9300 (code named Tukwila), built with a 65-nm feature size, became the first microprocessor to contain more than two billion transistors. Other manufacturers made similar progress: IBM released its eight-core POWER7 chip (1.2 billion transistors, 45 nm) in 2010; Advanced Micro Devices (AMD) introduced a pair of billion-plus transistor CPUs built with a 32-nm process in 2012; and Apple reached the 2 billion count with its 20-nm dual-core A8 processor (and 3 billion with the tri-core A8X) in 2014. Oracle (which acquired the former Sun Microsystems) broke the billion-transistor barrier with its SPARC T3 processor in 2010 and then surpassed all competitors with the 32-core SPARC M7, introduced in 2015 as the first microprocessor to contain 10 billion transistors.

As processors increased in transistor count and overall performance, main memory and secondary storage continued to steadily increase in capacity and speed. The same basic main memory technology that was used in most fifth-generation computers—Synchronous Dynamic Random Access Memory (SDRAM)—continued to be the dominant form of main memory through the sixth generation. Between 2003 and 2007 (corresponding to the "Windows XP generation" of laptop and desktop computing), it was common for such systems to have somewhere in the range of 256 to 1024 megabytes (or one fourth to one gigabyte) of second-generation double data rate (DDR2) SDRAM, which could transfer data at a peak speed of 3200 megabytes per second or better, depending on the specific modules used. Over the next few years RAM sizes gradually increased, such that by 2015 four gigabytes was typical of a very low-end PC, and most small systems came standard with 6–8 gigabytes or more. The introduction of DDR3 SDRAM in 2007 and DDR4 SDRAM in 2014 doubled and redoubled the peak data transfer rates to and from main memory. As far as secondary storage was concerned, magnetic hard disk drives continued to dominate the market, although hybrid drives (containing both a hard disk and a smaller amount of nonvolatile semiconductor flash memory) and even completely solid-state drives (SSDs, constructed solely of the more expensive but faster flash memory) increased in popularity over time. At the beginning of the sixth generation, small systems typically had a hard drive capacity of a few to a few dozen gigabytes, and the largest available drives were about 500 GB. Ten years later, one-terabyte (TB) drives were inexpensive and commonplace, and the most advanced drives offered 10 times that much storage.

The increase in main memory sizes for all sorts of systems from personal computers to servers and mainframes helped bring 64-bit CPUs (and operating systems) to the forefront during the sixth generation of computing systems. Although 64-bit supercomputer architectures debuted during the fourth generation and 64-bit servers began to appear during the fifth, it wasn't until 2003–2004 that first AMD and then Intel produced 64-bit versions of their ubiquitous x86 processors. In April 2005, Microsoft released a 64-bit version of its Windows XP Professional operating system, following up in January 2007 with 64-bit Windows Vista. These and subsequent operating systems would take advantage of the newest processors with their AMD64 and Intel 64 architectures, generically known in the industry as x86-64. While previous PCs and servers running 32-bit operating systems could only address $2^{32}$ bytes (4 gigabytes) of main memory, it henceforth became common for even inexpensive desktop and laptop systems to exceed that amount. Theoretically, 64-bit systems could address up to $2^{64}$ bytes (16 exabytes) of memory space, although (as of this writing) current implementations only use 48 of the 64 possible address bits for a total available memory space of 256 terabytes.

Although wireless local area network (WLAN) technology was first conceived during the third generation of computing (all the way back in 1971), and the first version of the Institute of Electrical and Electronics Engineers (IEEE) 802.11 standard was released in 1999 during the fifth generation, it was during the sixth generation that wireless digital communication technologies (including WLANs and cellular data transmission networks) became widely used. With the advent of third-generation (3G) cellular networks in the early 2000s and fourth-generation (4G) networks a few years later, in addition to the widespread availability of Wi-Fi WLAN "hotspots" or access points and the proliferation of touch-screen technology, an entirely new class of mobile computing systems, including tablet computers and smartphones, was born. At last, both computing power and digital communications capability could travel along with any user, almost anywhere he or she wanted to go.

Smartphones essentially replaced a pair of separate devices: a cellular telephone (which previously had been used only for voice communication and short text messages) and a personal digital assistant (a small, portable computing device used to keep track of appointments, to-do lists, etc., and, in some cases, access information over the Internet). BlackBerry's first smartphone was released in 2003. Within three years, use of the device had caught on so widely and addictively that the editorial staff of Webster's New World Dictionary chose "CrackBerry" as the winner of their Word of the Year contest. Apple's first iPhone was introduced in 2007, and the following year marked the debut of the HTC Dream—the first smartphone to use the Android mobile operating system. Not only could these devices access the World Wide Web and other Internet sources, act as Global Positioning Service (GPS) trackers, perform voice recognition or other artificial intelligence (AI) functions, and (of course) make telephone calls, their ability to run custom application programs ("apps") allowed almost limitless extensions to their functionality. Given the proliferation of ever more sophisticated apps, smartphone (and tablet) users demanded ever-greater computing power. With servers and even PCs making the leap to 64-bit processing, it was only a matter of time before they were joined by smaller, mobile devices. The first 64-bit Advanced RISC Machine (ARM) processor—the architecture used in most mobile computing devices—was announced in 2011. In 2013, Apple's iPhone 5S became the first 64-bit smartphone; the first Android devices to use 64-bit processors debuted the following year.

During the sixth generation, the concept of *cloud computing* became widely known and used in a variety of ways. Although there are probably as many different definitions of cloud computing as there are computer users, one common thread is that cloud-based approaches take advantage of wired and/or wireless data communication between local (often mobile) and remote computing systems. The precise location and nature of the system(s) providing cloud services, which could be computational, data storage, or both, is generally unknown and of no particular concern to the end user. Another common technique used in cloud computing is virtualization, where the resources as seen and accessed by a user may—but usually do not—correspond to a specific physical system. For example, a user may log into what appears to be a dedicated Windows or Linux server, but that capability may be only one of many services running on a remote machine that may be natively running a completely different operating system. By utilizing cloud-based resources rather than local computing hardware, users can gain a number of benefits, including reliability, scalability, and access to services from almost anywhere using a variety of different devices.

Although mobile devices and small computing systems made considerable advances during the sixth generation of computing, high-end systems, including servers and supercomputers, made great strides forward as well. One of the most important techniques to become popular during this time span was the use of Graphics Processing Units (GPUs) to supplement conventional CPUs in handling certain demanding computing jobs. GPUs, as their name suggests, are specialized processors optimized for the mathematically intensive process of creating images to be displayed to a human user—often in the context of a computer-based game, simulation, or virtual environment. Specialized graphics hardware has been in existence since at least the fourth generation of computing, and the term *GPU* was coined during the fifth generation (in 1999) by NVIDIA Corporation, which described its GeForce 256 processor as "the world's first GPU." However, it was several years later before the practice of using GPUs for nongraphical tasks (General-Purpose computing on Graphics Processing Units [GPGPU]) was widely adopted. As we shall explore further in Section 6.1.2, the same type of hardware parallelism that makes GPUs good at processing image data has also been found by researchers to be extremely efficient for processing other types of data, particularly performing scientific computations on highly parallel data sets composed of vectors or matrices. The development of proprietary programming languages, such as NVIDIA's Compute Unified Device Architecture (CUDA), and cross-platform languages, such as OpenCL, have allowed programmers to apply the power of GPUs to many computationally intensive problems. By November 2015, the fraction of the world's top 500 supercomputers that used GPU coprocessors/accelerators to boost their computational performance had risen to 21% (104/500).

Another new computing-related concept that came to the forefront during the sixth generation was *big data analytics*. The combined availability of large, inexpensive hard disk drives and fast, inexpensive networking technology enabled the construction of huge data storage repositories, which may take the form of direct-attached storage (DAS), network-attached storage (NAS), and/or storage area networks (SANs). These huge structures allowed for storage of tens or hundreds of terabytes or even petabytes of information, a.k.a. "big data" sets—often poorly structured or unstructured and therefore beyond the ability of traditional relational databases and software data analysis tools to handle. These massive treasure troves of data have been and are being generated constantly from a variety of sources, such as point-of-sale systems, web server logs, social media sites, sensor networks, and, of course, those very mobile devices that billions of us carry just about everywhere we go.

Of course, the point of collecting such vast data sets is to analyze them for useful information or insights—anything from a grocery store chain custom-selecting coupons to print for individual shoppers, to a political candidate's campaign team analyzing voting patterns, to financial analysts looking to predict stock market trends, to government agencies attempting to detect terrorist activity, to meteorologists investigating El Niño, and on and on. To enable timely and useful analysis of big data, new tools that leverage distributed processing power and storage have had to be developed. As this text was being written, one of the most popular platforms for big data analytics was *Hadoop*, the Apache Software Foundation's Java-based, open-source software framework (developed from Google's MapReduce programming model) for distributed data processing. Tools based on this and other technologies will continue to evolve as the data sets grow in size and complexity.

## 1.3 Types of computer systems

Many types of systems have been developed over the 75-plus year history of modern computing machines. In the previous section, we encountered several classifications of computers based on their size, speed, and intended applications. The first large computers, the descendants of which are still used in demanding business, big data analytics, and other applications, are referred to as *mainframe* computers. (The name dates back to the days when the processor took up an entire large frame, or mounting rack, stretching from the computer room's floor to the ceiling.) The somewhat less powerful (and considerably less expensive) machines that came to be used by small groups or individuals were smaller than mainframes and picked up the tag of *minicomputers*. (As previously mentioned, minicomputers as they were known in the 1960s, 70s, and 80s are now essentially extinct.) *Microcomputer* is a very general classification; literally, it refers to any machine using a microprocessor (or single-chip CPU) as its main processor. Because this is true of virtually all computers of the 21st century, the term has largely lost the ability it once had to distinguish between different classes of systems.

Common types of systems in today's computing world—all of which are microcomputers according to our definition of the term—include *personal computers* (both desktop and portable machines, including laptops, notebooks, and netbooks); scientific and engineering *workstations*, which include high-performance CPUs and GPUs to support applications such as CAD and data visualization; *mobile* computers, such as tablets, smartphones, smart watches, etc., which are designed to be easily carried or even worn by the user; *microcontrollers*, which are embedded inside other products, including appliances, consumer electronics, automobiles, and so on; *game consoles*, which are specialized machines designed to interact in real time with users and provide realistic virtual environments for game play; *digital signal processors* (DSPs), which are used to process digitized audio, video, and other analog information; and *supercomputers*, the most powerful machines of all, used to perform the intensive "number-crunching" needed to solve large-scale scientific and engineering problems.

As an alternative to classifying systems based on the descriptions above, it is also sometimes useful to group them not by size, speed, or application, but by their underlying architectural design. The most fundamental distinction is between systems that have only one central processing unit and those that use multiple (or *parallel*) processors. Individual CPUs may be described as having a Princeton or Harvard architecture; parallel systems may be classified in several ways depending on the number and nature of the processing units and the means used to connect them. Some experimental or special-purpose systems defy attempts at classification. We will briefly look at some of the architectural classifications of single and parallel systems in the following sections, then examine them in more depth in the chapters to follow.

## 1.3.1 Single processor systems

The original digital computer systems ran programs on a single CPU (or simply "processor"). This is still true of many systems today, especially those in which economy of implementation is more important than high computational performance. Although many enhancements have been made over the years to the original idea, almost every processor available today is descended from, and owes much of its architecture to, the original von Neumann architecture developed for the EDVAC and EDSAC back in the 1940s. Thus, in order to understand the operation of present and future architectures, it is important to first look back and consider the characteristics of the first practical modern computing system: the von Neumann machine.

The von Neumann architecture, also known as the Princeton architecture because John von Neumann was a researcher at Princeton University's Institute for Advanced Studies, was the first modern design for a computer based on the stored program concept originally developed by Babbage. The machine envisioned by von Neumann's team was very simple by today's standards, but its main features are readily recognizable by anyone who has studied the basic organization of most modern computer systems. The block diagram of the von Neumann computer shown in Figure 1.1 clearly shows input and output devices as well as the single memory used to store both data and program instructions. The control unit and arithmetic/logic unit (ALU) of the von Neumann machine are key parts of the CPU in modern microprocessors (internal registers were added later to provide faster storage for a limited amount of data).

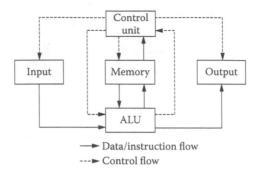

*Figure 1.1* Architecture of the von Neumann computer.

The goal of the original von Neumann machine was to numerically solve scientific and engineering problems involving differential equations, but it has proven remarkably adaptable to many other classes of problems from weather prediction to word processing. It is so versatile that the vast majority of computer systems today are quite similar, although much faster. The main factor distinguishing the von Neumann architecture from previous machines, and the primary reason for its success and adoption as the dominant computing paradigm, was the stored program concept. Because the machine receives its instructions from a (easily modified) program in memory rather than from hard wiring, the same hardware can easily perform a wide variety of tasks. The next four chapters take a much more detailed look at each of the major subsystems used in von Neumann–type machines. Memory is discussed in Chapter 2, the CPU in Chapters 3 and 4, and I/O in Chapter 5.

The *von Neumann machine cycle* is illustrated in Figure 1.2. This is the process that must be carried out for each instruction in a stored computer program. When all the steps are complete for a given instruction, the CPU is ready to process the next instruction. The CPU begins by fetching the instruction (reading it from memory). The instruction is then decoded— that is, the hardware inside the CPU interprets the bits and determines what operation needs to be done; the address (location) of the operand(s) (the data to be operated on) is also determined. The operands are fetched from registers, memory locations, or an input device and sent to the ALU, where the requested operation is performed. Finally, the results are stored in the specified location (or sent to an output device), and the processor is ready to start executing the next instruction. We will examine the details of the hardware required to carry out the von Neumann machine cycle when we discuss the basics of CPU design in Chapter 3.

The *Harvard architecture* is an alternative computer organization developed by Howard Aiken at Harvard University and used in the Mark-I and

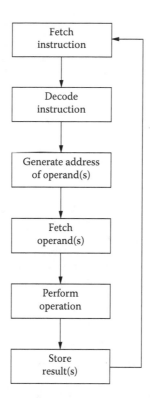

*Figure 1.2* The von Neumann machine cycle.

Mark-II machines. It aims to avoid the "von Neumann bottleneck" (the single path to memory for accessing both instructions and data) by providing separate memories and buses (access paths) for instructions and data (see Figure 1.3); thus, instructions may be fetched while data are being read or written. This is one of many examples in the field of computer systems

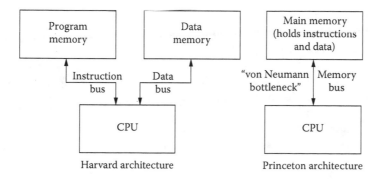

*Figure 1.3* Harvard architecture versus Princeton architecture.

design in which an increase in implementation cost and complexity may be justified in order to obtain a corresponding increase in performance.

Many modern systems use this type of structure, although it is very rare to see a "true" Harvard machine (with completely separate memories for code and data). Rather, in today's computing systems, it is common to employ a *modified Harvard architecture* in which main memory is "unified" (contains both code and data just like machines based on the Princeton architecture), but separate *cache* memories (discussed in Section 2.4) are provided for instructions and data. This model can achieve virtually the same performance advantage as the original Harvard architecture, but simplifies the overall design and operation of the system. (There are times when programs themselves must be manipulated as data, e.g., when compiling or assembling code; this would be much more cumbersome with completely separate memories.) With today's emphasis on raw microprocessor speed and the use of cache memories to bridge the CPU–memory speed gap, the modified Harvard architecture has become very widely used. It is particularly appropriate for modern, pipelined RISC (and post-RISC) architectures, which are discussed in Chapters 3 and 4.

## 1.3.2   Parallel processing systems

Systems that include more than one processor are considered to be *parallel* systems. Here the term is not used in its strict geometrical sense (coplanar lines that do not intersect), but rather to describe two or more pieces of hardware that work together, being simultaneously engaged in the same (or related) tasks. They are "parallel" in the sense of being independent but going in the same direction. Perhaps a more accurate description of such a system would be "concurrent" or "cooperative" processing, but the use of the term *parallel processing* is long established and unlikely to change. The first parallel processing systems were produced in the late 1950s; they included the Univac LARC, the Burroughs D825, and the IBM Sage machines.

There are many types of high-performance computer architectures, most of which are parallel to some greater or lesser extent. Some parallel systems have only a few processors, for example, two or four; others are "massively parallel," employing hundreds or thousands of CPUs. There are *multiprocessor* systems, in which several processors share access to main memory modules, and *cluster* (or *multicomputer*) systems, in which processors communicate by sending messages over a network rather than by sharing memory. (Modern *multicore* CPU chips can be considered single-chip multiprocessors.) Some high-performance systems (such as the Cray and NEC systems mentioned in Section 1.2.5) operate on *vectors* of data in a heavily pipelined CPU; a closely related class of machines, known as *array processors*, use a large set of processing elements to execute the same instruction on all elements of large arrays or matrices at once. All of these high-performance and parallel architectures are discussed in greater detail in Chapter 6.

### 1.3.3   Special architectures

Some present and possible future computer systems are not easily classified as Princeton or Harvard architectures, nor do they resemble any of the parallel systems mentioned in the previous section and described in Chapter 6. The uniqueness of such architectures means they are mostly used in special-purpose machines rather than for general computing applications. Some of these special architectures include *dataflow* machines (discussed in Section 7.1), which avoid the sequential programming used in von Neumann machines; *artificial neural networks* (Section 7.2), which are modeled on the human brain and nervous system; and *fuzzy logic* architectures (Section 7.3), which operate on logical values with a probability of being true or false rather than the absolute binary values 1 and 0.

Researchers are working to develop new computing devices based on the principles of quantum physics rather than Boolean algebra. These machines, if and when they prove to be practical, would use quantum effects to simultaneously perform vast numbers of calculations. Thus, quantum computers of the future may be orders of magnitude more powerful than the fastest supercomputers of today and may possibly render other computer architectures obsolete, at least for certain applications. Only time will tell whether the venerable von Neumann and Harvard architectures and their parallel descendants will persist or go the way of horse-drawn carriages. The possible characteristics of future machines based on quantum principles will be discussed more fully in Section 7.4.

## 1.4   Quality of computer systems

This section and the ones that follow deal with three concepts that are mentioned frequently with regard to computer systems and that influence their commercial success: *quality, cost,* and *performance*. Of course, high quality and high performance are good; no one wants to buy a system that is low quality or that does not perform well. Conversely, lower cost is good as long as it does not compromise quality or performance too much. But what do these terms really mean? The main goal of the rest of this chapter is to get the reader to think about what the terms *quality, cost,* and *performance* mean in the context of a computer system. What does it mean to say that one system has higher performance than another; by what yardstick do we measure performance? What, besides just the price tag on the system, could be considered cost factors? What does it mean to say that one architecture, or one physical implementation of an architecture, has higher quality than another? What do we mean by the quality of computer systems, and might different people have different definitions of it? What are some aspects of quality? It is difficult to get people to agree on what is a good computer, a fast computer, or even an inexpensive computer, but we will explore possible answers to these questions.

## 1.4.1    Generality and applicability

Every type of computer system one might design, buy, or use is designed with some application or applications in mind. Some computers—for example, most desktop or notebook personal computers—are designed with the idea of doing a variety of computing tasks at different times. You may use the system at one moment to browse the Web, then to balance your checkbook, then to draft a letter or memo, and then to run a simulation for a physics class or do a statistical analysis for a psychology experiment, or some other application. We say that the PC you use to do all these things is a *general-purpose* computer system or that it has high *generality*. Unfortunately, a computer that can do a variety of things reasonably well probably is not exceptionally good at any of these tasks. As the old saying goes, "jack of all trades, master of none."

Special-purpose machines, however, are designed to do one type of task very well. Often such machines are no better than average, and perhaps even very poor or inefficient, at doing other tasks. Microcontrollers, such as Freescale's ColdFire series, are very good for embedded control processors in microwave ovens, automobile transmissions, and the like but would be abysmal at floating-point number crunching or even word processing. IBM's mainframe computers are good for processing queries of large databases but not so good for real-time control. Cray supercomputers are excellent at large-scale scientific computations but probably wouldn't run an average-sized spreadsheet any better than an Apple MacBook Air (maybe not as well). If you need to do one task very well, however, and are willing to pay for performance, a special-purpose machine that is highly *applicable*—that is, tailored to your specific application—is often the best way to go.

By now the reader has probably concluded that generality is a somewhat problematic aspect of quality in that it is not always good, nor is it always bad. The breadth of applications to be run on a particular system determines how general its architecture and implementation should be. A more general machine is more flexible, providing reasonable performance in many scenarios. Generality affects cost: A machine that appeals to a wider audience of users is more marketable. Its design cost can be spread over more customers, and economies of scale make implementing it less expensive; thus, each machine can cost less. However, a more general architecture leads to a more complex implementation—one in which the designers try to do a little of everything. This may have a detrimental effect on quality as the more complex a system is, the more things can go wrong. The main trade-off is performance in a specific area versus performance on a general mix of applications. One should always keep this in mind when specifying or purchasing a system.

## 1.4.2   Ease of use

*Ease of use* is an aspect of quality that is self-explanatory except for the question of whose ease of use we are discussing. From an architectural standpoint, ease of use is referenced from the system programmer's point of view rather than that of an applications programmer or the end user of the system. In other words, ease of use and user-friendliness are two different concepts. The systems programmers who write operating systems, compilers, linkers, loaders, and so on, need to know and use the details of a computer architecture in order to extract the maximum performance from the system. If these people do their jobs well, the end users and even the applications programmers do not need to know a great deal about the details of the underlying architecture in order to realize good performance from the system. They can just write, or run, a high-level language program and let the compiler and operating system optimize for performance. However, the job the systems programmers are able to do, and thus the performance attained by the user's applications, is influenced by the machine's design and implementation. Some machines are easier to extract performance from than others. Does the architecture make it easy to optimize applications, or do applications have to fight the architecture to do what they need to do? Intel's legacy IA-32 (x86) architecture, for example, has been a challenge to systems programmers for more than 35 years. Excellent operating systems and compilers are available for it, but there is some question as to whether they would have been even better, and available sooner, if the x86 architecture had more ease of use.

## 1.4.3   Expandability

*Expandability* answers the question, "How easy is it to increase the capabilities of an architecture?" Over time, newer and bigger applications will always make new demands on computer architectures and the systems built using those architectures. Will system designers be able to respond to these demands within the confines of the architectural specifications? Can performance be scaled up to increase capability or down to save money? One might ask a number of expandability-related questions: Is the memory size specified explicitly, or is it left to the system designer? Can memory system performance enhancements, such as the use of interleaving, cache size/organization, and memory management strategy, be changed to fit changing circumstances, or are there hard architectural constraints? How easy or difficult (or impossible) is it to add CPUs in such a way that they can be utilized by programs?

Some architectures, by the way they are defined, are difficult to expand to higher levels of performance. In that case, either higher performance is difficult to realize, or compatibility (see Section 1.4.4) must be sacrificed. Other architectures are based on more flexible specifications that allow for a wide range of implementations with different performance characteristics. Oracle's (formerly Sun Microsystems') scalable processor architecture (SPARC) is one example of an architecture that has proven itself very expandable because it was designed that way. The architects' original idea, which panned out very well, was to have a wide variety of price and performance in computers available sharing a common architecture, and thus compatible with one another.

### 1.4.4   Compatibility

*Compatibility* is another important aspect of architectural quality. Strictly speaking, one could define compatibility as the ability of different computers to run the same machine language or object code programs. The physical implementation of two systems may be somewhat or even completely different; even some architectural elements, such as the number and type of buses, levels of the memory hierarchy, and exception handling techniques, may vary between them. The key point is that if the two machines have the same instruction set architecture (ISA), then they will be able to run the same software. (It may run much faster on one machine than the other, but both compatible machines understand the same machine language.)

As computer architectures evolve, they may exhibit different types of compatibility. Completely compatible computers are those that have identical instruction set architectures. Each machine understands and executes the exact same set of machine language instructions—no more, no less. This is usually because they are implemented by the same manufacturer at about the same time although sometimes one manufacturer goes out of its way to maintain compatibility with another's architecture. Advanced Micro Devices' (AMD's) "cloning" of Intel x86 compatible microprocessors is an example of essentially complete compatibility between different manufacturers' implementations of an architecture.

The most usual type of compatibility, seen as new features and enhancements are added to an architecture over time, is *upward* compatibility. Upwardly compatible computers implement the complete original architectural specification plus additional instructions, addressing or operational modes, and so on. A series of computers, each compatible with its predecessors, is known as a computer *family*. The members of the family are either completely or upwardly compatible. In upwardly compatible computer families, the newer machines can run the programs written (or compiled) for previous models without modification. However, these legacy programs do not take advantage of the features available only on the newer machines. An Intel Core-class machine, for example, will run executable programs compiled for a Pentium, 486, 386, or earlier processor, all the way back to the 8086/8088 CPUs of the late 1970s. Only programs optimized for the additional features of the newer processors will take advantage of all they have to offer, however. Earlier members of a family are not generally *downwardly* compatible with the later machines; in other words, you can run Intel 286 programs on a Pentium processor, but you could not run programs explicitly written for a Pentium on a 286-based machine.

*Forward* compatibility is upward compatibility carried to an extreme. Forward compatibility means that a later family of machines retains the ability (either through direct hardware implementation or emulation of the instruction set architecture) to run programs written and compiled for a previous family. Perhaps the best historical example of forward compatibility is the several families of mainframe computers built by IBM during the 1960s through the 2010s. The zEnterprise family machines being sold as this chapter is being written can still run, without recompilation, programs developed for the IBM System/360 and 370 families in the 1960s and early 1970s.

Is compatibility good or bad? As with generality, there is no universal answer to that question. Rather, it depends on one's circumstances and reasons for buying a particular system. If an organization is heavily dependent on the continuing use of legacy code that was written for an older architecture, compatibility is a highly desirable—perhaps even essential—attribute of a new system. However, compatibility has costs associated with it. Retaining the ability to execute legacy code while designing a new instruction set architecture significantly complicates the design process. For example, DEC's 32-bit VAX architecture was made more complex, and likely more expensive, by the decision to implement a PDP-11 compatibility mode. In addition to being more costly, a complex architecture is more likely to harbor hidden pitfalls or even outright flaws; simply put, the more computer engineers try to do in a given design, the more likely they are to make mistakes. Finally, compatibility may be at cross purposes with ease of use. Motorola's 68000 family architecture (which made a complete break with its earlier 6800 series processors) was widely considered to be "cleaner" than that of Intel's x86 chips, which retained a strong similarity to, if not complete compatibility with, its 8-bit 8080 and 8085 CPUs.

## 1.4.5   Reliability

In order to get useful work done by a computer system, it has to be up and functioning. Reliability, simply put, is the answer to the question, "Does the system stay up?" How often does it crash, and when it does, how difficult is it to get it going again? Our natural inclination, when studying this subject, is to concentrate on the reliability of the computer hardware. This would seem to be a question of implementation rather than architecture; after all, when considering the reliability of a car, we might reasonably care more about how well the assembly line workers who put the car together did their jobs than how well the automotive designers did theirs. However, reliability is also influenced by the architectural design. A well-designed car will be easier to assemble and to perform maintenance on, and a poorly designed one will be more subject to errors in those operations. (The better-designed vehicle will likely also be easier to drive.) A well-designed computer system will be implemented with parts that do their jobs and (both logically and physically) fit and work together well. The architectural design of its instruction set will also give rise to software that works well with the hardware, creating a seamless system that gets the job done without crashes. Ultimately, reliability is not just a feature or quality of the hardware, but of a complete system that includes software as well.

Reliability, although we mention it as a separate aspect of quality, is related to some of the other quality topics we have just discussed. Reliability (as perceived by the user) may be partly a function of compatibility. If two supposedly compatible architectures or machines are in reality not 100% compatible, some software may run properly on one machine but not the other; this will be perceived by the user as a defect. PCs using AMD processors, for example, might be judged inferior in reliability if, due to some subtle compatibility problem, they could not run an application that runs properly on a machine with an Intel processor even though the design and implementation of the AMD chip may be otherwise equal (or even superior) to Intel's. Ease of use can also affect reliability. If a machine's instruction set architecture is not clean and easy for systems programmers to work with, operating systems and compilers are more likely to contain subtle bugs that may show up down the line as an unstable system or application.

## 1.5   Success and failure of computer architectures and implementations

Why do some computer architectures succeed, sometimes remaining popular for years beyond the wildest expectations of their designers, and others fail, falling quickly out of use if they are adopted at all? The quality considerations we introduced in the previous section are part of the story, but by no means are they the only or even the major reasons why architectures succeed or fail. In this section, we will look at several factors that have an effect on the success of computer architectures.

## 1.5.1 Quality and the perception of quality

All of the quality factors we discussed may affect the success of a computer architecture. At least as important as the actual quality of an architecture (and its implementations) is the perception of quality on the part of those who buy and use computer systems. If users perceive the quality of one product as being higher than another, then all things being equal they will buy and use that first product. In many cases, they will even pay a premium for the system they believe has higher quality.

To return to our automotive analogy, some years ago consumer advocates (and eventually drivers) determined that certain vehicle manufacturers, for example, Toyota and Honda, were producing higher quality cars than some other manufacturers. This led other companies to follow their lead and produce higher quality cars and trucks also. Today, one could argue that the differences in quality between various makes of autos are much smaller, and not always in favor of the manufacturers that were once at the top of the list. However, those companies still have a widespread reputation for quality, and many people still perceive their products to be higher in quality than others. As a result, their cars continue to sell well, and drivers often pay more for those vehicles than comparable ones from other companies. The reality of present-day quality is, in this case, less important than the perception of quality from historical experience. The same sort of thinking is prevalent among buyers of computer systems and explains why people will pay more for, for example, an IBM system than a comparable machine from another manufacturer, or why some avoid PCs with AMD processors even though their Intel compatibility problems are many years in the past. The perception of quality, or the lack thereof, is a major factor in the success or failure of computer architectures.

## 1.5.2 Cost issues

If all things are equal, or nearly so, perceived quality is a major factor in determining whether people will buy a given computer system. In the real world, individuals and organizations have finite resources, so cost is always a consideration. When we think of cost, we often (somewhat naively) consider only the initial cost of purchasing a system. It makes more sense to consider the overall cost of continually operating the system over a period of time. Many factors contribute to the monetary cost of operating a computer system, and other cost factors may not directly equate to money but represent some other expenditure of limited resources. We will consider some important aspects of the cost of computer systems below.

*Hardware costs* are what most people immediately think of when considering the purchase of a computer system. Admittedly, a system with too high an initial cost can price itself out of consideration regardless of its quality or other desirable attributes (recall our previous mention of the Apple Lisa). However, computer hardware continues to become less expensive over time, as it has for the past 70-plus years. The initial cost of the hardware (and even the cost of subsequent hardware upgrades) is often dwarfed by the cost of support contracts, software licensing, maintenance, support, and other items over the life of the system.

*Software costs* are an important part of the overall cost of any computer system. Licensing fees and upgrades for system software and applications often exceed the initial hardware cost and sometimes must be periodically renewed, whereas once one has paid for the hardware it is typically owned outright. It would seem that software costs would be independent of the hardware architecture, but this is not always the case. Applications and system software for widely used architectures are often less expensive than the same software packages tailored to less popular systems. Thus, a system that has higher quality and performs better may not only have a higher hardware cost, but its software may, in some cases, be more expensive as well. Conversely, in some cases, higher hardware cost may be offset by available less expensive software, or vice versa.

It would be good if computer hardware and software never stopped working, but anyone who has used computer systems knows that systems inevitably crash (some more frequently than others). Thus, no comparison of computer system costs is complete without taking into consideration not only the initial purchase or licensing price for the hardware and software, but also the *maintenance and support costs*. Just as some automobiles cost more than others to maintain, some computer systems break down more frequently than others, and the cost of the replacement parts (and the labor to install them) varies greatly. Again, architectures that are widely used tend to have lower maintenance costs (for replacement and upgrade parts and sometimes labor, too) than those that are sold in smaller volume. Support contracts are often available from a system's manufacturer (or a third party), but these are often prohibitively expensive for individuals or small organizations.

Not all costs are monetary—at least, not directly. Any system that makes use of scarce resources incurs a cost. In many situations, cost considerations may include a system's power consumption, its physical mass and volume, or other factors. Any time a computer is embedded in a battery-operated device, for example, power consumption is critical, perhaps sufficiently so that a trade-off of performance for longer battery life is desirable. Any system that is to be placed in a vehicle adds weight and thus affects fuel consumption; it also takes up space that could otherwise be used for passengers or cargo.

A perfect example of a situation in which nonmonetary costs are of paramount concern is a system that must go on board a spacecraft. The overall monetary cost of a space mission is so high that the incremental price of a computer system may well be negligible. However, given the physics of launching the craft into orbit (or even interplanetary space), every gram of mass saved may be worth its Earth weight, or more, in gold. Minimizing the physical volume of the system may also be a major consideration, and with operation depending on electricity from solar cells, batteries, or other low-current sources, low power consumption may be even more important. If one has to launch a space mission to repair a faulty system (or worse, lose the entire craft because it cannot be reached to make repairs), reliability (or the lack thereof) may be the biggest cost factor of all.

### 1.5.3   *Architectural openness, market timing, and other issues*

Another important factor affecting the success of computer architectures is the *openness* of their specifications. In other words, are those specifications published and available so that others can build compatible hardware and software? Although the instruction set architecture is almost always widely publicized, the details of a particular hardware system's architecture and its implementations are sometimes held closely by the manufacturer to thwart competition. People want to be able to buy standard hardware and software cheaply, so although some "closed" or proprietary architectures have been very successful, the trend (especially in recent years) is for systems with open specifications to do well in the market. This principle was evidenced decades ago in consumer electronics by the commercial success of widely produced VCRs using the open VHS format, adopted by many manufacturers, over the technically superior Beta machines produced by only one or two manufacturers.

There is perhaps no better illustration of this phenomenon in the computer systems arena than the PC versus Macintosh personal computer wars that have been going on for more than 30 years. Hardly anyone will argue that the original IBM PC was as architecturally elegant, as technically advanced, or even as physically attractive as Apple's Macintosh. Most of the Apple specifications, however, were proprietary, while the PC's were open and available and became industry standards. As a result, PC-compatible "clones" were—and still are—made by a large number of manufacturers. Competition drove PC quality up and prices down, while Macs (and Macintosh parts) remained more expensive and less readily available for years. It also produced technical advances that allowed the performance of PCs to catch and even surpass that of Macs.

(The differences are much less pronounced since Apple adopted the Intel x86 architecture for Macs in 2006, replacing the PowerPC and 68000 family processors used in earlier models.) Other open architectures, such as Sun's SPARC architecture, have also done well in the marketplace. With the increasing acceptance of open source software by the computing community, open standards for hardware architectures may be even more important to their success in the future.

Stand-up comedians say that the most essential quality of good comedy is timing. The same could be said of many other areas of endeavor, including the marketing of computer systems. An architecture that comes to market before its time—or worse, after it—may not succeed even though it and its implementations are high in quality and low in cost. Conversely, inferior or difficult-to-use architectures have sometimes prevailed because they were in the right place at the right time. Many people have lamented IBM's choice of the Intel x86 architecture for its PCs—a choice that has made Intel the dominant manufacturer of microprocessors for more than 35 years. Intel's chief competitor in the late 1970s and early 1980s, when PCs first hit the market, was Motorola's 68000 family of CPUs. The 68000 architecture was arguably better than the 8086 and its successors in terms of expandability and ease of use (and also already had a built-in, compatible upgrade path to a fully 32-bit implementation), but it suffered from poor timing. Intel released its chips several months prior to Motorola's debut of the 68000. It was precisely during this time frame that IBM's engineers had to make the choice of a CPU for their upcoming PC.

Faced with the decision between Intel's available, working silicon or Motorola's promise of a better and more powerful chip in a few months, IBM chose to go with the "bird in the hand." Intel, which might not have survived (and certainly would have been much less prosperous) has been in the driver's seat of microprocessor development ever since. Motorola survived (its 68000 family chips were used in Sun's pre-SPARC workstations as well as the first several generations of Apple Macintosh computers, and their descendants, marketed under the Freescale Semiconductor banner until that company was bought out in 2015, are still widely used in embedded control applications) but never had another opportunity to dominate the market. Nor has Intel always been immune to bad timing: its iAPX 432 processor, developed during the late 1970s, was a design so far ahead of its time (and the technology available to implement its advanced ideas in silicon) that it failed from being produced too soon. Sometimes an architecture that works reasonably well and is available at the right time is better than a wonderful design that takes too long to produce and get to market or an overly ambitious design that is too costly to produce, demands too much of implementation technology, or is just plain unappreciated.

## 1.6   Measures of performance

Computer system performance depends on both the underlying architecture and its implementation. It is clear that both the architecture and
implementation must be optimized in order to get the maximum performance from a system, but what do we mean by performance, and how
do we measure it? Is it possible for two computer manufacturers to each
claim that their system has higher performance than the other and both
be telling the truth (or both be lying)? Probably so; there are many units
for quantifying computer system performance and many techniques for
measuring it. As is the case with quality, several different aspects of computer system performance may contribute more or less to the overall performance of a given system when running a given application. A system
that performs well on a particular application or benchmark may perform
poorly on others. Beware of general claims: What counts is how the system performs on your application.

### 1.6.1   CPU performance

CPU performance is the first and unfortunately the only, aspect of performance many people consider when comparing computer systems.
Of course, CPU performance is important, but memory and I/O performance can be just as important if not more so. It is important to consider
all aspects of performance when comparing systems. Even if the I/O and
memory performance of competing systems are essentially equal and
CPU performance is the deciding factor, there are many ways to measure
(and mis-measure) it. In this section, we examine some of the ways to
quantify, and thus compare, the performance of modern processors.

Most of the traditional measures of raw CPU performance begin with
the letter M, standing for the Greek prefix *mega* (or one million). It is common to identify and compare microprocessors based on their megahertz
rating, which is no more or less than the CPU's clock frequency. One
megahertz (MHz) is one million clock cycles per second. Most CPUs in
use today have clock speeds that are multiples of 1000 MHz or 1 gigahertz (GHz), which is 1 billion clock cycles per second. Clock cycle time or
period, $t$, is the reciprocal of clock frequency $f$ (mathematically, $t = 1/f$), so
a 100-MHz processor has a clock cycle time of 10 nanoseconds (ns), and a
1000-MHz (1-GHz) processor has a clock period of 1 ns.

All else being equal, a higher megahertz (or gigahertz) number means
that a given processor is faster. However, all else is rarely equal. Clock
speed in megahertz or gigahertz only tells us how many CPU cycles
occur each second; it tells us nothing about how much work the CPU gets
done each cycle or each second. Different architectures or even different
implementations of the same architecture may vary greatly with regard to

internal structure and breakdown of computational operations. Thus, not only is it not fair to compare, for example, an Intel Core i7 with an Oracle SPARC T5 based on clock speed alone, but it is nearly as unfair to compare the Core i7 with a Core i5 (or two Core i7 chips built on different micro-architectures) solely on that basis. Even when we consider two systems with the same type of processor, the system with the higher megahertz or gigahertz rating is not necessarily faster. The intelligent buyer will take a 3.2-GHz system with 16 GB of RAM over a 3.3-GHz machine with only 4 GB of RAM.

A somewhat better measure of processor performance is its MIPS (millions of instructions per second) rating. It is better to compare CPUs by instructions, rather than clock cycles, executed per second because, after all, it is instructions that do the work. Some CPUs have higher throughput (instructions completed per clock cycle) than others, so by considering MIPS instead of megahertz, we can get a slightly better picture of which system gets more work done. MIPS as a measure of performance is not perfect, however; it is only a fair comparison if the CPUs have the same (or a very similar) instruction set. For example, any contemporary Intel or AMD x86-64 architecture CPU executes essentially the same instruction set, so it is reasonable to compare any of these processors using their MIPS ratings. However, the instructions executed by a processor designed using another architecture, for example, SPARC or ARM, are very different from x86 instructions, so MIPS would not give a meaningful comparison of an AMD or Intel-based processor versus one of those other CPUs. (For this reason, one sarcastic, alternative definition of the MIPS acronym is "Meaningless Indication of Processor Speed.")

Another caveat to keep in mind when comparing systems by MIPS rating is to consider what types (and mixture) of instructions are being executed to achieve that rating. Manufacturers like to quote the "peak" performance of their machines, meaning the MIPS rate achieved under ideal conditions, while executing only those simple instructions that the machine can process the fastest. A machine may have a very high peak MIPS number, but if the code it is executing is all NOPs (no-operation instructions), then that number does not mean very much. If arithmetic computations are being tested, the peak MIPS rating may include only addition and subtraction operations, as multiplication and division usually take longer. This may not be representative of many real-world applications. A more realistic comparison would involve the MIPS rate sustained over time while executing a given mix of instructions, for example, a certain percentage each of memory access, branching, arithmetic, logic, and other types of instructions that would be more typical of real programs. Of course, it would be even better to compare the MIPS rate achieved while running one's actual program of interest, but that is not always feasible.

MIPS is normally a measure of millions of *integer* instructions that a processor can execute per second. Some programs, particularly scientific and engineering applications, place a much heavier premium on the ability to perform computations with real numbers, which are normally represented on computers in a floating-point format. If the system is to be used to run that type of code, it is much more appropriate to compare CPU (or *floating-point unit* [FPU]) performance by measuring millions of floating-point operations per second (MFLOPS) rather than MIPS. (Higher-performance systems may be rated in gigaflops [GFLOPS], billions of floating-point operations per second, or teraflops [TFLOPS], trillions of floating-point operations per second.) The same caveats mentioned with regard to MIPS measurements apply here. Beware of peak (M/G/T) FLOPS claims; they seldom reflect the machine's performance on any practical application, let alone the one you are interested in. Vector- or array-oriented machines may only be able to approach their theoretical maximum FLOPS rate when running highly vectorized code; they may perform orders of magnitude worse on scalar floating-point operations. The best comparison, if it is feasible, is to see what FLOPS rate can be sustained on the actual application(s) of interest, or at least some code with a similar mix of operations.

Many myths are associated with measures of CPU performance, foremost among them being that any single number—megahertz, MIPS, MFLOPS, or anything else—can tell you everything you need to know about which system is best. An even more fundamental misconception is that CPU performance alone, even if there were an exact way to measure it, is a valid comparison of computer systems. A good comparison must also consider memory and I/O performance. We will examine those topics next.

## 1.6.2   Memory system performance

Perhaps the most important factor affecting the overall performance of today's computer systems is the memory system. Because the CPU and I/O devices interact with main memory almost constantly, it can become a bottleneck for the entire system if not well designed and matched to the requirements of the rest of the machine. In general, the two most important characteristics of a memory system are its size and speed. It is important that secondary memory (typically disk drive[s]) be large enough to provide long-term storage for all programs and data, and likewise important that main memory be sufficiently large to keep most, if not all, of the programs and data that are likely to be in simultaneous use directly available to the processor without the need for time-consuming disk accesses. The main memory needs to be fast enough to keep up with the CPU and any other devices that need to read or write information. Although disk drives have no hope of reaching processor speeds, they need to be as fast as practicable, particularly in systems with large applications or data sets that must be frequently loaded. How do we quantify and compare the size and speed of computer memory systems? Let's take a look.

*Memory size* is the first, and often the only, measurement considered by novice computer users. To be sure, memory size is important, and even most naive users know that the more RAM or hard drive space one has, the better. Even people who don't know that a byte is a group of eight bits are usually aware that bytes, kilobytes (KB), megabytes (MB), and giga-bytes (GB) are the units of measurement for memory size (see Table 1.1). Disk drives, which are typically much larger than main memory, are often measured in terabytes (TB). In the future, as individual disk drives grow ever larger and networked arrays of storage devices become more and more common, users will have to become familiar with petabytes (PB) and exabytes (EB) as well.

*Memory speed* is at least as important as memory size. The pur-pose of memory is to store instructions and data for use by the CPU and to receive the results of its computations. A memory system that cannot keep up with the CPU will limit its performance. Ideally, the *cycle time* of memory (the time taken to complete one memory access and be ready for the next) should match the processor clock cycle time so that memory can be accessed by the CPU every cycle without wait-ing. (If the CPU needs to access an instruction plus an operand each clock cycle, a Harvard architecture can be used to avoid having to make memory twice as fast.) Synchronous memory devices, such as synchro-nous dynamic random access memory (SDRAM), typically are rated for their maximum compatible bus speed in megahertz or gigahertz rather than by cycle time, but it is easy to convert from one specification to the other by taking its reciprocal. *Access time* is another memory speed specification; it refers to the time required to read or write a single item in memory. Because many types of main memory devices require some

*Table 1.1* Units of memory storage

| Unit of storage | Common abbreviation | Size (decimal) based on SI prefixes | Nearest binary size (used for memory addressing) |
|---|---|---|---|
| Kilobyte | KB | 1 thousand ($10^3$) bytes | 1024 ($2^{10}$) bytes |
| Megabyte | MB | 1 million ($10^6$) bytes | 1,048,576 ($2^{20}$) bytes |
| Gigabyte | GB | 1 billion ($10^9$) bytes | 1,073,741,824 ($2^{30}$) bytes |
| Terabyte | TB | 1 trillion ($10^{12}$) bytes | $1.0995 \times 10^{12}$ ($2^{40}$) bytes |
| Petabyte | PB | 1 quadrillion ($10^{15}$) bytes | $1.1259 \times 10^{15}$ ($2^{50}$) bytes |
| Exabyte | EB | 1 quintillion ($10^{18}$) bytes | $1.1529 \times 10^{18}$ ($2^{60}$) bytes |

overhead or recovery period between accesses, cycle time may be some-what longer than access time.

Most semiconductor read-only memory (ROM) and dynamic random access memory (DRAM) devices typically used in computer main memory applications have cycle times significantly longer than those of the fastest CPUs. As we will see in Chapter 2, this speed gap between the CPU and main memory can be bridged to a considerable extent by using devices with shorter cycle times as a buffer, or cache, memory and/or by overlapping memory cycles in a technique known as memory interleaving.

Most semiconductor memories have constant access times (see the discussion of random access memories in Chapter 2). Secondary memory devices, such as disk and tape drives, have access times that are not only much longer, but variable. The time to read or write data from or to a disk, for example, includes several components: a small amount of controller overhead, the *seek time* required for the actuator to step the head in or out to the correct track, the *latency* or rotational delay required for the disk to get to the correct position to start accessing the information, and the *transfer time* required to actually read or write the data (mainly a function of rotational delay over the sector in question). The two predominant components of disk access time, namely the seek time and latency, vary considerably depending on the relative initial and final positions of the head and the disk, and thus usually are given as averages; best and worst case timings may also be specified. Average access times for disk storage are on the order of milliseconds, while access times for semiconductor memories are on the order of nanoseconds (roughly a million times faster). It is easy to see why disk drives are never used as main memory!

*Memory bandwidth*, or the amount of information that can be transferred to or from a memory system per unit of time, depends on both the speed of the memory devices and the width of the pathway between memory and the device(s) that need to access it. The cycle time of the memory devices (divided by the interleave factor, if appropriate) tells us how frequently we can transfer data to or from the memory. Taking the reciprocal of this time gives us the frequency of data transfer; for example, if we can do a transfer every 4 ns, then the frequency of transfers $f = 1/(4 \times 10^{-9}$ s$) = 250$ MHz or 250,000,000 transfers per second. To compute the bandwidth of the transfers, however, we need to know how much information is transferred at a time. If the bus only allows for 8-bit (or single byte) transfers, then the memory bandwidth would be 250 MB/s. If the memory system were constructed of the same type devices but organized such that 64 bits (8 bytes) of data could be read or written per cycle, then the memory bandwidth would be 2000 MB/s (2 GB/s).

### 1.6.3   I/O system performance

Just as there are many types of memory devices with different characteristics, there are many different types of input and output devices with
different purposes and properties. Some are used to interface with human
users, and others are used to exchange data with other computing hardware. The fundamental purpose of all I/O devices is the same: to move
data and programs into and out of the system. Because I/O devices need
to communicate data to and from the CPU or, in the case of large block
transfers, directly to and from main memory, their transfer rates (the number of I/O transfers completed per second) and bandwidth are important
performance characteristics.

I/O bandwidth (either for the entire I/O system or a particular I/O
device) is defined in essentially the same way as memory bandwidth: the
number of bytes (or kilobytes, megabytes, gigabytes, or terabytes) that can
be transferred per unit of time (per second or fraction thereof). The highest I/O bandwidth is usually achieved on direct transfers between main
memory and an I/O device, bypassing the processor. Bandwidth may be
specified as a peak rate (to be taken with the usual grain of salt assigned
to all peak performance figures) or, more realistically, as a sustained figure under given conditions (e.g., for a certain quantity of data or over a
certain period of time). The conditions are important because the effective
speed of I/O transfers often depends heavily on the size and duration of
transfers. A few large, block transfers of data are typically much more
efficient than a greater number of smaller transfers. However, very large
block transfers may overflow device buffers and cause the system to bog
down waiting on I/O. There is typically an optimum block size for data
transfers between a particular device and memory, which may or may not
correspond to the I/O characteristics of your application. Read the fine
print carefully.

I/O requirements vary greatly between applications—often much
more than do demands on the CPU and memory. Some applications
require almost no interaction with the outside world, and others (for
example, printer drivers and programs with extensive graphics output)
may be "I/O bound." Thus, although I/O performance is perhaps the most
difficult aspect of overall system performance to pin down, in some cases,
it may be the most important to investigate and understand.

### 1.6.4   Power performance

Everyone who purchases or builds a computer system wants its CPU(s),
memory, and I/O system to perform well with respect to the criteria
just discussed. However, performance always comes at a cost, and cost
has several aspects, as discussed in Section 1.5.2. In modern computer

systems, particularly at the high and low ends of the overall performance spectrum, power consumption (and the accompanying need to dissipate the heat generated by the operation of the system) has become one of the most significant costs. A wise design engineer will endeavor to minimize power consumption while targeting a system's computing performance goals, and a savvy buyer will not neglect to consider power requirements and thermal management when choosing a supercomputer or server—or, at the other end of the scale, a smartphone.

Power (P) is defined, from a physical standpoint, as the rate of doing work or the energy consumed per unit of time. The international unit of power is the watt (W), which is equivalent to one joule per second. In systems that operate on electrical power, which includes virtually all computers, the power drawn by a circuit depends on voltage (V, measured in volts [V]) and current (symbolized I, measured in amperes [A]). Note that the same SI order-of-magnitude prefixes used with other physical units apply to these quantities, so a very low-power circuit might operate on millivolts (mV, with "milli" denoting $10^{-3}$) while drawing microamperes ($\mu$A, where $\mu$ means $10^{-6}$) of current. Conversely, a large system might dissipate power in units of kilowatts or megawatts (thousands or millions of watts). In a direct current (DC) system (batteries and most computer power supplies deliver direct current to the components in a system), the relationship between these physical quantities is given by

$$P = VI$$

A current of one ampere flowing through a circuit with a potential of one volt will cause the dissipation of one watt of power. It can be seen from the above equation that, all else being equal, an increase in voltage or current (or both) will increase a system's power dissipation, and a decrease in either or both will reduce power dissipation. In the Complementary Metal Oxide Semiconductor (CMOS) circuit technology used to build microprocessor chips, current is mainly dependent on clock frequency; thus, overclocking increases the power required by a processor, and underclocking can be used to limit its power consumption. Reducing power by decreasing power supply voltage has been a continuing trend in microprocessor design. CPU chips of the 1970s and 80s generally required 5 volts for operation; by the 1990s, Intel's Pentium processors reduced this to 3.5 and then 3.3 volts; other manufacturers followed suit. Although most modern chips still use this 3.3–3.5 V range for input/output circuitry, operating voltages for the CPU cores themselves have been reduced even further. AMD's K6 processor (introduced in 1998) had a core voltage of just 2.2 volts; their latest (as of this writing) A- and FX-series chips as well as Intel's Core processors normally operate with $V_{CORE}$ in the range of 1.1–1.5 V although some versions can now run reliably at less than 1 volt.

It is in our interest to use such techniques to limit power consumption by the CPU (as well as other components in the system) from the standpoint of both providing energy to operate a given system (electricity is not free, and in mobile systems, battery life is a major concern) and removing the heat generated by its operation. From physics, we know that an electrical circuit neither creates nor destroys energy; it simply converts it from one form into another. Although in some cases, such as that of a light-emitting diode (LED) or audio speaker, electrical energy is converted into light or sound energy, in most computer systems, the vast majority of the electrical energy is converted into heat energy. Generally, this is done within some sort of confined space, whether it is the cramped interior of a smartphone case or the much larger volume of a server room. If we cannot adequately dissipate (disperse from the immediate environment) the heat produced by a computer system, its temperature will rise and eventually one or more components will overheat, causing it to fail.

High-performance, large-scale computers have consumed a great deal of electrical power ever since they were first built in the 1940s and 50s. The UNIVAC I, for example, operated at about 125 kilowatts; like many large systems that followed it, the UNIVAC required its own dedicated air-conditioning system that used chilled water and a powerful air blower, adding considerable cost to the overall system. In the decades since, as manufacturers have crammed ever-greater computing power into smaller and smaller spaces, the challenges of heat removal have only increased. By the 1970s, the Cray-1 supercomputer needed a patented liquid cooling system (using Freon refrigerant) to keep its logic circuits cool enough to operate. Its contemporary descendant, Titan (a Cray XK7 system housed at Oak Ridge National Laboratory, ranked number 2 on the November 2015 Top 500 list) draws a staggering 8.2 MW of power, but can be air-cooled as it is physically much larger than the Cray-1 (approximately 5000 square feet, or 465 square meters, in size). To be fair to Titan, however, it consumes only about 14.6 watts per each of its 560,640 computational cores (including CPUs and GPUs).

That last statement highlights the need for more "apples to apples" criteria for comparison between systems rather than just raw power consumption in watts. All else being equal, a larger and more computationally powerful system will tend to require more electrical power and generate more heat. So "watts per core" may be of more interest to a potential user (and by that standard, Titan is more than 7800 times as power-efficient as its 1970s ancestor). Perhaps an even better standard of comparison (because the purpose of a system is computation) would involve relating one of the computational performance measures described in Section 1.6.1 to a system's power requirements. Examples could include peak (or sustained) MIPS or MFLOPS per watt. For example, Titan's theoretical peak performance of 27,112.5 TFLOPS yields a maximum performance-to-power ratio of approximately 3.3 GFLOPS/W, and its measured performance of 17,590 TFLOPS on the LINPACK benchmark (to be described in the next section) gives the more realistic value of 2.14277 GFLOPS/W. Considering the Cray-1's paltry (by today's standards) ratings of about 1390 FLOPS/W peak and only 122 FLOPS/W while running LINPACK, we can see that computational power efficiency improved by considerably more than a millionfold over the 36-year time span between the introduction of the two machines.

Energy efficiency has become such an important consideration in modern computing that twice per year (since November 2007), a "Green500" list of the most efficient large computing systems has been published and released. As in the case of the biannual Top 500 list, each system is evaluated on LINPACK performance, but in this case, that value is divided (as we did above) by the system's power consumption. It is worth noting that the most powerful systems overall are not necessarily the most efficient and vice versa. Titan (#2 in the Top 500) is only number 63 on Green500; China's Tianhe-2 (#1 in total computational power) is even further down the Green500 list at number 90. The "greenest" machine on the November 2015 list, the Shoubu ExaScaler supercomputer at Japan's Institute of Physical and Chemical Research, had a rating of more than 7 GFLOPS/W (although it was only number 136 on the Top 500 list).

## 1.6.5 System benchmarks

We have noted that the best way to compare two or more computer systems is the most direct way possible: Run one's application of interest on each system and measure the time taken to compute a result (or perform typical tasks associated with the application). The system that gets that specific task or tasks done the fastest is the one that has the highest performance in the only sense that really matters. However, it is not always possible to run one's actual code on every system that merits consideration. One may not have physical access to the systems prior to making the decision, or it may not be possible to compile and run the program on every platform of interest, or one may just not have enough time to do so. One may even be specifying a system to run a future application that has not yet been developed. In this case, the next best thing is to identify a standard *benchmark* program, or suite of programs, that performs tasks similar to those of interest.

A benchmark is a program or set of programs chosen to be representative of a certain type of task. The idea is that if a system performs well on the benchmark, it will likely perform well on other applications with similar characteristics; if it performs poorly on the benchmark, it is not likely to be well suited for applications with similar demands.

Although we discussed CPU, memory, and I/O performance separately, it is extremely rare that any of these alone is the determining factor in performance on a real application. Real performance depends to some degree on the behavior of all three major subsystems, with the relative importance of each depending on the demands of a given application. Thus, many benchmark programs have been developed that exercise different parts of a system to different degrees and in different ways. Some benchmarks are very CPU-intensive, being dominated by the integer (or floating-point) performance of the system processor. Others are very memory-intensive, heavily I/O oriented, or well balanced between different types of operations. It is important to choose a benchmark that is representative of your intended application(s) if the results are to be meaningful. Using benchmark results to compare systems will never be quite as good as comparing performance directly on the application of interest, but if we choose the right benchmark, it is much better than using the manufacturers' peak MIPS or peak MFLOPS ratings, or other suspect measurements, to make the comparison.

A number of benchmarks have proven useful over the years and become classics, more or less de facto standards for evaluating the performance of systems intended for certain types of applications. Scientific application performance, for example, is often benchmarked using LINPACK, an adaptation of a Fortran linear algebra package (later translated into C) developed by Dr. Jack Dongarra of the University of Tennessee at Knoxville. The LINPACK benchmark solves a large system of simultaneous equations with single or double precision floating-point coefficients set up in large (100 by 100 or 1000 by 1000) matrices. There is also a version called HPLinpack (in which the HP stands for Highly Parallel) in which the size of the matrices can be adjusted in order to match the problem to the best capabilities of the machine under test. If one is interested in matrix manipulation of floating-point data, LINPACK provides a pretty good point of comparison. Another benchmark, Livermore Loops, is a set of 24 Fortran loops that operate on floating-point data sets. The length of the loops is varied so that the benchmark is run on short, medium, and long vectors. Introduced in 1970, the Livermore Loops benchmark has been used to evaluate the performance of several generations of supercomputers. Both LINPACK and Livermore Loops have historically been popular for evaluating vector-oriented machines (pipelined and/or highly parallel systems that perform operations on an entire vector, or row or column of a matrix, at once) because the matrix computations and loop operations can be vectorized by a good compiler to achieve top performance, so these benchmarks tend to show vector machines at their best. In modern systems, the highly parallel internal architecture of GPUs helps them excel on these same benchmarks.

If one is interested in floating-point performance in nonvector applications, particularly on machines that have scalar floating-point units, the venerable Whetstones benchmark may be useful. Developed in the early 1970s by Harold Curnow and Brian Wichmann, Whetstones was originally released in Algol and Fortran versions but was later translated into several other languages. The Whetstones benchmark produces system speed ratings in thousands of Whetstone instructions per second (KWIPS) or millions of Whetstone instructions per second (MWIPS). A similar benchmark without the emphasis on floating-point operations is known by the obvious pun Dhrystones. Originally written in Ada by Reinhold Weicker and later translated into Pascal and C, Dhrystones has been used since 1984 to measure performance on tasks using a mix of integer instructions. Dhrystones gives a better indication of general computing performance and has been particularly popular for evaluating UNIX systems. The benchmark produces performance ratings in Dhrystones per second; however, results are often converted to an approximate VAX MIPS rating by dividing a machine's rating by that of a DEC VAX 11/780, which was a popular minicomputer system (rated at 1 MIPS) at the time this benchmark was introduced.

Systems used in business transaction processing tend to make more intensive use of I/O than either floating-point or integer computations in the CPU. Thus, none of the previously mentioned benchmarks would be particularly appropriate for evaluating systems to be used in this type of application. Instead, systems are often compared using several benchmarks developed by the Transaction Processing Performance Council (TPC). The original TPC-A benchmark simulated online processing of debit and credit transactions in a banking environment; TPC-B was similar, but operated in batch mode instead of using a remote terminal emulator. Although these benchmarks are no longer in active use, their successor TPC-C (designed to address the shortcomings of TPC-A) models a business order processing and distributed warehouse environment and is still popular. Other currently (as of this writing) active TPC benchmarks include TPC-DI (data integration), TPC-DS (decision support/big data), TPC-E (online transaction processing for a brokerage firm), and TPC-H (decision support). None of these TPC benchmarks are particularly CPU intensive, but they do place heavier demands on a system's I/O and scheduling capabilities. Interestingly, the organization also has a TPC-Energy specification that outlines the approved method for inclusion of energy consumption metrics for systems running the various TPC benchmarks.

Perhaps the best known and most popular benchmark suites in recent years have been the ones developed by the Open Systems Group (OSG) of the Standard Performance Evaluation Corporation (SPEC), formerly known as the System Performance Evaluation Cooperative. The distinguishing feature of the SPEC benchmark suites is that they are composed of actual application code, not synthetic or artificial code (like Whetstones and Dhrystones) written just to exercise the system. The SPEC CPU benchmarks primarily evaluate the performance of the system processor and memory as well as the compiler; they are not very I/O intensive and make few demands of the operating system. SPEC CPU includes both an integer benchmark suite (SPECint) and a floating-point suite (SPECfp). Each suite consists of several programs (currently 19 floating-point applications written in Fortran, C, and C++ and 12 integer applications written in C and C++) with a result reported for each. The overall SPECfp or SPECint rating is a geometric mean of a given system's performances on each application.

Because SPEC requires system testers to report the results for individual applications as well as the composite ratings and because all the results are available to the public on SPEC's website (www.spec.org), anyone can compare systems of interest using the entire benchmark suite or any applicable subset of it. It is also worth noting that thorough documentation must be provided for each system tested. This includes not only the number, type, and clock frequency of system processor(s) and the amount and type of RAM installed, but also the details of cache memory, the exact operating system and compiler used (as well as the compiler optimizations switched on for each application), and the disk type and file system used. With all this information available, one can tell exactly what hardware and software are being evaluated and thus differentiate between apples-versus-apples and apples-versus-oranges comparisons.

The original SPEC CPU benchmarks were introduced in 1989; as computer systems have become more powerful, the SPEC suites have been updated with more challenging applications. This periodic revision prevents comparing systems with "toy" benchmarks. (Some of the original CPU89 programs would fit entirely in cache on modern systems.) SPEC CPU89 was followed by CPU92, CPU95, CPU2000, and the most recent generation, SPEC CPU2006. The development of its replacement, known for now as SPEC CPUv6 as it will be the sixth generation in the series, is underway. The rationale for periodically revising the benchmark suite was explained on SPEC's web page for CPU2000:

> Technology evolves at a breakneck pace. With this in mind, SPEC believes that computer benchmarks need to evolve as well. While the older benchmarks (SPEC CPU95) still provide a meaningful point of comparison, it is important to develop tests that can consider the changes in technology. SPEC CPU2000 is the next-generation industry-standardized CPU-intensive benchmark suite. SPEC designed CPU2000 to provide a comparative measure of compute intensive performance across the widest practical range

of hardware. The implementation resulted in source code benchmarks developed from real user applications. These benchmarks measure the performance of the processor, memory and compiler on the tested system.

Although SPEC's CPU2006 integer and floating-point CPU suites are its best known and most widely used benchmarks, SPEC has also developed other specialized test suites, such as SPECviewperf 12, which measures graphics performance; SPECjbb 2015, a benchmark for Java applications; SPEC SFS 2014, used to evaluate file servers; SPEC OMP 2012, which benchmarks the performance of shared-memory parallel processing systems; and several others. SPEC also created a package called SERT (Server Efficiency Rating Tool) that can be used to measure energy efficiency. All of these benchmarks embody SPEC's philosophy of being realistic, application-oriented, and portable to all platforms and operating systems, thus providing "a fair and useful set of metrics to differentiate candidate systems." Although no benchmark can perfectly predict performance on a particular real-world application, SPEC's openness, industry-wide acceptance, and continual development of new, relevant test suites make it likely that SPEC benchmarks will continue to help computing professionals choose systems for years to come.

## 1.7   Chapter wrap-up

An old saying goes, "The more things change, the more they stay the same." Perhaps no field exemplifies the truth of this aphorism more than computer systems design. Yes, much has changed about the way we build computers. Relays and vacuum tubes have given way to transistors and, ultimately, integrated circuits containing billions of tiny components. Magnetic drums, punched cards, and core memory have been replaced by high-speed hard drives, even faster flash drives, optical disk burners, and synchronous DRAM. Computer implementation technologies of all sorts have become orders of magnitude smaller, faster, and cheaper. Yet the basic ideas of computer architectural design have changed very little in decades. The original Princeton and Harvard architectures have been used with only relatively minor changes for more than 70 years. Index registers and addressing modes conceived in the 1950s are still in use today. The concept of virtual memory dates to the late 1950s, and cache memory has been around since the early 1960s. (Regardless of the time period, processors have always been faster than main memory.) Overlapping of operations in time by pipelining processors and interleaving memories are decades-old concepts. RISC architectures and superscalar execution? The CDC 6600 exemplified both concepts in 1964. Truly, although new implementation technologies arrive every year to dazzle us, there is little new under the sun, architecturally speaking. (A few notable exceptions will come to light later in this text.) This is why the study of historical computer architectures is still so valuable for computing professionals of the 21st century.

Another common thread, all through the modern history of computers, is that users have always sought the best and fastest computer system for their money. Obviously, the system that is currently the best or fastest for a given application (at a given price) has changed and will change, rapidly, over time. Even more fundamentally, the notions of what constitutes the best system, the means for establishing which is the fastest system, and the methods for establishing the overall cost of acquiring and operating a system have evolved over time. There are many aspects of system quality, and different ones are more or less important to different users. There are many ways to measure system speed that yield different results, some of which approximate the reality of a given situation more closely than others. Cost, too, has many aspects and modes of estimation (it is difficult to know the true cost of anything as complicated as a computer system), and although quality, performance, and cost are all important, other factors are sometimes just as crucial to the success or failure of an architecture in the marketplace (not to mention its desirability for a given application). In the end, each person who is tasked with purchasing or specifying a computer system is faced with a different set of circumstances that will likely dictate a different choice from that made by the next person. A thorough awareness of the history of computer architectures and their implementations, of important attributes of quality and cost, and of performance measures and evaluation techniques, will stand every computer professional in good stead throughout his or her career.

## REVIEW QUESTIONS

1. Explain in your own words the differences between computer systems architecture and implementation. How are these concepts distinct, yet interrelated? Give a historical example of how implementation technology has affected architectural design (or vice versa).
2. Describe the technologies used to implement computers of the first, second, third, fourth, fifth, and sixth generations. What were the main new architectural features that were introduced or popularized with each generation of machines? What advances in software went along with each new generation of hardware?
3. What characteristics do you think the next generation of computers (for example, 5 to 10 years from now) will display?

4. What was the main architectural difference between the two early computers ENIAC and EDVAC?
5. Why was the invention of solid-state electronics (in particular, the transistor) so important in the history of computer architecture?
6. Explain the origin of the term *core dump*.
7. What technological advances allowed the development of minicomputers, and what was the significance of this class of machines? How is a microcomputer different from a minicomputer?
8. How have the attributes of very high performance systems (a.k.a. supercomputers) changed over the third, fourth, fifth, and sixth generations of computing?
9. What is the most significant difference between computers of the past 10 to 15 years versus those of previous generations?
10. What is the principal performance limitation of a machine based on the von Neumann (Princeton) architecture? How does a Harvard architecture machine address this limitation?
11. Summarize in your own words the von Neumann machine cycle.
12. Does a computer system with high generality tend to have higher quality than other systems? Explain.
13. How does ease of use relate to user-friendliness?
14. The obvious benefit of maintaining upward and/or forward compatibility is the ability to continue to run legacy code. What are some of the disadvantages of compatibility?
15. Name at least two things (other than hardware purchase price, software licensing cost, maintenance, and support) that may be considered cost factors for a computer system.
16. Give as many reasons as you can why PC-compatible computers have a larger market share than Macs.
17. One computer system has a 3.2-GHz processor, and another has only a 2.7-GHz processor. Is it possible that the second system might outperform the first? Explain.
18. A computer system of interest has a CPU with a clock cycle time of 0.5 ns. Machine language instruction types for this system include integer addition/subtraction/logic instructions that require one clock cycle to be executed, data transfer instructions that average two clock cycles to be executed, control transfer instructions that average three clock cycles to be executed, floating-point arithmetic instructions that average five clock cycles to be executed, and input/output instructions that average two clock cycles to be executed.
    a. Suppose you are a marketing executive who wants to hype the performance of this system. Determine its peak MIPS rating for use in your advertisements.
    b. Suppose you have acquired this system and want to estimate its performance when running a particular program. You analyze

the compiled code for this program and determine that it consists of 40% data transfer instructions; 35% integer addition, subtraction, and logical instructions; 15% control transfer instructions; and 10% I/O instructions. What MIPS rating do you expect the system to achieve when running this program?

c. Suppose you are considering purchasing this system to run a variety of programs using mostly floating-point arithmetic. Of the widely used benchmark suites discussed in this chapter, which would be the best to use in comparing this system to others you are considering?

d. What does MFLOPS stand for? Estimate this system's MFLOPS rating. Justify your answer with reasoning and calculations.

19. Why does a hard disk that rotates at higher RPM generally outperform one that rotates at lower RPM? Under what circumstances might this not be the case?

20. A memory system can read or write a 64-bit value every 2 ns. Express its bandwidth in megabytes per second.

21. If a manufacturer's brochure states that a given system can perform I/O operations at 1500 MB/s, what questions would you like to ask the manufacturer's representative regarding this claim?

22. Fill in the blanks below with the most appropriate term or concept discussed in this chapter:

      _____ The actual, physical realization of a computer system as opposed to the conceptual or block-level design.

      _____ This was the first design for a programmable digital computer, but a working model was never completed.

      _____ This technological development was an important factor in moving from second-generation to third-generation computers.

      _____ This system is widely considered to have been the first supercomputer.

      _____ This early microcomputer kit was based on an 8-bit microprocessor; it introduced 10,000 hobbyists to (relatively) inexpensive personal computing.

      _____ This type of computer is embedded inside another electronic or mechanical device, such as a cellular telephone, microwave oven, or automobile transmission.

      _____ A type of computer system design in which the CPU uses separate memory buses for accessing instructions and data operands.

      _____ An architectural attribute that expresses the support provided for previous or other architectures by the current machine.

_____ A CPU performance index that measures the rate at which computations can be performed on real numbers rather than integers.

_____ A measure of memory or I/O performance that tells how much data can be transferred to or from a device per unit of time.

_____ A program or set of programs that are used as standardized means of comparing the performance of different computer systems.

# chapter two

# Computer memory systems

People who know a little bit about computer systems tend to compare machines based on processor speed or performance indices alone. As we saw in the previous chapter, this can be misleading because it considers only part of the picture. The design and implementation of a computer's memory system can have just as great, if not greater, impact on system performance as the design and implementation of the processor. Anyone who has tried to run multiple applications simultaneously under a modern operating system knows this. In this chapter, we examine important aspects of the design of memory systems that allow modern computers to function at peak performance without slowing down the CPU.

## 2.1 The memory hierarchy

Memory in modern computer systems is not one monolithic device or collection of similar devices. It is a collection of many different devices with different physical characteristics and modes of operation. Some of these devices are larger, and some are smaller (both in terms of physical size and storage capacity). Some are faster, and some are slower; some are cheaper, and some are more expensive. Why do we construct memory systems in this way? The answer is simple: because no memory device possesses all the characteristics we consider ideal. Every type of memory technology available has certain advantages or aspects in which it is superior to other technologies, but also some disadvantages or drawbacks that render it less than ideal, such that we cannot use it everywhere we need storage. By intelligently combining multiple types of memory devices in one system, we hope to obtain the advantages of each while minimizing their disadvantages. Before proceeding further, we should ask ourselves, "What would be the characteristics of an ideal memory device, assuming we could get one?" In the process of answering this question, we can gain a lot of insight into the design of computer memory systems.

### 2.1.1 Characteristics of an ideal memory

What would be the characteristics of an ideal memory device? As designers or at least potential users of a computer system, we could make a list of attributes we would like our memory system to have. Several come quickly to mind:

**Low cost:** Ideally, we would like memory to be free; failing that, we would like it to be as inexpensive as possible so that we can afford all we need. In order to make a fair comparison of cost between memory devices of different sizes, we generally refer to the price of memory per amount of storage. Once upon a time, it was common to refer to the cost of memory in terms of dollars per bit or byte of storage; now, it would be more appropriate to price memory in dollars per gigabyte or even dollars per terabyte. In any event, we would like this cost to be as low as possible while meeting any other requirements we might have.

**High speed:** Every type of memory has an associated *access time* (time to read or write a single piece of information) and *cycle time* (the time between repetitive reads or writes; sometimes, due to overhead or device recovery time, the cycle time is longer than the access time). Depending on the type of device, these times may be measured in milliseconds, microseconds, nanoseconds, or even picoseconds. The shorter the access and cycle times, the faster the device. Ideally, we would like to be able to store information, or access stored information, instantly (in zero time), but this is not possible in the real world. Practically, in order to keep our CPU busy rather than "sitting around" waiting, we need to be able to access information in memory in the same or less time that it takes to perform a computation. This way, while the current computation is being performed, we can store a previous result and obtain the operand for the next computation.

**High density:** An ideal memory device would have very high *information density*—that is to say, we would like to be able to store a great deal of information in a small physical space. We might refer to the number of gigabytes or terabytes that can be stored in a given area of circuit board space or, more properly, in a given volume (e.g., cubic inches or cubic centimeters). We cannot store an infinite amount of information in a finite amount of space or any finite amount of information in an infinitesimal space, but the closer a given memory technology can come to approximating this, the better we like it.

**Nonvolatile:** Many memory technologies are *volatile*; they require continuous application of power (usually electrical) in order to retain their contents. This is obviously undesirable, as power outages are an unavoidable fact of life (and according to Murphy's Law, they will always occur at the least desirable moment). Some types of memory, for example the dynamic RAM that is used as main memory in most computer systems, require not only continuous power but periodic refresh of the stored information. Such a memory is volatile in more ways than one. All else being equal, we would prefer to use a memory technology that is *nonvolatile*, meaning it maintains its stored information indefinitely in the absence of power and outside intervention.

**Read/write capable:** For maximum versatility, we would like to be able to store or retrieve information in memory at any time. Memory devices that allow the user to readily store and retrieve information are called *read/write memories* (RWMs). The less desirable alternative is a memory with fixed contents that can only be read; such a device is called a *read-only memory* (ROM). Of course, a write-only memory that allowed storage but not retrieval wouldn't make much sense; it would effectively be an information black hole. There are also some memory technologies that allow writes to occur, but in a way that is more costly (in terms of time, overhead, device life, or some other factor) than reads. We might refer to such a device, such as a flash memory, as a "read-mostly memory." Again, all else being equal, we would usually prefer a RWM over other types.

**Low power:** In an ideal world, memory would require no energy at all to operate; once again, that is not achievable with real technologies in the real world. Volatile memory devices require continuous application of power. Even nonvolatile memories, which can maintain their contents without power, require power for information to be read or written. Sometimes, this is a relatively minor consideration; in other applications, such as when heating is a problem or when a system must run off batteries, it is critical that our memory system consume as little power as possible. All else being equal, memory that consumes less power is always better (unless it is winter and your computer is doubling as a space heater).

**Durability:** We would like our memory system to last forever, or at least until the rest of the system is obsolete and we are ready to retire it. Based on historical data and knowledge of their manufacturing processes, memory device manufacturers may provide an estimate of the mean time between failures (MTBF) of their products. This is the average time a given part is supposed to last. (Keep in mind, however, that the life of any individual device may vary quite a bit from the average.) They may also express the expected lifetime of the product in other ways, for example, in terms of the total number of read and write operations it should be able to perform before failing. (For this information to be useful, one has to be able to estimate how frequently the device will be accessed during normal operations.) Durability may also be interpreted in terms of a device's ability to survive various forms of abuse, such as impact, temperature and humidity extremes, etc. In general, memory technologies that do not employ moving mechanical parts tend to last longer and survive more mistreatment than those that do.

**Removable:** In many instances, we consider it an advantage to be able to transport memory (and preferably its contents) from one computer system to another. This facilitates being able to share and back up information. In rare situations, for example, in which physical security of information (e.g., government or trade secrets) is extremely important, being able to remove memory devices may be considered undesirable. In most cases, however, it is a desirable feature, and in some cases, it is essential.

You can probably think of other desirable characteristics that a computer's memory system might ideally have, but even from the above list, it is obvious that no memory technology currently in use or likely to be developed in the near future has all of these ideal characteristics. In the next few pages, we explore some of the characteristics and limitations of popular memory technologies. Then, in the remainder of this chapter, we examine some of the techniques used in memory system design to maximize the advantages of each type of memory while minimizing or compensating for the disadvantages. The goal, of course, is a memory system that is fast, has high storage capacity, is readable and writable, maintains its contents under as many scenarios as possible, and yet is as inexpensive and convenient to use as possible.

## 2.1.2   *Characteristics of real memory devices*

Several types of memory devices are used in modern computer systems. The most popular types of memory are semiconductor chips (integrated circuits) and magnetic and optical media. There are several subtypes using each of these technologies, and each of these has some of the advantages mentioned in the previous section but also some disadvantages. As potential designers or at least users of computer systems, we need to be familiar with these memory technologies.

*Semiconductor memories* in general possess the advantage of speed. This is why the main memory space of virtually all modern computers is populated exclusively with semiconductor devices, and magnetic and optical devices are relegated to the role of secondary or tertiary (backup) storage. The CPU is built using semiconductor technology, and only a similar memory technology can keep up with processor speeds. In fact, not all semiconductor memories can operate at the full speed of most modern CPUs; this is why the vast majority of semiconductor main memory systems have an associated cache memory (see Section 2.4) made up of the very fastest memory devices.

The semiconductor memory technology with the highest information density is *dynamic random access memory* (DRAM). For this reason, because it is read/write memory, and because it has a relatively low cost per gigabyte, DRAM is used for the bulk of main memory in most computer systems. A DRAM device consists of a large array of capacitors (electrical devices capable of storing a charge). A charged capacitor is interpreted as storing a binary 1, and an uncharged capacitor indicates binary 0. Unfortunately, the capacitors in a DRAM device will discharge, or leak, over time; thus, to be able to continue to distinguish the 1s from the 0s and avoid losing stored information, the information must periodically be read and then rewritten. This process is called dynamic RAM *refresh*. It adds to the complexity of the memory control circuitry, but in general, this is a worthwhile trade-off due to the low cost and high storage density of DRAM.

Given the desired main memory size in most computer systems as compared to the amount of DRAM that can be fabricated on a single integrated circuit, DRAM is not usually sold as individual chips. Rather, several integrated circuits (ICs) are packaged together on a small printed circuit board module that plugs into the system board, or motherboard. These modules come in various forms, the most popular of which are known as *dual inline memory modules* (DIMMs) and *small outline dual inline memory modules* (SODIMMs). Some of these modules are faster (have lower access times and/or higher synchronous clock frequencies) than others, and different types plug into different size sockets (thus, it is important to buy the correct type for a given system), but they all use DRAM devices as the basic storage medium.

Although dynamic RAM offers relatively low-cost and high-density storage, in general it is not capable of keeping up with the full speed of today's microprocessors. Capacitors can be made very small and are easy to fabricate on silicon, but they take time to charge and discharge; this affects the access time for the device. The highest-speed semiconductor read/write memory technology is referred to as *static random access memory* (SRAM). In a SRAM device, the binary information is stored as the states of latches or flip-flops rather than capacitors. (In other words, SRAM is built in a very similar way to the storage registers inside a CPU.) SRAM is less dense than DRAM (it takes more silicon "real estate" to build a static RAM cell than a capacitor) and therefore is more expensive per amount of storage. SRAM, like DRAM, is a volatile technology that requires continuous application of electrical power to maintain its contents. However, because the bits are statically stored in latches, SRAM does not require periodic refresh. Contents are maintained indefinitely as long as power is applied. Compared to DRAM, SRAM circuitry requires more power for read/write operation, but some SRAMs, such as the Complementary Metal Oxide Semiconductor (CMOS) static RAM devices sometimes used to retain system settings, require very little current in standby mode and thus can maintain stored information for years under battery power.

Semiconductor read-only memories (ROMs), including *programmable read-only memories* (PROMs) and *erasable/programmable read-only memories* (EPROMs), are roughly comparable to SRAM in cost and density although they generally operate at DRAM speeds or slower. They are nonvolatile but have the major limitation of not being writable (although EPROMs can be reprogrammed in a separate circuit after erasure with ultraviolet light). Because they are not read/write memories, ROMs are only useful in limited applications, such as single-purpose embedded systems, video game cartridges, and the basic input/output system (BIOS) that contains the bootstrap code and low-level input/output (I/O) routines for most typical computer systems.

Semiconductor "read-mostly" memories include *electrically erasable programmable read-only memories* (EEPROMs) and their technological descendants, *flash memories*. These memories are nonvolatile, but unlike ROMs, they are rewritable in-circuit. Writes, however, can take significantly longer than reads to perform and in some cases must be done as "block" writes rather than individual memory locations. Also, these devices are more expensive than most other semiconductor memories and can only be rewritten a limited number (usually a few tens or hundreds of thousands) of times, so they are not suitable for populating the entire main memory space of a computer. Instead, read-mostly memories are typically used for special-purpose applications, such as digital cameras, portable thumb drives, hybrid drives, tablet computers, and smartphones.

*Magnetic memories* have been in use much longer than semiconductor memories—almost as long as there have been electronic computers. Mainframe computers of the 1950s often used rotating magnetic drums for storage; a few years later, magnetic *core memory* became the standard technology for main memory and remained so until it was replaced by integrated-circuit RAM and ROM in the 1970s. Magnetic core memory, like all magnetic memories, offered the advantage of nonvolatility (except in the presence of a strong external magnetic field). Access times were on the order of microseconds, however, and so this technology fell out of favor when faster semiconductor memories became cost-competitive. Another related (but slower) technology, *magnetic bubble memory*, was once thought ideal for long-term storage applications but could not compete with inexpensive disk drives, battery–backed-up SRAMS, and EEPROMs; it eventually died out. Ferroelectric RAM (FeRAM), another descendant of core memory, is still in production but has never caught on widely due to its much lower information storage density as compared with DRAM and flash memory.

Magnetic storage in most modern computer systems is in the form of disk and tape drives. Access times for magnetic disks are on the order of milliseconds or longer, so this technology is useful only for secondary storage, not main memory. Tape drives are even slower due to the frequent necessity of traversing a long physical distance down the tape in order to find the needed information. The chief advantages of magnetic memories, besides their nonvolatility, are very low cost per gigabyte of storage and extremely high information density (a hard drive can store a terabyte or more of data in a few cubic inches of space). Removable disks and tape cartridges (and some hard drives) also offer the advantage of portability.

Although magnetic memories are currently relegated to secondary storage applications, *magnetic RAM* (MRAM) is a developing memory technology that has the potential to eventually replace DRAM in main memory applications. MRAM operates on the principle of *magnetoresistance*, where an electric current is used to change the magnetic properties of a solid-state material. Pieces of this material are sandwiched between two perpendicular layers of wires. A bit is stored at each point where one wire crosses over another (see Figure 2.1). To write a bit, a current is passed through the wires; changing the polarity of the magnet changes the electrical resistance of the sandwiched material. Reading a bit is accomplished by passing a current through the wires connected to a sandwich and detecting its resistance; a high resistance is interpreted as a binary 1 and a low resistance as binary 0.

Because the bits are stored as magnetic fields rather than electrical charge, MRAM (like other magnetic memories) is nonvolatile. If it can achieve density, speed, and cost comparable to DRAM (no small feat, but a reasonable possibility), MRAM will enable the development of "instant-on" computers that retain the operating system, applications, and data in main memory even when the system is turned off. Several companies, including IBM, Honeywell, Everspin, and Cypress Semiconductor, have produced MRAM devices in limited quantities. However, perhaps due to continued high demand for DRAM and flash memory, manufacturers have been hesitant to commit resources (money and fabrication plants)

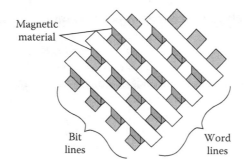

*Figure 2.1* Magnetic RAM construction.

to high-volume production of MRAM chips. If and when they are mass produced, MRAM devices could largely replace DRAM in computer main memory applications within a few years.

*Optical memories* are becoming more and more common—all the way down to low-end computer systems. Even inexpensive personal computers often have an optical drive that can at least read and often write various types of optical disks including *compact disks* (CDs), *digital versatile disks* (DVDs), and/or *Blu-ray disks* (BDs). Depending on type, an optical disk can store anywhere from several hundred megabytes of data (CD) to as much as 50 GB (Blu-ray) at a typical price of less than one dollar each. In addition to their low cost, optical disks offer most of the same advantages (portability, nonvolatility, and high density) as magnetic disks and also are immune to erasure by magnetic fields. They are much too slow to be used for main memory, however, and the writing process takes considerably longer than writing to a magnetic disk. Their most common uses are for distribution of software and digitally recorded audio/video and as an inexpensive form of backup/archival data storage.

### 2.1.3   Hierarchical memory systems

Having considered the characteristics of most of the available memory technologies, we can conclude that none of them are ideal. Each type of memory has certain advantages and disadvantages. It therefore makes sense to use a mixture of different types of devices in a system in order to try to trade off the advantages and disadvantages of each technology. We try to design the system to maximize the particular advantages of each type of memory while minimizing, or at least covering up, their disadvantages. In this way, the overall memory system can approximate our ideal system: large in capacity, dense, fast, read/write capable, and inexpensive with at least some parts being removable and the critical parts being nonvolatile. The typical solution is a computer system design in which a hierarchy of memory subsystems is made up of several types of devices. The general concept is depicted in Figure 2.2, and the specific names of the levels found in most modern computer systems are shown in Figure 2.3.

Notice that the upper levels of the hierarchy are the fastest (most closely matched to the speed of the computational hardware) but the smallest in terms of storage capacity. This is often due at least somewhat to space limitations, but it is mainly because the fastest memory technologies, such as SRAM, are the most expensive. As we move down the hierarchy, lower levels are composed of slower but cheaper and higher density components, so they have larger storage capacities. This varying capacity of each level is symbolized by drawing the diagram in the shape of a triangle.

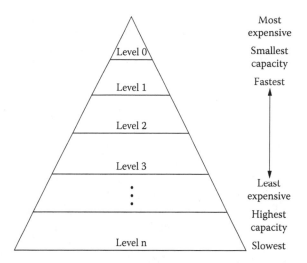

*Figure 2.2* Memory hierarchy (conceptual).

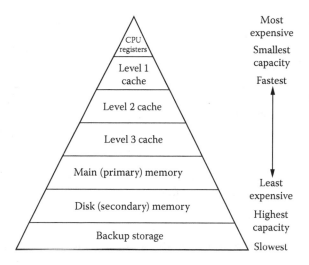

*Figure 2.3* Memory hierarchy (typical of modern computer systems).

Because the higher levels of the memory hierarchy have smaller capacities, it is impossible to keep all the information (program code and data) we need in these levels at one time. In practice, each higher level of the hierarchy contains only a subset of the information from the levels below it. The fundamental idea underlying the hierarchical memory concept is that we want to make as many accesses (as a percentage of the total) as we can to the upper levels of the hierarchy while only rarely having to access the lower levels, such that the resulting, overall memory system (taking into account all devices) approaches the speed of the highest levels while maintaining a capacity and cost per gigabyte approximating that of the lowest levels (the secondary storage devices). This requires a complex and well-thought-out design of which, for best acceptance, the details should be hidden from the end user. As much as possible, only the system designers should have to deal with the details of managing the memory system for optimal performance. However, if one is to be responsible for specifying computer systems whose performance is important or for developing code to run in such an environment, it is worthwhile to study the techniques used to optimize memory systems.

Optimizing the performance of memory systems has always been a big problem due to technological and cost limitations. For the vast majority of tasks, computer systems tend to require much more code and data storage than computational hardware; thus, it is generally not cost-effective to build the memory system using the same technology as the processor. Over the history of electronic computers, CPUs have increased in speed (or decreased their clock cycle times) more rapidly than memory devices. There has always been a performance gap between the CPU and main memory (and a much bigger gap between the CPU and secondary memory), and these gaps have only increased with time. Thus, design techniques that effectively close the gap between CPU and memory system performance are more important now than ever. The question has been, still is, and is likely to remain, "How do we fix things so that the memory system can keep up with the processor's demand for instructions and data?" The rest of this chapter examines several techniques that have been, and still are, used to help achieve this ever-challenging goal.

## 2.2   Main memory interleaving

We previously observed that the storage capacity of individual integrated circuit memory chips is such that a number of devices must be used together to achieve the desired total main memory size. This is unfortunate from a packaging and parts count standpoint, but does have some advantages in terms of fault tolerance (if one device fails, the others may still be usable) and flexibility of organization. In particular, constructing main memory from several smaller devices or sets of devices allows the designer to choose how the addressed locations are distributed among the devices. This distribution of memory addresses over a number of physically separate storage locations is referred to as *interleaving*. Given a particular pattern of memory references, the type of interleaving used can affect the performance of the memory system. We will examine alternative interleaving strategies and their performance implications.

## 2.2.1   *High-order interleaving*

Most introductory digital logic and computer organization texts contain a description of *high-order interleaving*. This is the simplest and most common way to organize a computer's main memory when constructing it from a number of smaller devices. A simple example would be the design (see Figure 2.4) of a 64-KB memory using four 16K × 8 RAM devices.

A memory with 64K (actually 65,536 or $2^{16}$) addressable locations requires 16 binary address lines to uniquely identify a given location. In this example, each individual device contains $2^{14} = 16,384$ locations and thus has 14 address lines. The low-order 14 address bits from the CPU are connected to all four devices in common, and the high-order two address bits are connected to an address decoder to generate the four chip select (CS) inputs. Because the decoder outputs are mutually exclusive, only one of the four memory devices will be enabled at a time. This device will respond to its address inputs and the read/write control signal by performing the desired operation on one of its $2^{14}$ byte-wide storage locations. The data to be read or written will be transferred via the data bus.

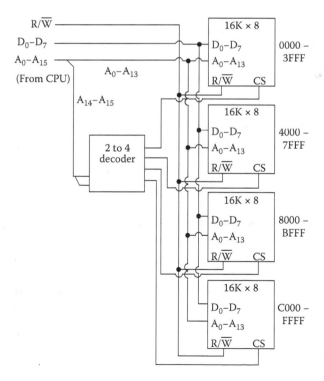

*Figure 2.4* Simple memory system using high-order interleaving.

The operation of the memory system would not be materially altered if smaller devices were used. If "narrower" devices (say, 16K × 4 or 16K × 1) were available, we would simply replace each 16K × 8 device with a bank of multiple devices, and each smaller device would be connected to a subset of the data bus lines. If we were to use "shallower" devices, such as 8K × 8 memory chips, each chip would require fewer of the low-order address lines (in this case 13 instead of 14), and we would need a larger address decoder (3 to 8 instead of 2 to 4) to generate the additional chip selects from the high-order address lines. The basic theory of operation would still be the same.

The distribution of memory addresses over the several devices (or banks of devices) in this high-order interleaved system is such that consecutively numbered memory locations are in the same device, except when crossing a 16K boundary. In other words, device 0 contains memory locations 0 through 16,383 (0000000000000000 through 0011111111111111 binary, or 0000 through 3FFF hexadecimal). Device 1 contains locations 16,384 through 32,767, device 2 contains locations 32,768 through 49,151, and device 3 contains locations 49,152 through 65,535.

This high-order interleaved memory organization is simple, easy to understand, requires few external parts (just one decoder), and offers the advantage that if one of the devices fails, the others can remain operational and provide a large amount of contiguously addressed memory. In our example, if device 0 or 3 fails, we would still have 48 KB of contiguous memory space, and if device 1 or 2 fails, we would have one working 32-KB block of memory and one 16-KB block. It also has the beneficial side effect that if the memory system is to be dual- or multiported (accessible from more than one bus, as in a system with multiple processors) and if the necessary hardware is added to support this, then much of the time accesses may occur to separate banks simultaneously, thus multiplying the effective memory bandwidth.

The disadvantage of high-order interleaving (when used with a single data/address bus) is that, at any given time, all but one (three fourths in our example) of our memory devices (or banks of devices) are idle. This one device or group of devices will respond to a read or write request in its specified access time. The memory system as a whole will be only as fast as any one device. We might ask ourselves if there is some way to improve on this situation; the following discussion of low-order interleaving will reveal how this can be done under certain circumstances.

## 2.2.2    *Low-order interleaving*

High-order memory interleaving is so common—the default organization for most main memory systems—that most textbooks do not even refer to it as a form of interleaving. What most computer architecture texts refer to as an interleaved memory system is the type of interleaving used to improve bandwidth to a single processor (or any other device capable of reading and writing memory). This is known as *low-order interleaving.*

The idea of low-order interleaving is as simple as the concept of high-order interleaving. In both cases, we have a larger main memory system constructed from a number of smaller devices. The difference is in how we map the memory addresses across the different devices or groups of devices. Let us return to our example in which we designed a 64-KB memory using four 16K × 8 RAM devices, with one apparently minor but significant change (see Figure 2.5): Instead of connecting the low-order 14 address bits from the CPU to all four devices in common, we connect the higher-order 14 bits; and instead of connecting the high-order two address bits to the external decoder, we generate the four chip select inputs by decoding the two lowest-order address bits. The decoder outputs are still mutually exclusive, so still only one of the four memory devices will be enabled at a time. What have we accomplished by doing this?

The important difference between this example and the previous one is in the permutation of memory addresses over the several devices. There are still a total of 65,536 memory locations equally divided over the four

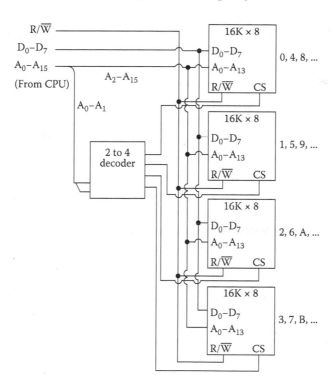

*Figure 2.5* Simple memory system using low-order interleaving.

chips, but now consecutively numbered memory locations are always in different devices. The addresses are assigned in rotation, such that device 0 contains memory locations 0, 4, 8, 12, ... through 65,532 (all the ones whose binary addresses end in 00). Device 1 contains all the locations with binary addresses ending in 01 (1, 5, 9, 13, ..., 65,533). Devices 2 and 3, respectively, contain all the locations with addresses ending in binary 10 and 11. Thus, if we access sequentially numbered memory locations (which is a more frequent occurrence than one might think), the accesses will be distributed over all four devices on a rotating basis.

The big advantage of this organization is that, given a fast enough bus and some extra hardware (to allow separate latching of the addresses and transfer of data for each of the devices or banks of devices), it is possible to have several, in this case up to four, memory accesses in progress at the same time. The likelihood of being able to take advantage of this low-order interleaving scheme is rather high because computer systems frequently access sequentially numbered memory locations consecutively. For example, program instructions are stored and executed sequentially except when that order is modified by control transfer instructions. Block I/O transfers (see Chapter 5) are normally done to or from sequential locations in a memory buffer. Many data structures, such as arrays, lists, and strings, are stored consecutively in memory. Even scalar variables are often grouped together by compilers into a contiguous block of memory.

In a low-order interleaved system, any time we access consecutively numbered memory locations for reading or writing, each successive access is to a different device or bank of devices. This allows a significant performance improvement over high-order interleaving because it is not necessary to wait for the current memory access to complete before starting the next one. Suppose that in our example we want to read memory locations 0 through 63 in succession. We initiate a read operation to location 0, which is in device 0; say the cycle time for the memory device is $t$ nanoseconds. After $t/4$ ns have passed, we initiate a read operation to location 1. (We can do this because this location is in device 1, which is currently idle.) Another $t/4$ ns later, we start a read operation on location 2, which is in device 2; after another $t/4$ ns, we start a read of location 3, which is in device 3. At this point, we have four memory accesses in progress simultaneously. After four $t/4$ intervals, the data from the read of location 0 are placed on the bus and transferred to the CPU. Device 0 is now free again, and we can initiate the read of location 4 from that same device. In another $t/4$ ns, we will transfer the contents of location 1 and start reading location 5; $t/4$ ns later, we will transfer the contents of location 2 and start the read of location 6; and so on, rotating among the four devices until we transfer the contents of all 64 memory locations. By overlapping memory accesses and keeping all four devices busy at the same time, we will get the entire job done in approximately one quarter the time that would have been required if the system had used high-order interleaving.

It is not necessary that the locations to be accessed be sequentially numbered to realize a performance benefit from a low-order interleaved main memory. Any access pattern that is relatively prime with the interleaving factor will benefit just as much. For example, in our four-way interleaved system (the number of *ways* is the interleaving factor, generally a power of two because addresses are binary numbers), if we were to access locations 0, 5, 10, 15, ... or 2, 9, 16, 23, ..., we could still get the full speed-up effect and have an average cycle time of $t/4$.

If we tried to access every second memory location (e.g., locations 3, 5, 7, 9, ...), we would lose some, but not all, of the potential speed-up. The accesses would be spread over two (but not all four) of the devices, so our average steady-state cycle time would be $t/2$ (twice that of the best case scenario, but still half that of the high-order interleaved system). The worst case scenario would occur if we tried to access every fourth memory location (0, 4, 8, 12, ...), or every eighth, or any interval composed of an integer multiple of the interleaving factor (four in this case). If this occurs, we will continually be accessing the same device, and low-order interleaving will give us no benefits at all. The effective cycle time will revert to $t$, that of an individual device.

The obvious benefit of a low-order main memory interleave is that, when transferring data to or from a single device (for example, the CPU) we can achieve a speed-up approaching $n$ (where $n$ is the interleaving factor). In the best case (sequential access), an $n$-way low-order interleave using devices with a cycle time of $t$ can give us the same performance as a noninterleaved or high-order interleaved memory built using devices with a cycle time of $t/n$ (which would likely be much more expensive). For example, an eight-way low-order interleave of 10 ns DRAMs could, under ideal conditions, approximate the performance of much costlier 1.25 ns SRAMs. Even in "real computing" in which not all accesses are sequential, we can often achieve enough of a performance increase for low-order interleaving to be worthwhile.

Low-order interleaving must have some costs or disadvantages or else it would be used universally. The most obvious disadvantage is an increase in hardware cost and complexity. A high-order interleaved system (or a noninterleaved system built from a monolithic device) can have a very simple, inexpensive bus interface because only one memory access is in progress at a time. When low-order interleaving is used, it becomes necessary to multiplex the addresses and data values for up to $n$ simultaneous transactions across the same bus. This requires very fast bus connections and associated hardware (decoders, latches, transceivers, etc.) as these have to do $n$ times the work in the same amount of time. The additional hardware required, even if built using very fast components, has some propagation delay that may cut into the potential speed-up. The alternative would be to make the bus $n$ times as wide as the size of an individual, addressable memory location (e.g., 32 bits wide in our example of a four-way interleave of 8-bit devices). This, of course, also increases the cost of implementation.

One other potential disadvantage of low-order memory interleaving in systems with multiple processors (or other devices that might need to access memory) is that the memory system is designed to maximize the bandwidth of transfers to or from a single device. In other words, if one processor is taking advantage of accessing sequentially numbered memory locations, it is using up the full bandwidth of all the memory devices and there is no opportunity for any other processor (or I/O controller, etc.) to access memory without halting the first. A potential remedy for this situation, if main memory is quite large with respect to the size of the individual devices, would be to use both high- and low-order interleaving in the same system. The memory addresses would be divided into not two, but three logical parts (see Figure 2.6); both the upper and lower bits would be externally decoded. The upper bits would select an address range composed of sets of devices; the low-order bits would choose a device or set of devices, permuted by address, within this larger set; the middle bits would be decoded internally by the devices to select a particular location. This combined interleaving scheme is the most complex and costly to implement but can be worthwhile in systems in which fast access to a large memory space is needed.

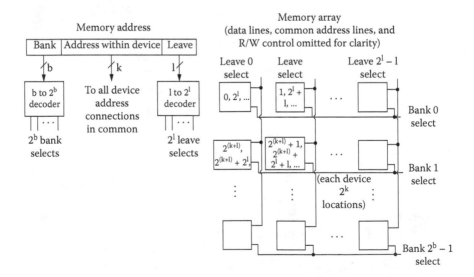

*Figure 2.6* Memory system using both high- and low-order interleaving.

## 2.3    Logical organization of computer memory

The previous section on interleaving introduced the basic main memory design common to most computer systems and showed a way that, under certain conditions, access to the main memory can be made faster. The underlying assumption of our discussion of main memory was the random access property, although we have yet to discuss what "random access" means or what other types of memory organization might be possible. We know that the bulk of main memory in most computer systems is semiconductor RAM (though portions of memory that are required to be nonvolatile may be constructed of ROM, flash, or other devices). However, certain types of computer memory (including some that may be very important to system performance) are not random access in their logical organization. Two other important types of memories are known as sequential access memories and associative memories. We discuss the differences between these logical organizations of memory in the following section.

### 2.3.1    Random access memories

Anyone who has worked with computers to any significant extent knows that main memory, for the most part, is made up of RAM. Few people, however, consider what the term really means. Computer programs do not really access memory at random, but according to some programmed sequence in order to carry out a given task. When they access memory, they do so by generating a number, called the *address*, of the location to be read or written. The important property of a RAM is that all locations are created equal when it comes to reading or writing. In other words, if a memory location is to be read, any arbitrarily (or even randomly) chosen location can be read in the same amount of time. Likewise, any location in a writable RAM, no matter what its address, can be written in the same amount of time.

From this definition, it is clear that semiconductor DRAMs and SRAMs are not the only random access memories in computer systems. Semiconductor ROMs (and associated technologies such as PROM, EPROM, and EEPROM), flash memories, and some other devices have the property of equal read access time for all locations and thus may correctly be referred to as RAMs. In fact, a more correct term for semiconductor RAM is *read/write memory* to distinguish it from read-only or read-mostly memories, many of which are also random access in their organization. However, the use of the term *RAM* as a synonym for read/write memory has become so entrenched that using the correct terminology is more apt to cause confusion than enlightenment.

In any RAM, whether it is a DRAM, SRAM, ROM, or some other type of device, each memory location is identified by a unique, numerical (specifically binary) address. An addressed location may consist of an individual bit, although more usually addresses are assigned to bytes (groups of 8 bits) or words (groups of a specified number of bits, depending on the particular architecture). Because of this, the term *absolutely addressed memory* is often used as a synonym for RAM. Strictly speaking, some types of memory (such as magnetic bubble memories and charge-coupled devices) are absolutely addressed but not truly random access, but as these technologies have generally fallen out of favor, the distinction has mostly been lost.

All of the RAMs we have discussed so far (except the ones with addresses for individual bits) are accessed by what we call a *word slice:* All the bits in a given numbered word are accessed at the same time. As shown in Figure 2.7, we present the address $i$ of a word and can then read or write all the bits of word $i$ simultaneously. There is no mechanism for reading or writing bits from different words in one operation. This is usually fine; however, some particular computer applications (graphics and certain types of array processing come to mind) can benefit by being able to access information by *bit slice*—that is to say, we may want to read or write bit $j$ of all or some defined subset of the memory locations (see Figure 2.8).

We could create a bit-slice-only memory easily enough by rearranging the connections to a regular RAM; however, we would then no longer be able to access it by word slice. If we needed to be able to access information by bit slice or word slice, we could construct what is called an *orthogonal memory* (Figure 2.9). *Orthogonal* is a term in geometry meaning perpendicular; the name describes our perception of bit slices and word slices as being logically perpendicular to each other, though they may or may not be physically arranged that way on an integrated circuit. Orthogonal memories are not seen very often in general-purpose computers but they have been used in special-purpose machines, such as the Goodyear Aerospace STARAN computer (an array processor developed in the early 1970s). Our main purpose in mentioning them is to point out that special

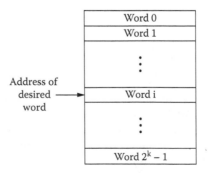

*Figure 2.7* Memory access by word slice.

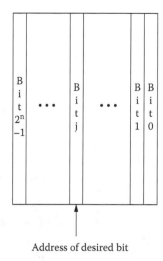

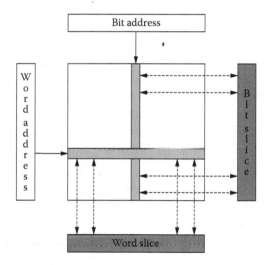

*Figure 2.8* Memory access by bit slice.

*Figure 2.9* Orthogonal memory.

problems sometimes need special solutions and that there are other ways, besides word slice, to address a RAM.

## 2.3.2   Sequential access memories

A second, frequently used, type of memory system is called a *sequential access memory* (SAM). The classic example of a sequential access memory is a magnetic tape (or, if one is old enough to remember, a punched paper tape). Information is stored by recording it on a physical medium that travels past a read/write mechanism or *head*. In order to read or write information in a particular location, the tape must physically move past the head. It is obvious that such a tape is not a random access memory; a location closer to the present position of the read/write head can be accessed more quickly than one that is far away. If the head is currently at position $n$ and we want to access location $n + 5$, for example, we must first advance sequentially past locations $n + 1$, $n + 2$, $n + 3$, and $n + 4$. In other words, we must move +5 positions from the current location. If instead we wanted to access location $n - 50$, we would have to move 50 positions down the tape in the opposite direction. With other types of sequential access memory, access may be sequential in more than one dimension. In the case of magnetic and optical disks, for example, both the radial distance the head must be stepped in or out from the center spindle, and the angular distance around the head's path, must be specified and traversed to access the desired information.

In a sense, sequential access memories are also addressed, but in a different way from RAMs. Instead of finding the desired item using its *absolute address* (its unique binary identifier), the important concept in a sequentially organized memory is the *relative address* of the information, which tells us not specifically where it is but rather how far it is from our current position in a particular direction. Using absolute addressing is analogous to telling "Scotty" of *Star Trek* fame to beam someone to the building at 122 Main Street; relative addressing is like living on Main Street and giving someone directions to walk to the sixth house to the north. Either approach, if properly followed, will get the person to the correct building, but the absolute addressing used for the transporter beam is not dependent on one's present location and will get us to the destination in the same amount of time regardless of the address without having to walk past every building in between. When relative addressing is used in a SAM, not only the location number, but also the access time, is proportional to the distance between the current and desired locations.

Because of the uniformity of addressing and access times, RAMs can easily be interfaced in a synchronous or asynchronous fashion, as the designer prefers. Sequential access memories, practically speaking, can only use an asynchronous interface because synchronous transfers of data would always have to allow for the worst case access time, which may be extremely long. Thus, disk and tape drives never interface directly to the CPU, but rather connect indirectly through a drive controller. Because of their simplicity and flexibility in interfacing with a (synchronous) CPU, RAMs are preferred by system designers and are essential for main memory. However, the advantages of magnetic and optical disk memories in terms of cost, storage density, and nonvolatility ensure that sequential access memories will be used, at least in secondary storage applications, for quite some time to come.

### 2.3.3   Associative memories

*Associative memories* are a third type of memory organization—radically different from the two just discussed. The operation of an associative memory is best summarized by referring to it by its alternate name, *content addressable memory* (CAM). Both random access and sequential access memories identify stored information by its location, either in an absolute or relative sense. Associative memories identify stored information by the actual content that is stored (or at least some subset of it). Rather than provide an absolute or relative address for a memory location and telling the hardware to store an item there or asking it what is in that location, we specify the contents we are looking for and in effect ask the memory system, "Got any of those?" A lighthearted description of associative memory is that it is the "Go Fish" approach to memory access.

The astute reader will probably have already posed the question, "If we already know the contents of a memory location, why would we need to look in memory for them? Wouldn't that be a waste of time?" Indeed, in most cases, if we know the contents of an 8-bit memory location are supposed to be, for example, 01101101, it does not do us much good just to verify that is the case. The real power and utility of an associative memory is the ability to match on a selected part of the contents, which we do know, in order to obtain the related information that we seek. In other words, we might find it more useful to ask the memory whether it contains any entries with bit 0 equal to 1 and bit 3 equal to zero, or to provide us with the first entry that starts with the bit pattern 011, or some other partial contents. This is directly analogous to a software database application that allows us to look up all of a customer's information if we know his or her name or telephone number. However, an associative memory does the same thing in hardware and thus is much faster than a software search.

In making an associative query (to use a database term) of a memory system, we need to identify three things. First, we need to provide an *argument*, or search term—in other words, the word we are trying to match the memory contents against. We also need to specify a *mask* or *key* that identifies which bit positions of the argument to check for a match on and which to ignore. Finally, we need some sort of control over conflict resolution, or at least a way to detect conflicts (multiple matches). After all, any associative search may produce no matches, a unique match, or several matches. When extracting information from the memory, knowing which of these events has occurred is often significant. If we desire to update the stored information, it is very important to detect the lack of a match so that the write operation can be aborted, and it is probably just as important to detect multiple matches so that we can determine which location(s) to update.

Figure 2.10 shows a block diagram of an associative memory array. Notice that there is an *argument register* A that holds the item to be searched for and a *key register* K in which bits equal to 1 indicate positions to check for a match and zeroes denote positions to be ignored. The results of the search are found in the *match register* M, which contains one bit for each word in the associative array. If the logical OR of all the match register bits is zero, no match was found; if it is one, at least one match was found. Examining the individual bits of M will allow us to determine how many matches occurred and in what location(s).

The construction of the registers A, K, and M is straightforward, but how can we construct the associative array itself? The memory cells could be constructed of capacitors (such as DRAM), flip-flops (such as SRAM), or some other technology. Because the main purpose of associative memory is to be able to perform a high-speed search of stored information, we will assume that each bit of data is stored in a D flip-flop or similar device. The mechanism for reading and writing these bits is the same as it would be in any static RAM: To store a bit, we place it on the D input and clock the device, and to read a stored bit, we simply look at the state of the Q output. However, additional logic is required in order to perform the search and check for matches. This logic will decrease the density of the memory cells, increase power consumption, and add considerably to the cost per

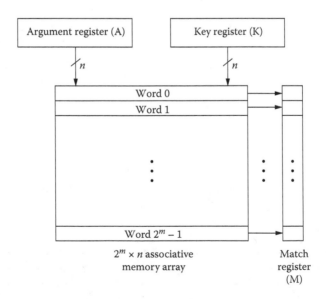

*Figure 2.10* Associative memory block diagram.

bit of fabricating the memory but (in some cases, at least) may be worth it in terms of speeding up the search for information.

Figure 2.11 shows logic that could be used for an individual associative memory cell. $Q_i$ stores the state of the $i$th bit of a word (obviously, to store all $n$ bits of a word will require $n$ flip-flops, and an associative array with $2^m$ words will require $2^m \times n$ flip-flops). $A_i$ is the corresponding bit of the argument to be searched for; it must be compared to the $Q_i$ bit in every word simultaneously. The equivalence (a.k.a. Exclusive-NOR or XNOR) gate outputs a logic 1 if the stored bit matches the corresponding bit of the argument. This gate's output is logically ORed with the inverse of the corresponding key bit $K_i$ to indicate a match $m_i$ in this bit position. This is because if $K_i = 0$, this bit position is a "don't care" for the purposes of matching, so we do not want a mismatch between $A_i$ and $Q_i$ to disqualify the word from matching. Bit $m_i$ will be 1 if either $A_i = Q_i$ or $K_i = 0$. All of the individual bit position match bits $m_i$ for the word can then be ANDed together to detect a match between the argument and that word and generate the corresponding bit to be stored in the match register M. If at least one bit of M is 1, the selected bits (according to the key) of the argument are contained in memory.

The advantage of going to all the effort and expense of building an associative memory is search speed. All the bits of all the words in the memory are compared to the argument bits simultaneously (in parallel). Rather than perform a search sequentially in software by examining one word after another, we have effectively built the search function into the hardware. Finding a match in any word, whether the first, the last, or anywhere in between (or even in multiple words), takes the same (brief) amount of time. Contrast this with a software search algorithm that takes a variable, and generally much longer, time to complete.

As with many design choices in computing, the choice between a parallel search using CAM and a sequential search through the contents of a RAM boils down to a trade-off of competing, mutually exclusive criteria. RAM is much more information-dense, cheaper, and less complex to build, not to mention useful for a much wider range of applications, but it takes a long time to search. CAM gives much better performance for a particular application (search) but offers little, if any, assistance to most

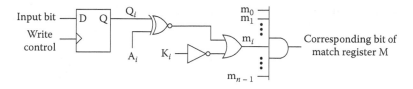

*Figure 2.11* Associative memory cell logic.

other computing functions and is much higher in cost and complexity for the same storage capacity. Its use as main memory would only be economically justified in systems tailored to very specialized applications. However, as will be seen in the next section, general-purpose machines can benefit from using a small amount of associative memory in a particular way to improve overall main memory performance.

## 2.4   Cache memory

Low-order interleaving, as discussed in Section 2.2, is one way to try to improve the performance of a computer's main memory. As we saw, however, interleaving has its limitations, and the performance improvement that can be realized is highly dependent on the precise ordering of memory references. The technique of main memory *caching* is a somewhat more general way of improving main memory performance that we will examine in this section.

A *cache memory* is a high-speed buffer memory that is logically placed between the CPU and main memory. (It may be physically located on the same integrated circuit as the processor core, nearby in a separate chip on the system board, or in both places.) Its purpose is to hold data and/ or instructions that are most likely to be needed by the CPU in the near future so that they may be accessed as rapidly as possible—ideally, at the full speed of the CPU with no "wait states," which are usually necessary if data are to be read from or written to main memory. The idea is that if the needed data or instructions can usually be found in the faster cache memory, then that is so many times that the processor will not have to wait on the slower main memory. The concept of cache memory goes back at least to the early 1960s, when magnetic core memories (fast for the time) were used as buffers between the CPU and main storage, which may have been a rotating magnetic drum.

The word *cache* comes from the French verb *cacher*, which means "to hide." The operation of the cache is transparent to or, in effect, hidden from the programmer. With no effort on his or her part, the programmer's code (or at least portions of it) runs "invisibly" from cache, and main memory appears to be faster than it really is. This does not mean that no effort is required to design and manage the cache; it just means that the effort is expended in the design of the hardware rather than in programming. We examine aspects of this in the next sections.

### 2.4.1   Locality of reference

Typically, due to cost factors (modern cache is built from more expensive SRAM rather than the DRAM used for main memory), cache is much smaller in size than main memory. For example, a system with 4 GB of main memory might have only 4 MB (or less) cache. One might ask how much good this would do. Because cache is only 1/1024 the size of main memory, it would appear that it would be of almost negligible benefit. Indeed, if memory references for code and data were uniformly distributed throughout the address space, we would only expect one access in every 1024 to occur to the faster cache memory. In this case, the additional expense of a cache could hardly be justified.

Fortunately for the performance of computer memory systems, computer programs do not access memory at random. Instead, most programs confine the vast majority of their memory references for instructions and data to small areas of memory, at least over any given limited stretch of time. This observed, nearly universal behavior of computer programs exemplifies the principle of *locality of reference*. Simply put, this principle states that programs tend to access code and data that have recently been accessed, or which are near code or data that have recently been accessed. This principle explains why a relatively small cache memory can have a large impact on memory system performance. It is OK to have a cache that is 0.1% or less of the size of main memory as long as we make sure it contains the *right* 0.1% of the information—that which is most likely to be used in the near future. Determining which 0.1% to keep in the cache at any given time is the challenging, but fortunately not impossible, task.

Aspects of locality of reference include *temporal, spatial,* and *sequential* locality. Temporal (time-related) locality says that if a given memory location is accessed once, there is a high probability of its being accessed again within a short time. Spatial locality means that locations near a recently accessed location are also likely to be referenced. Sequential locality is a specific form of spatial locality; it tells us that memory locations whose addresses sequentially follow a referenced location are extremely likely to be accessed in the very near future. Low-order interleaved memory takes advantage of sequential locality, but cache memory is a more general technique that can benefit from all three forms of locality.

These aspects of locality of reference are readily illustrated by a variety of common programming practices and data structures. Code, for example, is normally executed sequentially. The widespread use of program loops and subroutines (a.k.a. functions, procedures, methods) contributes to temporal locality. Vectors, arrays, strings, tables, stacks, queues, and other common data structures are almost always stored in contiguous memory locations and are commonly referenced within program loops. Even scalar data items are normally grouped together in a common block or data segment by compilers. Obviously, different programs and data sets exhibit different types and amounts of locality; fortunately, almost all exhibit locality to a considerable degree. It is this property of locality that makes cache memory work as a technique for improving the performance of main memory systems.

## 2.4.2   Hits, misses, and performance

Because of the locality property, even though cache may be much smaller than main memory, it is far more likely than it would otherwise seem that it will contain the needed information at any given point in time. In the hypothetical case (4 GB main memory, 4 MB cache) presented in the previous section, the cache is less than 0.1% of the size of main memory, yet for most programs, the needed instruction or data item may be found in the cache memory 90% or more of the time. This parameter—the probability of avoiding a main memory access by finding the desired information in cache—is known as the *hit ratio* of the system. (A *cache hit* is any reference to a location that is currently resident in cache, and a *cache miss* is any reference to a location that is not cached.) Hit ratio may be calculated by a simple formula:

Hit ratio = $p_h$ = number of hits/total number of main memory accesses

or, equivalently,

$p_h$ = number of hits/(number of hits + number of misses)

The hit ratio may be expressed as a decimal fraction in the range of 0 to 1 or, equivalently (by multiplying by 100), as a percentage. For example, if a given program required a total of 142,000 memory accesses for code and data but (due to locality) 129,000 were hits and only 13,000 were misses, then the hit ratio would be

$p_h$ = 129,000/(129,000 + 13,000) = 129,000/142,000 = 0.9085 = 90.85%

We can correspondingly define the *miss ratio* (the fraction of memory references not satisfied by cache) as $(1 - p_h)$. The hit ratio is never really a constant. It will vary from system to system depending on the amount of cache present and the details of its design. It will also vary on the same system depending on what program is currently running and the properties of the data set it is working with. Even within a given run, hit ratio is a dynamic parameter as the contents of cache change over time. In the example above, most of the 13,000 misses probably occurred early in the run before the cache filled up with useful data and instructions, so the initial hit ratio was probably quite low. Later on, with a full cache, the hit ratio may have been much higher than 0.9085. (It still would have varied somewhat as different routines containing different loops and other control structures were encountered.) The overall value computed was just the average hit ratio over some span of time, but that is sufficient for us to estimate the effect of cache on performance.

With a high hit ratio (close to 1.0 or 100%), dramatic gains in performance are possible. The speed ratio between cache and main memory may easily be 10:1 or greater. Cache ideally operates at the full speed of the CPU (in other words, we can access cache in a single processor clock cycle), and main memory access typically takes much longer. Let us say that in the preceding example ($p_h = 0.9085$) the main memory access time is 10 ns, and the cache can be accessed in 0.5 ns (a 20:1 speed ratio). What is the effective time required to access memory, on average, over all references? We can compute this simply using the following formula for a weighted average:

$$t_{a \text{ effective}} = t_{a \text{ cache}} \times (p_h) + t_{a \text{ main}} \times (1 - p_h)$$

$$t_{a \text{ effective}} = (0.5 \text{ ns})(0.9085) + (10 \text{ ns})(0.0915) = 0.45425 \text{ ns} + 0.915 \text{ ns} = 1.36925 \text{ ns}$$

which is much closer to the speed of the cache than it is to the speed of the main memory. The cache itself is 20 times the speed of the main memory; the combined system with cache and main memory is about 7.3 times as fast as the main memory alone. That small, fast cache memory has bought us a lot of performance, and this is a fairly conservative example; it is not uncommon for cache hit ratios to be in the range of 97% to 98% in practice. A hit ratio of 0.98 would have brought the access time down to just 0.69 ns, or approximately 14.5 times the speed of main memory. By spending a relatively small amount on 4 MB of fast memory, we have achieved nearly as much improvement in performance as we would have realized by populating the entire main memory space (4 GB) with the faster devices, but at a small fraction of the cost. "Such a deal I have for you," a system designer might say. "A magic memory upgrade kit! Just plug it in and watch performance soar!"

Of course, as we will shortly see, it's not quite that simple. We cannot just plug these extra SRAM devices in on the CPU-memory bus and expect them to work like magic, automatically choosing the right information to keep while discarding items that will not be needed any time soon. The operation of the cache must be intelligently controlled in order to take advantage of locality and achieve a high hit ratio. We need circuitry to decide such things as which main memory locations to load into cache, where to load them into cache, and what information already in the cache must be displaced in order to make room for new information that we want to bring in. The cache control circuitry must also be able to handle issues such as writes to locations that are cacheable. It must be able to detect whether a location that is written is in cache, update that location, and make sure the corresponding main memory location is updated. We will explore the details and ramifications of these control issues next.

Before we begin our discussion of the details of cache design, we need to make one observation regarding typical cache operation. Although programs normally interact with memory in the form of reading or writing individual bytes or words, transfers of data or instructions into or out of cache typically are done with less granularity. To put it more simply, most caches are designed such that a block of data, rather than a single byte or word, is loaded or displaced at a time. The unit of information that is moved between main memory and cache is referred to as a *refill line* or simply a *line*. Depending on system characteristics, line size may range from just a few (e.g., 8 or 16) bytes to a fairly substantial chunk of memory, perhaps as large as 256 bytes.

Cache may be partitioned into refill lines rather than individual bytes or words for several reasons. First, given the typical size and performance characteristics of buses between the CPU and main memory, it is usually more efficient to perform a small number of larger transfers than a large number of small ones. Also, because of the locality principle (which is the basis of cache operation anyway), if a program needs a given byte or word now, it will probably need the surrounding ones soon, so it makes sense to fetch them all from main memory at once. Finally (this is especially true for the fully associative cache organization described in the next section), it is generally less expensive to build a cache with a small number of large entries than vice versa. Thus, although line size can (in the simplest case) equal one byte or one word, it is usually somewhat larger and (like the size of most memories with binary addresses) virtually always an integer power of two.

### 2.4.3   Mapping strategies

The most notable job of the cache controller, and the one with probably the most significance in regard to hit ratio, is the task of *mapping* the main memory locations into cache locations. The system needs to be able to know where in cache a given main memory location is, so that it can be retrieved rapidly in the event of a hit. Even more basically, the system needs to be able to detect a hit versus a miss—to quickly determine whether or not a given main memory location is cached. Three strategies are widely used for performing this mapping of main memory addresses to cache locations. These mapping strategies, also referred to as *cache organizations*, are known as associative mapping, direct mapping, and set-associative mapping. Each has certain advantages and disadvantages that we will explore.

*Associative mapping* is a cache organization that takes advantage of the properties of associative (or content-addressable) memories that we studied in the previous section. Associative mapping, often referred to as *fully associative* mapping to distinguish it from the set-associative mapping to be discussed later, is the most flexible mapping scheme. Because of this, all other things being equal, it will have the highest hit ratio and thus improve performance more than the other mapping strategies. However, because it relies on a CAM to store information, a fully associative cache is the most expensive type to build.

Each entry, or refill line, in a fully associative cache is composed of two parts: an address *tag*, which is the information to be matched on associatively, and one or more data or instruction bytes/words that are a copy of the correspondingly addressed line in main memory. If a line is an individual memory location, the tag is the complete memory address of that location; if, as is more usually the case, a line contains several memory locations, the tag consists of the high-order address bits that are common to all locations in that line. For example, if the (byte-addressable) main memory uses 32-bit addresses and each refill line contains $2^6 = 64$ bytes, then the associative tags will be the upper $(32 - 6) = 26$ address bits. Figure 2.12 shows the general layout of the fully associative cache.

Note that the tag storage is all CAM, but the cached information that goes with each tag is not needed in the matching process, so it can be stored in "plain old" static RAM. Each tag is logically associated with one, and only one, line of information stored in the SRAM.

The power of a fully associative cache organization is that when a main memory address is referenced, it is quick and easy to determine whether it is a cache hit or a cache miss. The upper bits of the main memory address are checked against all the cache tags simultaneously. We never cache the same main memory location in more than one place, so

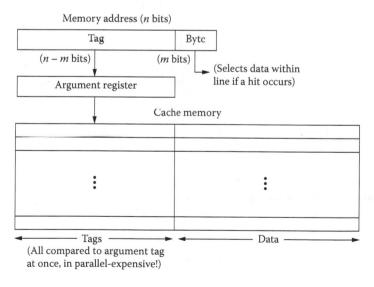

*Figure 2.12* Fully associative cache.

there will be either one match or none. If no tag matches the supplied tag, the access is a miss, and main memory must be referenced (note that we will then place this line into the cache, displacing another line if necessary, so that subsequent references to it will result in hits). If one tag is a match, then a hit has occurred, and the lower main memory address bits will identify which byte or word within the line is to be accessed. In this case (which, of course, is what we hope for) the main memory access can be omitted.

Not only is the check for a hit very fast because all the tags are checked at once, but this cache organization is the most flexible because there are no limitations on where in the cache any given information from main memory may be mapped. Any line from main memory may reside in any line of cache. Thus, any combination of main memory contents can reside in cache at any time, limited only by the total size of the cache. Because of this flexibility, hit ratios tend to be high for a given cache size; however, the need for a CAM to hold the tags makes this a costly strategy to implement.

*Direct mapping* is the opposite extreme of fully associative mapping. The idea is simple: Because the cost of a fully associative cache is dominated by the cost of the matching hardware required to associatively compare all the many large tags at once, we could achieve considerable savings by reducing the number of comparisons and/or the size of the tags. A direct mapping does both of these things by sacrificing flexibility. Instead of an entry (a given item from main memory) being able to reside anywhere in the cache, it is constrained to be in one particular line if it is in cache at all. The particular line into which it may be mapped is determined by part of its main memory address, referred to as its *index*.

Because it is only possible for a given item to be in one place, we only have to compare the tag portion of the main memory address with one stored tag (the one with the same index) to determine whether or not a hit has occurred. Also, because some of the memory address bits are used for indexing, although line size is independent of the mapping strategy (and thus could be the same in a direct-mapped cache as it might be in a fully associative cache), fewer bits are required for the tags. If main memory is $2^n$ bytes and the line size is $2^m$ bytes, then the tags in a fully associative cache are $n - m$ bits in size regardless of the number of lines in the cache. However, in a direct-mapped cache containing $2^k$ lines, $k$ bits are used for the index, so the tags are only $n - k - m$ bits long.

Figure 2.13 shows an example of a direct-mapped cache. Suppose a system is to have 4 GB ($2^{32}$ bytes) of main memory and 2 MB ($2^{21}$ bytes) of cache, which is to be organized as $2^{15} = 32{,}768$ lines of $2^6 = 64$ bytes each. The 32 address bits would be divided into 11 tag, 15 index, and 6 byte bits as shown. To check for a cache hit on any given main memory access, the index bits are used to uniquely choose one of the 32,768 stored tags.

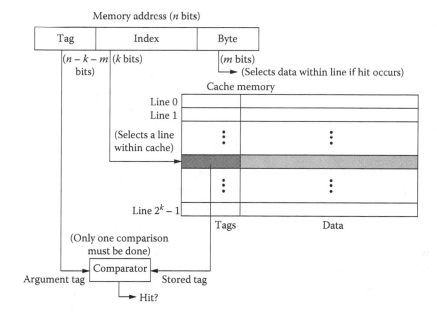

*Figure 2.13* Direct-mapped cache.

That tag (and that tag only) is checked against the tag bits of the supplied address. If they match, the access is a hit, and the six low-order bits can be used to identify which location within the line is to be accessed. If they do not match, the access is a miss, and main memory must be accessed.

As an example, let's say the CPU wants to read main memory location $0040018A_{16}$ ($0000000001000000000000110001010_2$). Although this address is a monolithic binary value as far as the CPU is concerned, for the purposes of the cache, it is treated as three separate values: a tag of $00000000010_2$, an index of $000000000000110_2$, and a byte address within a given line of $001010_2$. Because the index is $00000000110_2$, or decimal 6, the tag in position 6 is accessed and checked. Because this tag is equal to $00000000010_2$, the access is a hit, and the data at position $001010_2$ (10 decimal) within that line are forwarded to the CPU. If the contents of the tag in position 6 had been anything other than $00000000010_2$, the access would have been a miss, and a main memory read would have been required to get the data. In that case, the previous contents of cache line 6 would be replaced by the contents of main memory locations $00400180_{16}$ through $004001BF_{16}$ (the 64-byte block including the referenced location) because, by the principle of locality of reference, those memory contents would likely be needed in the near future.

The key to cost savings using a direct mapping is that no associative memory is required. Both the tags and data are stored in a fast, but otherwise ordinary, static RAM, which is more expensive than DRAM but significantly cheaper than a corresponding amount of CAM. Only one small comparator circuit (just 11 bits in our example) is needed to check a single tag for a match. The trade-off is a potentially lower hit ratio and thus lower overall system performance given the same size cache. Why is this the case? Let us look at the last example. Suppose the program was processing two arrays: one stored beginning at location $00400180_{16}$, including the location accessed in the example, and another stored beginning at location $01C00180_{16}$. The tag portion of the address for this second array would be $00000001110_2$ instead of $00000000010_2$, but it would have an index of $000000000000110_2$ just like the first array. With a fully associative organization, caching both arrays at the same time would be no problem; any subset of main memory contents can be cached at the same time. If the direct mapping of the example is used, these two arrays are mutually exclusive when it comes to caching. Any time elements of either array are loaded into cache, they can only go in line 6, displacing elements of the other array if they have already been loaded.

We should point out here that there are three types of cache misses, classified by the reason that the needed data are not currently in the cache. The first type is a *compulsory miss*, also known as a "first reference miss" or a "cold start miss." The first time a program makes reference to an item (instruction or operand) in a given line-sized block of memory, it will not have been loaded into cache yet, and so a miss will occur. Subsequent references to the same block will result in hits unless that line has been evicted from cache in the meantime. The second type is a *capacity miss*. Because cache is finite in size (and always smaller than main memory), it is possible that the program code and/or data set will be too large to completely fit in the cache. Once it fills up, some lines that are still in use may have to be evicted to make room for newly referenced lines. The misses that subsequently occur for no other reason than because the cache is full are capacity misses. Any cache, regardless of organization, will suffer from both compulsory and capacity misses. However, a fully associative cache is immune from the third type: *conflict misses*. Conflict misses occur when two or more blocks of main memory map to the same spot in the cache (and, of course, cannot occupy it at the same time). Anytime one of the conflicting blocks (i.e., ones with the same index but different tags) is referenced, it will cause the eviction of another that is currently cached; a subsequent reference to that (no longer cached) region of memory will cause a conflict miss.

It is easy to see that if the program frequently needs to access both of the arrays in the above scenario, conflict misses will become frequent, and performance will be significantly degraded. Obviously, the system hit ratio will be lower than it would be under other circumstances, so the apparent access time for main memory as seen by the CPU will increase.

There may also be a hidden performance cost if other operations, such as I/O, are going on in the background. Even though the cache "fills" after each miss may be transparent to the CPU, they still consume main memory bandwidth; thus, other devices that may be trying to access main memory will have to wait.

This contention for a line of cache by multiple items in main memory is not always a problem, but the laws of probability (not to mention Murphy's Law) tell us that it is likely to happen at least some of the time. For this reason, a direct-mapped cache may not be the best choice due to performance concerns. Conversely, due to cost constraints, a fully associative cache may not be feasible. A popular solution that attempts to realize most of the benefits of these two strategies while minimizing their disadvantages is discussed next.

*Set-associative mapping* is a compromise between the two extremes of fully associative and direct-mapped cache organization. In this type of cache, a particular item is constrained to be in one of a small subset of the lines in the cache rather than in one line only (direct) or in any line (fully associative). A set-associative cache can perhaps be most conveniently thought of as a group or set of multiple direct-mapped caches operating in parallel. For each possible index, there are now two or more associated lines, each with its own stored tag and associated data from different areas of main memory. These different lines from main memory will be ones that would contend with each other for placement in a direct-mapped cache, but, due to the duplication of the hardware, they can coexist in the set-associative cache.

The hardware for a set-associative cache is somewhat more complex than what is required for a direct mapping. For example, instead of one comparator being used to check the supplied address against the single tag with a given index, the set-associative cache will require two, four, or more comparators. (The cache is said to be as many "ways" associative as there are different places to cache a given piece of information, so a cache with four parallel sets and thus four comparators would be called a four-way set-associative cache.) Also, when all of the lines with a given index are full and another line needs to be loaded, some sort of algorithm must be employed to determine which of the existing lines to displace. (In a direct-mapped cache there is only one option and thus no similar choice to make.) On the other hand, the hardware for a set-associative cache is not nearly as complex or expensive as that required for a fully associative cache.

Let us go back to the direct-mapped example and reorganize it as a two-way set-associative cache. We still assume that there is 4 GB of main memory space, so 32-bit addressing is used. The 2 MB total cache size is the same, but now we partition it into two subdivisions of 1 MB and treat each as though it were an independent direct-mapped cache. Assuming that each line is still 64 bytes, this means there are $2^{14} = 16,384$ lines in each half of the cache. There will thus be 14 index bits instead of 15, and the tags will consist of the upper 12 address bits rather than 11 (see Figure 2.14). On each memory reference, two 12-bit tags, rather than one 11-bit tag, must be checked. The two hypothetical arrays we spoke of, at main memory addresses $00400180_{16}$ and $01C00180_{16}$, still both map to line 6. Because there are two separate lines with index 6, however, it is possible for these items to coexist in cache (though accessing a third area of main memory with the same index would necessarily displace one of them).

If we were concerned about this latter possibility, we could divide the cache into four 512 KB partitions (ways) and use a four-way set-associative design with a 13-bit index and four 13-bit tags to check on each access. By extension, it is not difficult to envision how an eight-way, 16-way, or even more associative cache could be designed. These more highly associative organizations are rarely used as they add cost and complexity, and it is rare for more than two to four widely separated areas in memory to be in frequent use at the same time. Thus, considering benefits versus cost, a two- or four-way set-associative organization is often the best overall design for a data or mixed instruction/data cache. (When instructions are cached separately, a direct mapping may do virtually as well as a set-associative one, as program flow tends to be more localized than data access.)

By way of summing up, we might point out that the set-associative cache is the most general of the three organizational concepts. A

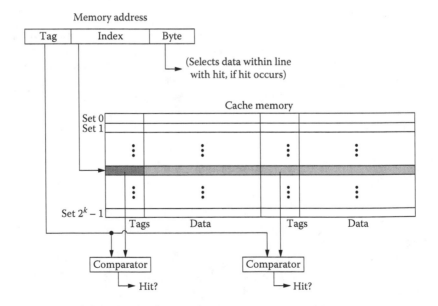

*Figure 2.14* Two-way set-associative cache.

direct-mapped cache is really just a one-way set-associative cache by another name, and a fully associative cache with $2^k$ lines is the same as a $2^k$-way set-associative cache with one line per way. The direct-mapped cache will have the fewest tag bits and the most index bits; as associativity increases for the same cache size, the number of index bits decreases and the number of tag bits increases. Ultimately, in the fully associative case, there are 0 index bits, and the tags are the maximum size, which is the entire main memory address minus the number of bits that identify a specific location within a line.

## 2.4.4   Cache write policies

Cache operation using any of the three mappings just discussed is fairly straightforward as long as memory accesses are limited to reads. This may be possible in the case of an instruction-only cache, as self-modifying code is not a particularly good idea anyway. However, when we are designing the data cache in a system with a split (modified Harvard) cache architecture, or a unified (Princeton or mixed instruction/data) cache, we always have to allow for writes to cached locations. Writes complicate the design process because of what must happen when a line is displaced from cache. If the contents of that line have only been read, they can simply be overwritten by the new data. But if any location in that line has been modified, we must make sure main memory reflects the updated contents before loading the new data in their place. This can be done in different ways, which we will discuss below.

*Write-through cache* is the simplest approach to keeping main memory contents consistent with the cache. Every time we write to a location in cache, we also perform a write to the corresponding main memory location. These two writes can usually be started at the same time, but the main memory write takes longer and thus determines the overall time for the write operation.

One advantage of this method is that it is relatively simple to build into the hardware. In addition, because of the write-throughs, main memory always has the most current data, identical to the contents of the cache. This is not particularly important from the CPU's standpoint because subsequent reads of the same location will hit the cache and get the most recent data anyway. However, if there are other devices (for example, Direct Memory Access [DMA] controllers or I/O processors) in the system, always having main memory updated with the latest data can simplify things quite a bit.

The obvious disadvantage of a write-through cache is that an item may be accessed (for reading or writing) several times while it is in cache. (Indeed, we are depending on this to happen if we are to realize a performance benefit from using cache.) We may spend the time required to write a value to main memory, then turn around and read it again or write it again or both—possibly a number of times—before it is finally displaced from the cache. All the extraneous writes (all of them, that is, except the last one, which commits the final value to main memory) exact a performance penalty; it is as though we never hit on a write because all writes access main memory. Although both reads and writes can nominally be hits in the sense of referencing a cached location, only read hits are beneficial in terms of performance. The *effective* hit ratio will be lower than the actual hit ratio; exactly how much lower depends on the behavior of a particular program, but the difference can be significant.

*Write-back cache* is more complex to implement but can improve performance if writes are done frequently. On a write that hits the cache, only the cache location is updated. Main memory is only written when a line that has been modified is displaced from the cache to make room for a new line to be loaded. Implementing this policy requires that we add a bit to each tag to indicate whether or not the associated line has been written. This bit is often called the "inconsistent bit" (or, more colorfully, the "dirty bit"—a cache location that has been written to may be called a "dirty cell" or "dirty word"). If this bit is 0, then this line has only been read, and the cache control hardware will know that a new line can simply be loaded over the existing information. If the dirty bit is 1, then that line (or at least any modified locations within it) must be copied or "written back" to main memory before it is overwritten by new information. If this were not done, it would be as though the write had never occurred; main memory would never be updated, and the system would operate incorrectly.

The advantages and disadvantages of a write-back strategy are the converse of those for a write-through cache. Using the write-back approach will generally maximize performance because write hits can be nearly as beneficial as read hits. If we write to a cached location 10 times, for example, only one write to main memory is required. Nine of the 10 write hits did not require a main memory access and thus took no more time than read hits.

With this approach, data in main memory can be "stale"—that is, we are not guaranteed that what is in main memory matches the most recent activity of the CPU. This is a potential problem that must be detected and dealt with if other devices in the system access memory. In addition, the logic required to do write-backs is more complex than that required to perform write-throughs. We not only need an extra bit added to every tag to keep track of updated lines; we need logic to examine these bits and initiate a line write-back operation if needed. To do a write-back, the controller must either hold up the read operation for filling the new line until the write-back is complete, or it must buffer the displaced information in a temporary location and write it back after the line fill is done.

## 2.4.5   Cache replacement strategies

Speaking of displaced information, when designing a cache, we also need to build into the control hardware a means of choosing which entry is to be displaced when the cache is full and a miss occurs, meaning that we need to bring in the line containing the desired information. In a direct-mapped cache, there is only one place a given item can go, so this is simple, but in a set-associative or fully associative cache, there are multiple potential places to load a given line, so we must have some way of deciding among them. The major criterion for this, as in other aspects of cache design, is that the algorithm should be simple so that it can be implemented in hardware with very little delay. In order to maximize the hit ratio, we would also like the algorithm to be effective; that is, to choose for replacement a line that will not be used again for a long time. Doing this perfectly would require foreknowledge of future memory access patterns, which, of course, our hardware cannot have unless it is psychic. However, there are several possibilities that may approximate an ideal replacement strategy well enough for our purposes.

One possible replacement strategy is a *least frequently used* (LFU) algorithm. That is, we choose for replacement the line that has done us the least good (received the fewest hits) so far, reasoning that it would be likely to remain the least frequently used line if allowed to remain in cache. Lines that have been frequently hit are "rewarded" by being allowed to remain in cache, where we hope they will continue to be valuable in the future. One potential problem with this approach is that lines that have been loaded very recently might not yet have a high usage count and so might be displaced even though they have the potential to be used more in the future. The main problem, however, is with the complexity of the required hardware. LFU requires a counter to be built for each entry (line) in the cache in order to keep up with how many times it has been accessed; these count values must be compared (and the chosen one reset to zero) each time we need to replace a line. Because of this hardware complexity, LFU is not very practical as a cache replacement algorithm.

Other replacement algorithms that may achieve results similar to LFU with somewhat less hardware complexity include *least recently used* (LRU) and *first-in, first-out* (FIFO). The LRU algorithm replaces the line that was hit the longest time ago, regardless of how many times it has been used. FIFO replaces the "oldest" line—that is, the one that has been in cache the longest. Each of these approaches, in its own way, attempts to replace the entry that has the least temporal locality associated with it in hopes that it will not be needed again soon.

Some studies have shown that once cache gets to the sizes that are common in modern computers, performance is not particularly sensitive to the particular replacement algorithm used. Therefore, to keep the hardware as simple and fast as possible, some cache designers have chosen to use a very basic *round-robin* algorithm (in which candidacy for replacement is simply rotated among the cache lines) or even a *random* replacement strategy, in which some arbitrary string of bits is used to identify the line to be replaced. Any algorithm that is simple and has little effect on performance over the long term is a viable candidate. Designers typically make this choice, as well as other design decisions such as the degree of associativity, by running simulations of cache behavior with different memory reference sequences taken from logged program runs.

## 2.4.6   Cache initialization

One more topic that should be addressed before we conclude our discussion of cache memory is cache initialization. Once a program has been running for a while and the cache is full, its operation is fairly straightforward, but how do we handle filling up the cache to start with, for example, after a reset or when a new process begins to run? More to the point, on these occasions, how do we make sure that invalid data are not read from the cache by the CPU?

It is important to realize that like any RAM, cache memory always contains something, whether that something is meaningful data and instructions or "garbage." When the system is reset, for example, the cache will either contain residual information from before the reset or (if power was interrupted) a more or less random collection of 0s and 1s. In either case, the contents of the cache are invalid, and we need to keep addresses generated by the CPU from accidentally matching one of the (random) tags and feeding it random garbage data (or worse, instructions).

The simplest and most usual approach used to reinitialize and validate the cache uses another bit associated with each tag (similar to the dirty bit used in a write-back cache), which we call the "valid bit." When a reset occurs, the cache control hardware clears the valid bit of every line in the cache. A tag match is not considered to be a hit unless the associated valid bit is set, so this initially forces all accesses to be (compulsory) misses. As misses occur and valid lines are brought into cache from main memory, the cache controller sets the corresponding valid bits to 1. Any tag match on a line with a valid bit = 1 is a legitimate hit, and the cached information can be used. Eventually, all valid bits will be set; at that point, the cache is full and will remain so until something happens to clear some or all of the valid bits again.

Many architectures support not only invalidating the cache in hardware on a system reset, but also under supervisor software control (by the operating system). This allows all or part of the cache to be "flushed" (for purposes of protection) on an interrupt, context switch, or any time the operating system deems it necessary without a machine reset having to occur. Other related cache control enhancements may include mechanisms to "freeze" the contents of the cache or to lock certain entries in place and prevent them from being evicted. The goal of all such design features is to make sure that, as much as possible, the cache is kept full of valid, useful data that will contribute to a high hit ratio and maximize memory system performance.

## 2.5 Memory management and virtual memory

The two major "speed gaps" in most modern computer memory systems are the gap between the CPU and main memory speeds and the gap between main and secondary storage speeds. We have seen how techniques like cache memory and low-order interleaving can help main memory appear faster than it really is and thus bridge the speed gap between the processor and main memory. In this section, we learn about approaches that are used to make the main memory appear much larger than it is—more like the size of the slower secondary memory. (Alternatively, we could say that we are making the large secondary memory space appear to be as fast as the main memory and directly addressable.) In this way, the overall system will appear, from the processor's (and the programmer's) point of view, to have one large, fast, homogeneous memory space rather than the hierarchy of different types of devices of which it is actually composed.

### 2.5.1 Why virtual memory?

Computer programs and their associated data sets may be very large. In many cases, the code and data are larger than the amount of main memory physically present in a given system. We may need to run the same program on a variety of systems, some of which have more memory than others. Because most modern general-purpose computers have operating systems that support multitasking, other programs may be (and usually are) resident in memory at the same time, taking up part of the available space. It is generally impossible to know in advance which programs will be loaded at a particular time and thus how much memory will be available for a given program when it runs.

It is possible to divide a large program or data set into smaller chunks or "overlays" and load the needed parts under program control, unloading other parts as they are no longer needed. Each time the program needs to load or unload information, it must explicitly interact with the operating system to request or relinquish memory. This is a workable scheme, but it puts the burden on the application programmer to manage his or her own memory usage. It would be preferable from a programmer's point of view to be able to assume that there will always be enough memory available to load the entire application at once and let the operating system handle the problems if this is not the case. To make this happen, it is necessary for the addresses used by the program to be independent of the addresses used by the main memory hardware. This is the basic idea behind memory management using the *virtual memory* approach.

## 2.5.2   *Virtual memory basics*

In a system using virtual memory, each program has its own *virtual address space* (sometimes referred to as a *logical address space*) within which all memory references are contained. This space is not unlimited (no memory system using addresses with a finite number of bits can provide an infinite amount of storage), but the size of the virtual addresses is chosen such that the address space provided exceeds the demands of any application likely to be run on the system. In the past, 32-bit virtual addressing (which provided a virtual address space of 4 GB) was common. More recently, as applications have gotten larger and a number of systems have approached or exceeded 4 GB of RAM, larger virtual address spaces have become common. A 48-bit address allows a program 256 terabytes (TB) of virtual space, and a 64-bit address provides for a currently unimaginable 16 exabytes (EB). For the foreseeable future, 64-bit virtual addressing should be adequate (remember, however, that this was once said of 16- and 32-bit addressing as well). The purpose of this large address space is to give the programmer (and the compiler) the illusion of a huge main memory exclusively "owned" by his or her program and thus free the programmer from the burden of memory management.

Each program running on a system with virtual memory has its own large, private address space for referencing code and data. However, several such programs may have to share the same, probably much smaller, physical main memory space that uses its own addresses. Thus, before any actual memory reference can proceed, there must be a translation of the virtual address referenced by the program into a physical address where the information actually resides. This process is symbolized in Figure 2.15.

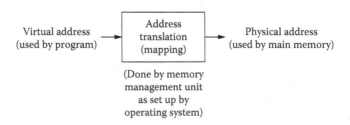

**Figure 2.15** Address translation in a system with virtual memory.

Virtual-to-physical address translation is done by a combination of hardware and (operating system) software. Doing everything in hardware would be expensive and inflexible, and doing most or all of the translation in software would be intolerably slow. Typically, translation on a cycle-by-cycle basis is handled by a hardware component called the *memory management unit* (MMU), which can operate autonomously as long as translations are routine. When a problem arises, such as a reference to a virtual location that is not currently loaded in main memory, the MMU signals an exception so that the operating system can intervene and solve the problem.

Even when translations are routine and accomplished completely in the MMU hardware, there is some "overhead," or inefficiency, involved. Any physical device, including the MMU, has an associated propagation delay that adds to the delay of the main memory devices in determining the memory cycle time. This must be compensated for by using faster (and more expensive) memory devices or by restricting the bus speed to allow for the greater delay. When the operating system has to intervene, the overhead is increased considerably as the currently running program must be suspended and then resumed after the problem is corrected (for example, by loading needed data into main memory from a disk drive). It takes time for the operating system to determine what needs to be done and then do it. However, the hardware and software overhead inherent to virtual memory systems has been found to be worthwhile, as it is compensated for by reductions in programming complexity and programmer effort.

Virtual memory systems are generally classified as either *paged* or *segmented*, although it is possible to combine attributes of both in a single system. We examine the attributes, advantages, and disadvantages of each approach in the next sections.

## 2.5.3 Paged virtual memory

A paged virtual memory system is also known as a *demand-paged* virtual memory system because the pages are loaded on demand, or as they are requested. A paged system is hardware-oriented in the sense that the size of the *pages* (the blocks that are moved between main and secondary memory) is fixed, based on a hardware consideration: the granularity of secondary storage (disk sector size). Main memory is divided into *page frames* of a constant size, which is either equal to or an integer multiple of the sector size. Different architectures and operating systems may have different page sizes, generally in the range of 512 bytes to 16 KB (with 4 to 8 KB being typical), but once this is chosen, it is normally a constant for a given system. (Some modern systems allow for "huge pages," which may be megabytes in size, to coexist with normal-sized pages.)

A virtual address in a system that uses paging can be divided into two parts as shown in Figure 2.16. The high-order bits may be considered the virtual page number, and the low-order bits represent the offset into a page. The number of bits in the offset, of course, depends on the page size. If pages are 4 KB ($2^{12}$ bytes), for example, the offset consists of the 12 least significant address bits. Because the pages are always loaded on fixed page frame boundaries, the physical offset is the same as the virtual offset; only the virtual page number needs to be translated into a physical page frame number. This process, which uses a lookup table in memory, is illustrated in Figure 2.17. A page table base register points to the base of a lookup table; the virtual page number is used as an index into this table to obtain the translation information.

To avoid having to maintain a single, huge page table for each running process, the virtual page number may be (and usually is) divided into two or more bit fields. (See Figure 2.18 for an example with three fields.) This allows the lookup to be done in a stepwise fashion, in which the higher-level tables contain pointers to the start of lower-level page tables, and the next bit field is used to index into the next lower-level table. The lowest-level table lookup completes the translation. This multiple-level lookup procedure takes longer than a single-stage lookup, but the resulting tables are smaller and easier to deal with.

The information obtained from the lowest-level page table is called a *page table entry*. Each entry includes the page frame number that tells where in the physical main memory the page is located, if it is currently loaded. Other information will include a *validity bit*, or *presence bit*, which tells whether or not the page exists in main memory, and a *dirty bit* (similar to that kept for a cache line), which tells whether the page has been modified while in memory. Other page attribute bits may include *protection bits*, which govern who or what may access the page and for what

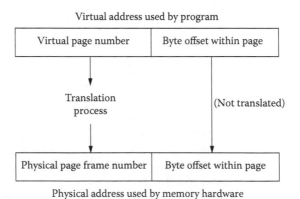

*Figure 2.16* Address translation in a system with demand-paged virtual memory.

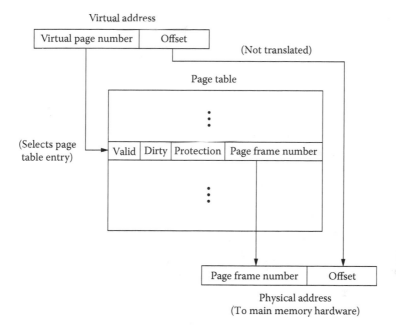

*Figure 2.17* Table lookup for address translation in a paged system.

purposes (read only, read/write, etc.). Assuming the reference is to a page that is present in main memory and that the program has a right to access, the page frame number bits are concatenated with the untranslated offset bits to form the physical address of the item in memory.

Sometimes, because the entire program is not loaded into main memory at once, a reference is made to a page that is not present in main memory. This situation is known as a *page fault*. The memory access cannot be completed, and the MMU interrupts the operating system to ask for help. The operating system must locate the requested page in secondary memory, find an available page frame in main memory (displacing a previously loaded page if memory is full), communicate with the disk controller to cause the page to be loaded, and then restart the program that caused the page fault. To keep the entire system from stalling while the disk is accessed, the operating system will generally transfer control to another process. If this second process has some pages already loaded in main memory, it may be able to run (and thus keep the CPU busy) while the first process is waiting for its page to load. If the second process also encounters a page fault (or has to wait for I/O, etc.), then a third process will be run, and so on.

When a page fault occurs and main memory is full, a previously loaded page must be displaced in order to make room for the new page to

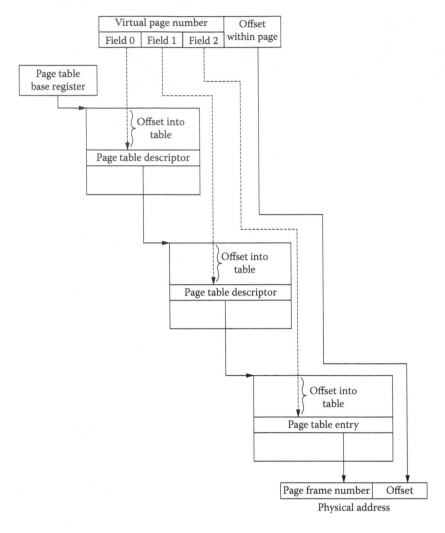

*Figure 2.18* Multiple-level table lookup in paged system.

be loaded. The good news is that paged systems are not subject to external fragmentation; because all the pages are the same size, equal to the size of a main memory page frame, there is never a problem achieving a fit for the page being loaded. However, pages do suffer from internal fragmentation; even if only a few bytes of memory are needed, they take up an entire page frame in memory. (As DRAMs have become larger and less expensive, internal fragmentation is less of a concern than it used to be.) If the displaced page has been modified (if its dirty bit is set), it first must be copied back to disk. Otherwise, it can simply be overwritten with the new page.

Page replacement policies are typically similar to those used for replacement of lines in a cache; however, because pages are replaced much less frequently than cache lines and because the replacement algorithm can be implemented in software rather than hardware, the scheme can be made a bit more sophisticated if desired. FIFO and LRU replacement schemes are common. The main concern in choosing a page to be replaced is to maximize the likelihood of it not being needed again soon. If it is, the system may find itself in the situation of the same page, or the same small number of pages, being repeatedly loaded and displaced. This condition, known as page *thrashing*, results in a large number of page faults. This is a much more costly (in time) situation than the analogous behavior of a cache as page faults take on the order of milliseconds to process, while cache misses cost only a few nanoseconds.

Another potential complicating factor exists in a system with paged virtual memory in which the processor uses a complex instruction set computer (CISC) architecture. As we will discuss in Sections 3.1.6 and 4.4, CISC processors often have machine language instructions that perform operations on vectors or strings that occupy many contiguous memory locations. It is possible that the vector or string being processed may overlap a page boundary; thus, a page fault may occur in the middle of the instruction's execution with part of the operand having been processed and the rest still to be processed after it is loaded into memory. Such a page fault is known as a *delayed page fault*. To handle it, the processor must either be able to handle an exception occurring in the middle of an instruction by later *restarting* the instruction from the point at which the fault occurred, or it must be able to undo, or *roll back,* the effect of the faulting instruction and then re-execute the entire instruction after the needed page is loaded. Both of these mechanisms are nontrivial and significantly complicate the design of the processor. Alternatively, the MMU could precheck all locations that will be accessed by a given instruction to see if any of them will cause a page fault, but this would complicate the design of the MMU and require its designers to have specific knowledge of the CPU architecture. Whatever the solution, the delayed page fault problem shows that no part of a computer system can be designed in a vacuum. CPU design affects memory system design and vice versa. The wise reader will keep this in mind throughout his or her course of study.

## 2.5.4 Segmented virtual memory

Another widely used virtual memory technique is called *segmentation*. A *demand-segmented* memory system maps memory in variable-length segments rather than fixed-size pages. Although it obviously requires hardware for implementation, segmentation is software-oriented in the sense that the length of the segments is determined by the structure of the code or data they contain rather than by hardware constraints, such as disk sector size. (There is always some maximum segment size due to hardware limitations, but it is typically much larger than the size of a page in a demand-paged system.) Because segments can vary in size, main memory is not divided into frames; segments can be loaded anywhere there is sufficient free memory for them to fit. Fragmentation problems are exactly the reverse of those encountered with paging. Because segments can vary in size, internal fragmentation is never a problem. However, when a segment is loaded, it is necessary to check the size of available memory areas (or other segments that might be displaced) to determine where the requested segment will fit. Invariably, over time there will arise some areas of main memory that do not provide a good fit for segments being loaded and thus remain unused; this is known as external fragmentation. Reclaiming these areas of memory involves relocating segments that have already been loaded, which uses up processor time and makes segmentation somewhat less efficient than paging.

A logical address in a system that uses segmentation can be divided into two parts as shown in Figure 2.19. The high-order bits may be considered the segment number, and the low-order bits represent an offset into the segment. The maximum size of a segment determines the number of bits required for the offset; smaller segments will not use the full addressing range. If the maximum segment size is 256 KB ($2^{18}$ bytes), the 18 least significant address bits are reserved for addressing within a segment.

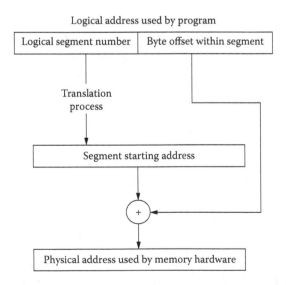

*Figure 2.19* Address translation in a system with demand-segmented virtual memory.

The address translation process, illustrated in Figure 2.20, uses a lookup table in memory, which contains similar information to that found in page tables. A segment table base register points to the base of a lookup table; the logical segment number is used as an index into the table to obtain the *segment table entry* containing the translation information, validity bit, dirty bit, protection bits, etc. As in a paged system, it is possible—and often more efficient—to use a hierarchy of smaller segment tables rather than one large table.

The variable size of segments, as opposed to the fixed size of pages, gives rise to a significant difference between the translation processes. Because segments can be loaded beginning at any address in main memory rather than only on fixed page frame boundaries, the offset cannot simply be concatenated with the translated address. Rather than producing a physical page frame number that provides only the upper bits of the physical address, the segment table lookup produces a complete main memory address that represents the starting location for the segment. The offset within the segment must be added to, rather than simply concatenated with, this address to produce the correct physical address corresponding to the logical address generated by the program.

The occurrence of *segment faults* is analogous to that of page faults in a paged virtual memory system. If the memory access cannot be completed because the requested segment is not loaded in main memory, the

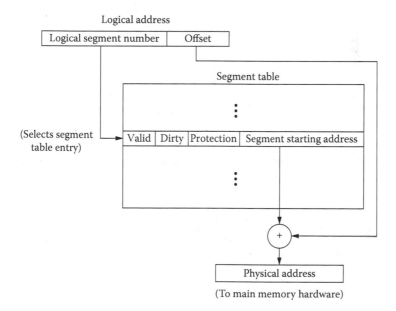

*Figure 2.20* Table lookup for address translation in a segmented system.

MMU alerts the operating system, which locates the segment in secondary memory, finds available space in main memory (a more complicated task than in a paged system), loads it, and then restarts the program that caused the segment fault. Segment replacement algorithms are generally similar to those used in paged systems, but are complicated by the necessity to find not merely an area of memory that has not recently been used, but one into which the requested segment will fit. As in a paged system, if the displaced information has been modified (if its dirty bit is set), it first must be copied back to disk. Otherwise, it can simply be overwritten with the new segment. As segments can be considerably larger than pages, segment faults tend to occur somewhat less frequently than page faults. However, because there is usually more information to load on a segment fault in addition to more overhead, segment faults are more costly, in time, than page faults.

## 2.5.5   Segmentation with paging

Segmentation as a virtual memory technique offers certain advantages over paging as far as the software is concerned. Because segments naturally shrink or grow to accommodate the size and structure of the code and data they contain, they better reflect its organization. Protection and other attributes can more readily be customized for small or large quantities of information as appropriate. Compared with paging, however, segmentation suffers from external fragmentation and other inefficiencies due to mismatches between the characteristics of hardware and software. It is possible to combine segmentation with paging in a single system in order to achieve most of the advantages of both systems. In such a system, segments can still vary in size (up to some maximum) but not arbitrarily. Instead of being sized to the level of individual bytes or words, segments are composed of one or more pages that are a fixed size. Because main memory is divided into fixed-size page frames, segmentation with paging avoids the external fragmentation problems normally associated with segmentation. By keeping the page size fairly small, internal fragmentation (which in any case is a less serious problem in modern systems) can be minimized.

   In a system using segmentation with paging, virtual addresses are divided into (at least) three fields (see Figure 2.21). The upper part of the address is considered the segment number and is used to index into a segment table. From this table, the system obtains the starting address of a page table; the next set of bits, the page number, is used to index into the page table for that segment. The page table entry contains the same information and is used in the same way it would be used in a purely paged memory management scheme. In particular, the page frame number obtained from the page table is concatenated with the offset within the page to form the physical main memory address of the requested

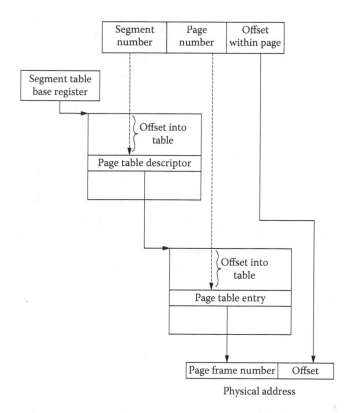

*Figure 2.21* Address translation in a system using segmentation with paging.

information. Thus, at the hardware level, this scheme behaves like a paged system, and to the software, it resembles a segmented implementation.

## 2.5.6   *The MMU and TLB*

Page and segment tables can be quite large and thus have historically been maintained in main memory. They are maintained by the operating system, not by user code, so they must be in an area of main memory that is off limits to user programs. The MMU, which may be on a separate integrated circuit but (in modern systems) is usually fabricated on the same silicon chip as the CPU, typically contains a page (or segment) table base register that points to the beginning address of the top-level table for a given process. (It can be reinitialized by the operating system when a task switch occurs.) It thus serves as a starting point for the indexing of all tables in main memory.

The obvious problem with doing a "table walk" (traversing at least one, and possibly several, levels of lookup tables to obtain a virtual-to-physical address translation) through main memory is that it is a slow process. The program needs to reference memory, which takes a certain cycle time, $t_c$. If a three-level table walk is required to obtain the physical address of the desired location, the time to access this virtual location will be at least $4t_c$. Even if only a single table lookup is required, the memory cycle time will at least double. This added delay will be intolerable in any system in which memory performance is critical.

To avoid having to access memory multiple times for each reference to a virtual memory location, modern MMUs incorporate a feature known as a *translation lookaside buffer* (TLB). The TLB, sometimes known by other names such as an *address translation cache* (ATC), is a cache specially designed to hold recently used page or segment table entries. Because most accesses to a virtual page or segment are soon followed by a number of additional accesses to the same area of memory, the TLB will be "hit" the vast majority of the time, thus dispensing with the main memory table walk for address translation. The needed information is instead obtained from the TLB with only a very small delay. Because the working set of pages or segments for most programs is reasonably small and because only the translations (rather than the memory contents) are cached, it doesn't take a very large TLB to have a very beneficial effect on performance.

## 2.5.7   Cache and virtual memory

Cache and virtual memory are two of the most widely used techniques to improve the performance of memory in modern computer systems. There are many similarities and a few important differences between these two techniques. Some important issues should be taken into consideration when, as is now usually the case, cache and virtual memory are used in the same system. We examine these issues as we conclude our study of memory systems.

Virtual memory and cache are similar in several ways. Most fundamentally, both cache and virtual memory exist for the ultimate purpose of providing instructions and data to the CPU. Because they are used with hierarchical memory systems, both of these techniques involve a larger, slower memory and a smaller, faster memory. The goal of each is to maximize the advantages of two levels of the hierarchy, approximating the speed of the smaller memory while making use of the capacity of the larger. To make this happen, both the cache controller and the memory management unit use hardware to map addresses. Both paged and segmented virtual memory systems, as well as cache, operate on a demand basis—that is, they all replace older information in the faster memory with newer information as it is requested by the CPU, not by trying to predict usage in advance. Both cache and paged virtual memory systems transfer fixed-size blocks of data between the faster and slower memories. (Segmented virtual memory implementations also transfer blocks of data, but they are variable in size.) The principle of locality of reference is important to the operation of both cache and virtual memory systems. Because of locality, the vast majority of accesses are satisfied by referring to the smaller, faster memory. When the needed information is not found in the smaller memory, both techniques suffer a significant performance penalty.

The significant differences between cache and virtual memory implementations are due to their different places within the hierarchy and the relative speeds of the levels involved. Although both transfer blocks of information between levels, the size of the cache blocks (refill lines) is generally significantly smaller than the size of the blocks (segments or pages) transferred between main memory and disk in a system with virtual memory. Refill lines are generally only a few, up to a maximum of 128 to 256, bytes in size; pages are often 4 KB or more; and segments may be even larger. At least partially due to this size discrepancy, segment or page faults tend to be much rarer than cache misses. Cache hit ratios are typically in the range of 90 to 98%, while referenced segments or pages are found in main memory well in excess of 99.9% of the time. This is fortunate because the time penalty for a page or segment fault is much greater than that incurred by a cache miss. Often only a few to a few dozen clock cycles (wait states) may be required to access main memory on a cache miss, during which time the CPU can simply idle and then immediately resume executing instructions; at most, only a few nanoseconds are wasted. By contrast, a segment or page fault requires one or more disk accesses, which may take several milliseconds each (an eternity to a modern CPU), plus the overhead of a task switch in order to run another process while the faulting one waits. Because of the speed required at the uppermost levels of the memory hierarchy, cache control is done solely in hardware; management of virtual memory is done partly in hardware (by the MMU) and partly in software (by the operating system).

The only purpose of cache is to increase the apparent speed of main memory; virtual memory has several functions. The most important of these is to provide each program with the appearance of a large main memory (often the size of secondary memory or greater) for its exclusive use. Secondary goals include support for multiprogramming (multiple programs resident in memory at the same time), program relocation, and memory space protection (so that one program does not access memory belonging to another). All of these goals are realized at some cost in performance, but ideally without slowing main memory access too much.

We have treated virtual memory and cache design as two separate topics, but neither exists in a vacuum in modern computer system design. All but the simplest systems incorporate one or more cache memories as part of an overall memory system including virtual memory management. The cache controller and MMU must interact on a cycle-by-cycle basis; they are usually designed together, and both usually reside on the same integrated circuit as the CPU itself. What are some of the design issues that must be dealt with when designing a virtual memory system that includes cache memory?

The first and most fundamental cache design choice in a system with virtual memory is whether to cache information based on virtual or physical addresses. In other words, do we use part of the untranslated (virtual) address as the tag (and index if the cache is not fully associative) required to locate an item in cache, or do we translate the virtual address first and determine the cache tag and index from the physical address? There are advantages and disadvantages either way.

The main advantage of a virtually addressed cache is speed. Because the cache controller does not have to wait for the MMU to translate the supplied address before checking for a hit, the needed information can be accessed more quickly when a hit occurs. Misses can likewise be detected and the required main memory access started almost immediately. (The address translation in the MMU can proceed in parallel with the cache controller's determination of hit versus miss, with translation being aborted in the event of a hit.) Another advantage is consistency of cache behavior. Because cache access patterns depend on the virtual addresses used by the program and because these are the same from one run to the next, identical runs will result in identical cache usage patterns and thus identical hit ratios.

Physically addressed caches, however, may cause programs to exhibit performance variations between otherwise identical runs. This is because the operating system may load the same program at different physical addresses depending on extraneous factors, such as other programs already in memory. This means the cache tags and indices for different runs may be different and may lead to a different pattern of hits, misses, and line replacements. In a set-associative or (especially) a direct-mapped cache, different addresses can result in different patterns of contention and significantly affect the hit ratio (and overall performance). In addition to variations in performance, physically addressed caches are not quite as fast in the best case because the address translation must be completed before the cache lookup can begin.

Do these disadvantages mean that physically addressed caches are never preferable? Not at all: In some situations, they may be preferable to virtually addressed caches, or even necessary. Because all the cache tags and indices are based on a single, physical address space rather than a separate virtual address space for each process, information can be left in a physically addressed cache when a task switch occurs. In a virtually addressed cache, we would have to worry about address $n$ from one process matching the cached address $n$ of a different process, so the cache would have to be "flushed" or completely invalidated on each change of context. (Alternatively, process IDs could be stored along with the cache tags, with a hit being recognized only if both match; however, this would increase implementation cost.) Conversely, because identical virtual addresses referenced by different processes will map to different physical locations unless data are being intentionally shared, there is no need to flush the physically addressed cache. This property may give a performance advantage in a multithreaded, multitasking system in which task switches are frequent. Physical cache addressing may be necessary in some applications, particularly when an off-chip, level 2 or 3 cache is being designed. If the MMU is fabricated on the same package with the CPU, then address translation takes place before off-chip hardware ever "sees" the address, and there is no alternative to a physical cache mapping.

## 2.6 Chapter wrap-up

Many nuances of memory system design are beyond the scope of this book. As with any other highly specialized craft, one learns memory design best by actually doing it, and current knowledge has a way of quickly becoming obsolete. As technology changes, approaches that were once in favor become less attractive and vice versa. However, the basic principles of a hierarchical memory system design have been the same for at least the past 40 or 50 years, and they are likely to remain valid for some time. Although the most intricate details of memory system design are constantly changing (and perhaps best left to specialists), an appreciation of the basic principles behind these designs as explained in this chapter is important to any professional in the computing field. Whether you are an application programmer wanting to extract the maximum performance from your code, a systems guru trying to find the most efficient way to manage multitasking on a particular architecture, or a technical manager looking to purchase the best-performing system for your department's applications, memory system performance is critical to your goals. The better you understand how computer memory systems work and why they are designed the way they are, the more informed—and likely, more successful—will be your decisions.

### REVIEW QUESTIONS

1. Consider the various aspects of an ideal computer memory discussed in Section 2.1.1 and the characteristics of available memory devices discussed in Section 2.1.2. Fill in the columns of the table below with the following types of memory devices, in order from most desirable to least desirable: magnetic hard disk, semiconductor DRAM, CD-R, DVD-RW, semiconductor ROM, DVD-R, semiconductor flash memory, magnetic floppy disk, CD-RW, semiconductor static RAM, and semiconductor EPROM.

| Cost/bit (lower is better) | Speed (higher is better) | Information density (higher is better) | Volatility (non-volatile is better) | Readable/ writable? (both is usually better) | Power consumption (lower is better) | Durability (more durable is better) | Removable/portable? (more portable is usually better) |
| --- | --- | --- | --- | --- | --- | --- | --- |

2. Describe in your own words what a hierarchical memory system is and why it is used in the vast majority of modern computer systems.

3. What is the fundamental, underlying reason that low-order main memory interleaving and/or cache memories are needed and used in virtually all high-performance computer systems?

4. A main memory system is designed using 15-ns RAM devices using a four-way low-order interleave.

   a. What would be the effective time per main memory access under ideal conditions?

   b. What would constitute ideal conditions? (In other words, under what circumstances could the access time you just calculated be achieved?)

   c. What would constitute worst-case conditions? (In other words, under what circumstances would memory accesses be the slowest?) What would the access time be in this worst-case scenario? If ideal conditions exist 80% of the time and worst-case conditions occur 20% of the time, what would be the average time required per memory access?

   d. When ideal conditions exist, we would like the processor to be able to access memory every clock cycle with no wait states (that is, without any cycles wasted waiting for memory to respond). Given this requirement, what is the highest processor bus clock frequency that can be used with this memory system?

   e. Other than increased hardware cost and complexity, are there any potential disadvantages of using a low-order interleaved memory design? If so, discuss one such disadvantage and the circumstances under which it might be significant.

5. Is it correct to refer to a typical semiconductor integrated circuit ROM as a random access memory? Why or why not? Name and describe two other logical organizations of computer memory that are not random access.

6. Assume that a given system's main memory has an access time of 6.0 ns, and its cache has an access time of 1.2 ns (five times as fast). What would the hit ratio need to be in order for the effective memory access time to be 1.5 ns (four times as fast as main memory)?

7. A particular program runs on a system with cache memory. The program makes a total of 250,000 memory references; 235,000 of these are to cached locations.

   a. What is the hit ratio in this case?

   b. If the cache can be accessed in 1.0 ns but the main memory requires 7.5 ns for an access to take place, what is the average time required by this program for a memory access, assuming all accesses are reads?

    c.  What would be the answer to (b) if a write-through policy is used and 75% of memory accesses are reads?

8. Is hit ratio a dynamic or static performance parameter in a typical computer memory system? Explain your answer.

9. What are the advantages of a set-associative cache organization as opposed to a direct-mapped or fully associative mapping strategy?

10. A computer has 64 MB of byte-addressable main memory. A proposal is made to design a 1 MB cache memory with a refill line (block) size of 64 bytes.

    a.  Show how the memory address bits would be allocated for a direct-mapped cache organization.

    b.  Repeat (a) for a four-way set-associative cache organization.

    c.  Repeat (a) for a fully associative cache organization.

    d.  Given the direct-mapped organization and ignoring any extra bits that might be needed (valid bit, dirty bit, etc.), what would be the overall size ("depth" by "width") of the memory used to implement the cache? What type of memory devices would be used to implement the cache (be as specific as possible)?

    e.  Which line(s) of the direct-mapped cache could main memory location $1E0027A_{16}$ map into? (Give the line number[s], which will be in the range of 0 to $[n-1]$ if there are $n$ lines in the cache.) Give the memory address (in hexadecimal) of another location that could not reside in cache at the same time as this one (if such a location exists).

11. Define and describe virtual memory. What are its purposes, and what are the advantages and disadvantages of virtual memory systems?

12. Name and describe the two principal approaches to implementing virtual memory systems. How are they similar, and how do they differ? Can they be combined, and if so, how?

13. What is the purpose of having multiple levels of page or segment tables rather than a single table for looking up address translations? What are the disadvantages, if any, of this scheme?

14. A process running on a system with demand-paged virtual memory generates the following reference string (sequence of requested pages): 4, 3, 6, 1, 5, 1, 3, 6, 4, 2, 2, 3. The operating system allocates each process a maximum of four page frames at a time. What will be the number of page faults for this process under each of the following page replacement policies?

    a.  LRU

    b.  FIFO

    c.  LFU (with FIFO as tiebreaker)

15. In what ways are cache memory and virtual memory similar? In what ways are they different?

16. In systems that make use of both virtual memory and cache, what are the advantages of a virtually addressed cache? Does a physically addressed cache have any advantages of its own, and if so, what are they? Describe a situation in which one of these approaches would have to be used because the other would not be feasible.

17. Fill in the blanks below with the most appropriate term or concept discussed in this chapter:

_____ A characteristic of a memory device that refers to the amount of information that can be stored in a given physical space or volume.

_____ A semiconductor memory device made up of a large array of capacitors; its contents must be periodically refreshed in order to keep them from being lost.

_____ A developing memory technology that operates on the principle of magnetoresistance; it may allow the development of "instant-on" computer systems.

_____ A type of semiconductor memory device, the contents of which cannot be overwritten during normal operation but can be erased using ultraviolet light.

_____ This type of memory device is also known as a CAM.

_____ A register in an associative memory that contains the item to be searched for.

_____ The principle that allows hierarchical storage systems to function at close to the speed of the faster, smaller level(s).

_____ This occurs when a needed instruction or operand is not found in cache, so a main memory access is required.

_____ The unit of information that is transferred between a cache and main memory.

_____ The portion of a memory address that determines whether a cache line contains the needed information.

_____ The most flexible but most expensive cache organization, in which a block of information from main memory can reside anywhere in the cache.

_____ A policy whereby writes to cached locations update main memory only when the line is displaced.

_____ This is set or cleared to indicate whether a given cache line has been initialized with "good" information or contains "garbage" because it is not yet initialized.

_____ A hardware unit that handles the details of address translation in a system with virtual memory.

_____ This occurs when a program makes reference to a logical segment of memory that is not physically present in main memory.

_____ A type of cache used to hold virtual-to-physical address translation information.

_____ This is set to indicate that the contents of a faster memory subsystem have been modified and need to be copied to the slower memory when they are displaced.

_____ This can occur during the execution of a string or vector instruction when part of the operand is present in physical main memory and the rest is not.

# chapter three

# Basics of the central processing unit

The central processing unit (CPU) is the brain of any computer system based on the von Neumann (Princeton) or Harvard architectures introduced in Chapter 1. Parallel machines have many such brains, but normally each of them is based on the same principles used to design the CPU in a *uniprocessor* (single CPU) system. A typical CPU has three major parts: the arithmetic/logic unit (ALU), which performs calculations; internal registers, which provide temporary storage for data to be used in calculations; and the control unit, which directs and sequences all operations of the ALU and registers as well as the rest of the machine. (A block diagram of a simple CPU is shown in Figure 3.1.) The control unit that is responsible for carrying out the sequential execution of the stored program in memory is the hallmark of the von Neumann–type machine, using the registers and the arithmetic and logical circuits (together known as the *datapath*) to do the work. The design of the control unit and datapath have a major impact on the performance of the processor and its suitability for various types of applications. CPU design is a critical component of overall system design. In this chapter, we look at important basic aspects of the design of a typical general-purpose processor; in the following chapter, we will go beyond the basics and look at modern techniques for improving CPU performance.

## 3.1 The instruction set

One of the most important features of any machine's architectural design, yet one of the least appreciated by many computing professionals, is its instruction set architecture (ISA). The ISA determines how all software must be structured at the machine level. The hardware only knows how to execute machine language instructions, but because almost all software is now developed in high-level languages rather than assembly or machine language, many people pay little or no attention to what type of instruction set a machine supports or how it compares to those used in other machines. As long as the system will compile and run a C++ program or support a Java virtual machine, is its native machine language really all that important? If all you are concerned with is getting a given program to run, probably not. But if you are interested in making a system perform

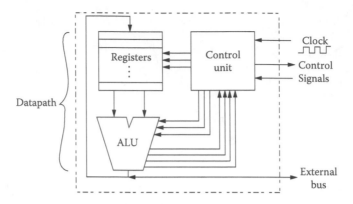

**Figure 3.1** Block diagram of a simple CPU.

to the best of its abilities, then it behooves you to know what those abilities really are, and the only ability of a CPU that really matters is its ability to execute machine instructions. What those instructions are, what they do, and how they support your high-level task can have a great deal to do with how quickly and efficiently that task will be done. Thus, it is worthwhile to study the similarities and differences between ISAs in order to be able to pick the best system for a given application. We examine important features of computer ISAs in this section.

### 3.1.1  *Machine language instructions*

In order to appreciate the features of machine ISAs, it is important to realize what a computer program is. You may think of it as a C, C++, Fortran, or Java listing, but all that high-level syntax is left behind in producing an executable file. In the end, regardless of how it was originally specified, a computer program is a sequence of binary machine language instructions to be performed in order (that order, of course, often being alterable based on the outcome of tests performed and decisions made while the program is running). Each machine language instruction is no more and no less than a collection of bits that are decoded by logic inside the processor's control unit; each combination of bits has a unique meaning to the control unit, telling it to perform an operation on some item of data, move data from one place to another, input or output data from or to a device, alter the flow of program execution, or carry out some other task that the machine's hardware is capable of doing.

Machine language instructions may be fixed or variable in length and are normally divided into "fields," or groups of bits, each of which is decoded to tell the control unit about a particular aspect of the instruction. A simple machine might have 16-bit instructions divided into, for example, five fields as shown in Figure 3.2. (This example is taken from the instruction set of DEC's PDP-11 minicomputer of the early 1970s.)

The first set of bits comprise the *operation code* field (*op code* for short). They define the operation to be carried out. In this case, because there are four op code bits, the computer can have at most $2^4 = 16$ instructions (assuming this is the only instruction format). The control unit would use a 4 to 16 decoder to uniquely identify which operation to perform based on the op code for a particular instruction.

The remaining 12 bits specify two operands to be used with the instruction. Each operand is identified by two three-bit fields: one to specify a register and one to specify an *addressing mode* (or way of determining the operand's location given the contents of the register in question). Having three bits to specify a register means that the CPU is limited to having only $2^3 = 8$ internal registers (or, at least, that only eight can be used for the purpose of addressing operands). Likewise, allocating three bits to identify an addressing mode means there can be no more than eight such modes. One of them might be register direct, meaning the operand is in the specified register; another might be register indirect, meaning the register contains not the operand itself but a pointer to the operand in memory. (We will examine addressing modes more thoroughly in Section 3.1.3.) Each of the three-bit fields will be interpreted by a 3 to 8 decoder, the outputs of which will be used by the control unit in the process of determining the locations of the operands so that they can be accessed. Thus, within the 16 bits of the instruction, the CPU finds all the information it needs to determine what it is to do next.

A machine may have one or several formats for machine language instructions. (A processor with variable-length instructions will have multiple machine language formats, but there may also be multiple formats for a given instruction size. Oracle's SPARC architecture, for example, has three 32-bit instruction formats.) Each format may have different-sized bit fields and have different meanings for the fields. Figure 3.3 shows an example of an architecture that has fixed-length, 32-bit instructions with three formats. Notice how the two leftmost bits are used to identify the format of the remaining 30 bits. This would correspond to a two-level decoding scheme in which the outputs of a 2 to 4 decoder driven by the two format bits would determine which set of secondary decoders would be used to interpret the remaining bits. This type of arrangement

| Op code | Reg. 1 | Mode 1 | Reg. 2 | Mode 2 |
|---------|--------|--------|--------|--------|
| (4) | (3) | (3) | (3) | (3) |

*Figure 3.2* Simple machine instruction format with five bit fields.

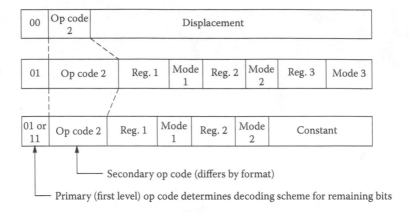

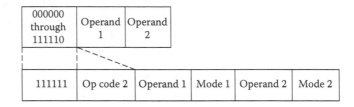

*Figure 3.3* Architecture with multiple instruction formats of the same length.

*Figure 3.4* Architecture with multiple instruction formats of different lengths.

makes the control unit a little more complex but allows the machine to have a greater variety of instructions than would be possible with just one instruction format.

Because the types of instructions and number of operands required for each sometimes vary considerably, it is often more space-efficient to encode some types of instructions in fewer bits while others take up more bits. Having instructions of variable lengths complicates the process of fetching and decoding instructions (the first two steps in the von Neumann execution cycle; see Figure 1.2) but may be justified if keeping executable code size small is important. Figure 3.4 shows an example of an instruction set with some 16-bit instructions and some 32-bit instructions. One particular op code (111111) from the shorter format is used to tell the CPU that the next 16 bits are to be interpreted as the second part of the current instruction, rather than the next instruction.

### 3.1.2   *Functional categories of instructions*

Computer architectures vary considerably in their design details and intended applications. Some are intended for scientific computing, some for business applications, some for networking, some for embedded control, and so on. Some are designed to maximize performance, and some are intended to minimize cost, power consumption, or other expenditure of resources. Some are reduced instruction set computers (RISCs), some are complex instruction set computers (CISCs), and some are somewhere in between. Because of these differences, computer ISAs vary widely in the types of operations they support. However, there is a great deal of commonality between the instruction sets of almost every Princeton or Harvard architecture machine, going all the way back to the first generation. Although the specifics vary greatly, virtually every computer architecture implements instructions to carry out a few general categories of tasks. Let us look at the major classifications.

*Data transfer instructions* are the most common instructions found in computer programs. To help operations proceed as quickly as possible, we like to work with operands in the processor's internal registers. (Some architectures—particularly RISCs—require this.) However, there are a limited number of registers available, so programs spend much of their time moving data around to get values where they are needed and store them back into memory when they are no longer needed. Instructions of this type include transfers from one internal register to another, from main memory to a register, or from a register to a memory location. Typical assembly language mnemonics for data transfer instructions include MOVE, LOAD, STORE, XCHG, PUSH, POP, and so on.

*Computational instructions* are usually the second most numerous type in a program. These are the instructions that "do the work"—that is, perform manipulations (arithmetic, logical, shift, or other operations) of data. Some texts separate computational instructions into the categories of arithmetic operations and logical operations, but this is an arbitrary distinction. All of these instructions, regardless of the specific operation they implement, involve passing binary data through the processor's functional hardware (usually referred to as the ALU) and using some combination of logic gates to operate on it, producing a result (and, in most cases, a set of condition codes or flags that indicate the nature of the result: positive/negative, zero/nonzero, carry/no carry, overflow/no overflow, etc.). Typical assembly language mnemonics for computational instructions include ADD, SUB, MUL, AND, OR, NOT, SHL, SHR, etc.

*Control transfer instructions* are usually the third most frequent type encountered in machine-level code. These are the instructions that allow the normally sequential flow of execution to be altered, either unconditionally or conditionally. When an unconditional transfer of control is encountered, or when a conditional transfer of control succeeds, the next instruction executed is not the next sequential instruction in memory, as it normally would be in a von Neumann machine. Instead, the processor goes to a new location in memory (specified either as an absolute binary address or relative to the location of the current instruction) and continues executing code at that location. Control transfer instructions are important because they allow programs to be logically organized in small blocks of code (variously known as procedures, functions, methods, subroutines, etc.) that can call or be called by other blocks of code. This makes programs easier to understand and modify than if they were written as huge blocks. Even more importantly, conditional control transfer instructions allow programs to make decisions based on input and the results of prior calculations. They make it possible for high-level control structures, such as IF and SWITCH statements, loops, and so on, to be implemented at the machine level. Without these types of instructions, programs would have to process all data the same way, making code much less versatile and useful. Typical mnemonics for control transfer instructions include JMP, BNZ, BGT, JSR, CALL, RET, etc.

The remaining, less commonly used instructions found in most computer architectures may be lumped together or divided into groups depending on one's taste and the specifics of a given instruction set. These instructions may include *input/output* (I/O), *system*, and *miscellaneous instructions*. I/O instructions (typified by the Intel x86's IN and OUT) are essentially self-explanatory; they provide for the transfer of data to and from I/O device interfaces. Not all architectures have a distinct category of I/O instructions; they are only present on machines that employ *separate I/O* (to be discussed in Chapter 5) rather than *memory-mapped I/O*. Machines with memory-mapped I/O use regular memory data transfer instructions to move data to or from I/O ports.

System instructions are special operations involving management of the processor and its operating environment. As such, they are often *privileged* instructions that can only be executed when the system is in "supervisor" or "system" mode (in which the operating system runs). Attempts by user code to execute such instructions would then result in an *exception* (see Chapter 5) alerting the OS of the privilege violation. System instructions include such functions as virtual memory management, enabling and freezing the cache, masking and unmasking interrupts, halting the system, and so on. Such tasks need to be done in many systems, but not all programs should be allowed to do them.

Finally, most systems have a few instructions that cannot easily be categorized in one of the preceding classifications. This may be because they perform some special function unique to a given architecture or because they perform no function at all. We may refer to these as miscellaneous instructions. Perhaps the most ubiquitous such instruction is NOP, or no operation, which is used to kill time while the processor waits for some other hardware (or a human user). An instruction with no functionality is difficult to categorize.

All, or almost all, of the functional categories of instructions presented in this section are common to every type of computer designed around a Princeton or Harvard architecture. What differs considerably from one architecture to another is how many, and which, instructions of each type are provided. Some architectures have very rich instruction sets with specific machine instructions for just about every task a programmer would want to accomplish, and others have a minimalist instruction set with complex operations left to be accomplished in software by combining several machine instructions. The Motorola 68000, for example, has machine instructions to perform signed and unsigned integer multiplication and division. The original SPARC architecture provided only basic building blocks, such as addition, subtraction, and shifting operations, leaving multiplication and division to be implemented as library subroutines. (The more recent UltraSPARC, or SPARC v9, architecture does provide for built-in multiply and divide instructions.)

The point one should take from this variety of instruction sets is that hardware and software are equivalent, or interchangeable. The only machine instructions that are required are either a NAND or NOR instruction to do all computations (remember, those two are the only universal logic functions from which all other Boolean functions can be derived) plus some method for transferring data from one place to another and some way of performing a conditional transfer of control (the condition can be forced to true to allow unconditional transfers). One or two shift instructions would be nice, too, but we could probably get by without them if we had to. Once these most basic, required hardware capabilities are in place, the amount of additional functionality that is provided by hardware versus software is completely up to the designer. At various points in the history of computing, the prevailing wisdom (and economic factors) have influenced that choice more in the direction of hardware and, at other times, more in the direction of software. Because hardware and software perform the same functions and because technological and economic factors are continually changing, these choices will continue to be revisited in the future.

## 3.1.3  Instruction addressing modes

Regardless of the specific machine instructions implemented in a given computer architecture, it is certain that many of them will need to access operands to be used in computations. These operands may reside in CPU registers or main memory locations. In order to perform the desired operation, the processor must be able to locate the operands, wherever they may be. The means within a machine instruction that allow the programmer (or compiler) to specify the location of an operand are referred to as *addressing modes*. Some architectures provide a wide variety of addressing modes, and others provide only a few. Let us review some of the more commonly used addressing modes and see how they specify the location of an operand.

*Immediate addressing* embeds an operand into the machine language instruction itself (see Figure 3.5). In other words, one of the bit fields of the instruction contains the binary value to be operated upon. The operand is available immediately because the operand fetch phase of instruction execution is done concurrently with the first phase, instruction fetch.

This type of addressing is good for quick access to constants that are known when the program is written but is not useful for access to variables. It is also costly in terms of instruction set design because the constant takes up bits in the machine language instruction format that could be used for other things. Either the size of the instruction may have to be increased to accommodate a reasonable range of immediate constants, or the range of values that may be encoded must be restricted to fit within the constraints imposed by instruction size and the need to have bit fields for other purposes. These limitations have not proven to be a real deterrent; almost all computer ISAs include some form of immediate addressing.

*Direct addressing* (sometimes referred to as absolute addressing) refers to a mode in which the location of the operand (specifically, its address in main memory) is given explicitly in the instruction (see Figure 3.6). Operand access using direct addressing is slower than using immediate addressing because the information obtained by fetching the instruction does not include the operand itself, but only its address; an additional memory access must be made to read or write the operand. (Because the operand resides in memory rather than in the instruction itself, it is not read-only as is normally the case with immediate addressing.) In other respects, direct addressing has some of the same limitations as immediate addressing; as with the immediate constant, the memory address embedded in the instruction takes up some number of bits (quite a few if the machine has a large addressing range). This requires the instruction format to be large or limits the range of locations that can be accessed using this mode. Direct addressing is useful for referencing scalar variables but

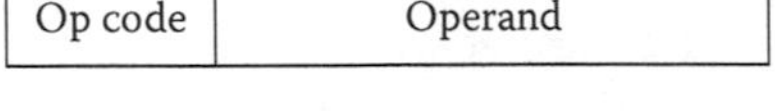

| Op code | Operand |
| --- | --- |

**Figure 3.5** Machine language instruction format using immediate addressing.

| Op code | Address of operand |
| --- | --- |

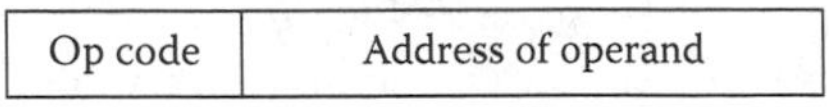

**Figure 3.6** Machine language instruction format using direct addressing.

has limited utility for strings, arrays, and other data structures that take up multiple memory locations.

*Register addressing*, where the operand resides in a CPU register, is logically equivalent to direct addressing because it also explicitly specifies the operand's location. For this reason, it is also sometimes known as *register direct addressing*. The principal advantage of register addressing is quick access to the operand (as there is no need to read or write memory). However, the number of registers accessible to the program is limited. One common optimization is placing the most frequently accessed variables in registers, while those that are referenced less often reside in memory.

*Indirect addressing* is one logical step further removed from specifying the operand itself. Instead of the instruction containing the operand itself (immediate) or the address of the operand (direct), it contains the specification of a pointer to the operand. In other words, it tells the CPU where to find the operand's address. The pointer to the operand can be found in a specified register (*register indirect addressing*, see Figure 3.7) or, in some architectures, in a specified memory location (*memory indirect addressing*, see Figure 3.8).

The principal advantage of indirect addressing is that not only can the value of the operand be changed at run time (making it a variable instead of a constant), but because it resides in a modifiable register or memory location, the address of the operand can also be modified "on the fly." This means that indirect addressing readily lends itself to the processing of arrays, strings, or any type of data structure stored in contiguous memory locations. By incrementing or decrementing the pointer by the size of an individual element, one can readily access the next or previous element in the structure. This coding structure can be placed inside a loop for convenient processing of the entire data structure. The only major drawback of

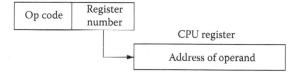

*Figure 3.7* Register indirect addressing.

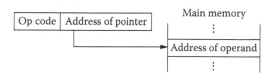

*Figure 3.8* Memory indirect addressing.

indirect addressing is that determining the operand's address takes time. The CPU must locate the pointer, obtain its contents, and then use those contents as the address for reading or writing the operand. If the pointer is in a memory location (memory indirect), then memory must be accessed at least three times: once to read the instruction, a second time to read the pointer, and a third time to read or write the data. (If the operand is both a source of data and the destination of the result, a fourth access would be needed.) This is slow and tends to complicate the design of the control unit. For this reason, many architectures implement indirect addressing only as register indirect, requiring pointers to be loaded into registers before use.

*Indexed addressing*, also known as displacement addressing, works similarly to register indirect addressing but has the additional feature of a second, constant value embedded in the instruction that is added to the contents of the pointer register to form the effective address of the operand in memory (see Figure 3.9). One common use of this mode involves encoding the starting address of an array as the constant displacement and using the register to hold the array index or offset of the particular element being processed—hence the designation of indexed addressing.

Most computer architectures provide at least a basic form of indexed addressing similar to that just described. Some also have more complex implementations in which the operand address is calculated as the sum of the contents of two or more registers, possibly in addition to a constant displacement. (One example is the *based indexed* addressing mode built into the Intel x86 architecture.) Such a mode, if provided, complicates the design of the processor somewhat but provides a way for the assembly language programmer (or compiler) to easily access elements of two-dimensional arrays or matrices stored in memory.

*Program counter relative addressing* is used in many computer architectures for handling control transfers, such as conditional branches and

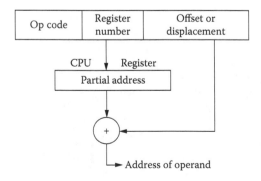

*Figure 3.9* Indexed addressing.

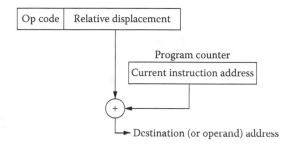

*Figure 3.10* PC relative addressing.

subroutine calls. The idea is that an offset (usually signed to allow forward or backward branching) is added to the current program counter (PC) value to form the destination address. This value is copied back to the PC, effectively transferring control to that address (see Figure 3.10).

The advantage of using PC relative addressing, rather than simply specifying an absolute address as the destination, is that the resulting code is position independent—that is, it can be loaded anywhere in memory and still execute properly. The absolute address of a called routine or branch target may have changed, but it is still in the same place relative to the instruction that transfers control to it. For this reason, most computer architectures provide at least some subset of control transfer instructions that operate this way. Some architectures (the Motorola 680x0 family is a notable example) also implement PC relative addressing for instructions that access memory for operands. In this way, blocks of data can be relocated along with code in memory and still appear to the program logic to be in the same place.

*Stack addressing* involves references to a last-in, first-out (LIFO) data structure in memory. Nearly all architectures have "push" and "pop" instructions, or their equivalents, that facilitate storing items on and retrieving them from a stack in memory. These instructions use a given register as a stack pointer to identify the current "top of stack" location in memory, automatically decrementing or incrementing the pointer as items are pushed or popped. Most architectures only use stack addressing for data transfers, but some machines have computational instructions that retrieve an operand(s) and store a result on a memory stack. This approach has the same problems of complexity and sluggishness that are inherent to any manipulation of data in memory, but it does reduce the need for CPU registers to hold operands.

## 3.1.4 Number of operands per instruction

One of the most significant choices made in the development of a computer architecture, and one of the most obvious (evident as soon as one

looks at a line of assembly code), is the determination of how many operands each instruction will take. This decision is important because it
balances ease of programming against the size of machine instructions.
Simply put, the more operands an instruction accesses, the more bits are
required to specify the locations of those operands.

From a programming point of view, the ideal number of operands
per instruction is probably three. Most arithmetic and logical operations,
including addition, subtraction, multiplication, logical AND, OR, NAND,
NOR, XOR, etc. take two source operands and produce one result (or destination operand). For complete flexibility, one would like to be able to independently specify the locations (in registers or memory) of both source
operands and the destination. Some architectures thus implement *three-
operand instructions*, but many do not. Reasons include the complexity of
instructions (especially if some or all of the operands can be in memory
locations) and the need for larger machine language instructions to have
enough bits to locate all the operands.

Many architectures, including some of the most popular, have adopted
the compromise of *two-operand instructions*. This maintains a reasonable
degree of flexibility while keeping the instructions somewhat shorter than
they would be with three operands. The rationale is that many frequently
used operations, such as negation of a signed number, logical NOT, and
so on, require only two operands anyway. When two source operands
are needed, one of them can double as a destination operand, being overwritten by the result of the operation. In many cases, at least one of the
operands is only an intermediate result and may safely be discarded; if
it is necessary to preserve its value, one extra data transfer instruction
will be required prior to the operation. If the architects determine that an
occasional extra data transfer is worth the savings in cost and complexity
from having only two operands per instruction, then this is a reasonable
approach.

*One-operand instructions* were once fairly common but are not seen in
many contemporary architectures. In order for a single operand specification to be workable for operations that require two source operands, the
other operand must reside in a known location. This is typically a special
register in the CPU called the *accumulator*. The accumulator normally provides one of the operands for every arithmetic and logical operation and
also receives the result; thus, the only operand that must be located is the
second source operand. This is a very simple organizational model that
was quite popular in the days when processors could have few internal
registers. However, because a single register is involved in every computational instruction, the program must spend a great deal of time (and
extra machine instructions) moving operands into, and results out of,
that register. The resulting "accumulator bottleneck" tends to limit performance. For this reason, accumulator-based architectures are seldom

found outside of embedded control processors where simplicity and low cost are more important considerations than speed.

*Zero-operand instructions* would seem to be an oxymoron. How can an architecture have operational instructions without operands? It cannot, of course; as in the case of an accumulator machine, the operand locations must be known implicitly. A machine could have two accumulators for operands and leave all results in one of the accumulators. This would be workable but could result in even more of a bottleneck than using one-operand instructions.

More practically, zero-operand instructions could be used in a machine with a *stack architecture* (one in which all computational operands come from the top of a stack in memory and all results are left on the stack as described at the end of Section 3.1.3). A limited number of architectures have employed this approach, which has the virtue of keeping the machine instructions as simple and short as possible (consisting only of an op code). Compiler design is also simplified because a stack architecture dispenses with the problem of register allocation. However, stack architectures suffer from a significant performance penalty because operands must be fetched from and stored back to memory for each computation. (A register-based machine allows intermediate results and repeatedly used values to be kept in the processor for quicker access until a final answer is calculated and returned to memory.) Because of this disadvantage, stack architectures are even less common than accumulator architectures.

### 3.1.5 Memory-register versus load-store architectures

Because of the limitations of stack and accumulator-based architectures, most modern CPUs are designed around a set of several (typically 8, 16, 32, or more) general-purpose internal registers. This allows at least some of the operands used in computations to be available without the need for a memory access. An important question to be answered by the designer of an instruction set is, "Do we merely want to *allow* the programmer (or compiler) to use register operands in computations, or will we *require* all computations to use register operands only?" The implications of this choice have more far-reaching effects than one might think.

A *memory–register* architecture is one in which many, if not all, computations may be performed using data in memory and in registers. Many popular architectures, such as the Intel x86 and Motorola 680x0 families, fall into this category. The Intel processor has a considerable number of two-operand instructions written in assembly language as "ADD destination, source." (Subtraction and logical operations use the same format.) Both operands are used in the computation, with the first being overwritten by the result. It is allowable for both operands to be CPU registers; it is also possible for either one (but not both) to be main memory locations, specified using any addressing mode the architecture supports. So "ADD EAX, EBX," "ADD EAX, [EBX]," and "ADD [EBX], EAX" are all valid forms of the addition instruction. (EAX and EBX are the names of CPU registers, and square brackets indicate register indirect addressing.) Both register–register and memory–register computations are supported. Most x86 instructions do not provide for this, but some other architectures allow all of an instruction's operands to be in memory locations (a *memory–memory* architecture).

The advantages of a memory–register (or memory–memory) architecture are fairly obvious. First of all, assembly language coding is simplified. All of the machine's addressing modes are available to identify an operand for any instruction. Computational instructions such as ADD behave just like data transfer instructions such as MOV. (Fortunately, Intel is better at CPU design than it is at spelling.) The assembly language programmer and the high-level language compiler have a lot of flexibility in how they perform various tasks. To the extent that operands can reside in memory, the architecture need not provide as many registers; this can save time on an interrupt or context switch because there is less CPU state information to be saved and restored. Finally, executable code size tends to be smaller. A given program can contain fewer instructions because the need to move values from memory into registers before performing computations is reduced (or eliminated if both operands can reside in memory).

The disadvantages of a memory–register architecture (which are further exacerbated in a memory–memory architecture) are a bit less obvious, which helps explain why such architectures were dominant for many years and why they continue to be widely used today. The main disadvantage of allowing computational instructions to access memory is that it makes the control unit design more complex. Any instruction may need to access memory one, two, three, or even four or more times before it is completed. This complexity can readily be handled in a *microprogrammed* design (see Section 3.3.3), but this generally incurs a speed penalty. Because the number of bits required to specify a memory location is generally much greater than the number required to uniquely identify a register, instructions with a variable number of memory operands must either be variable in length (complicating the fetch/decode process) or all must be the size of the longest instruction format (wasting bits and making programs take up more memory). Finally, the more memory accesses an instruction requires, the more clock cycles it will take to execute. Ideally, we would like instructions to execute in as few clock cycles as possible; and even more to the point, we would like all of them to execute in the same number of clock cycles if possible. Variability in the number of clock cycles per instruction makes it more difficult to *pipeline* instruction execution (see Section 4.3), and pipelining is an important technique for increasing CPU performance. Although pipelined, high-speed implementations of several memory–register architectures do exist, they achieve good CPU performance in spite of their instruction set design rather than because of it.

The alternative to a memory–register architecture is known as a *load–store* architecture. Notable examples of the load–store philosophy include the MIPS and SPARC families of processors. In a load–store architecture, only the data transfer instructions (typically named "load" for reading data from memory and "store" for writing data to memory) are able to access variables in memory. All arithmetic and logic instructions operate only on data in registers (or possibly immediate constants); they leave the results in registers as well. Thus, a typical computational instruction might be written in assembly language as ADD R1, R2, R3, and a data transfer might appear as LOAD [R4], R5. Combinations of the two, such as ADD R1, [R2], R3, are not allowed.

The advantages and disadvantages of a load–store architecture are essentially the converse of those of a memory–register architecture. Assembly language programming requires a bit more effort, and because data transfer operations are divorced from computational instructions, programs tend to require more machine instructions to perform the same task. (The equivalent of the x86 instruction ADD [EBX], EAX would require three instructions: one to load the first operand from memory, one to do the addition, and one to store the result back into the memory location.) Because all operands must be in registers, more registers must be provided or performance will suffer; however, more registers are more difficult to manage and more time-consuming to save and restore when that becomes necessary.

Load–store machines have advantages to counterbalance these disadvantages. Because register addressing requires smaller bit fields (five bits are sufficient to choose one of 32 registers, while memory addresses are typically much longer), instructions can be smaller (and, perhaps more importantly, consistent in size). Three-operand instructions, which are desirable from a programming point of view but would be unwieldy if some or all of the operands could be in memory, become practical in a load–store architecture. Because all computational instructions access memory only once (the instruction fetch) and all data transfer instructions require exactly two memory accesses (instruction fetch plus operand load or store), the control logic is less complex and can be implemented in *hardwired* fashion (see Section 3.3.2), which may result in a shorter CPU clock cycle. The arithmetic and logic instructions, which never access memory after they are fetched, are simple to pipeline; because data memory accesses are done independently of computations, they do not interfere with the pipelining as much as they otherwise might. Do these advantages outweigh the disadvantages? Different CPU manufacturers have different answers; however, load–store architectures are becoming increasingly popular.

## 3.1.6   CISC and RISC instruction sets

When we discuss computer instruction sets, one of the fundamental distinctions we make is whether a given machine is a CISC or a RISC. Typical CISC and RISC architectures and implementations differ in many ways, which we discuss more fully in Section 4.4, but as their names imply, one of the most important differences between machines based on each of these competing philosophies is the nature of their instruction sets. CISC architectures were dominant in the 1960s and 1970s (many computing veterans consider DEC's VAX to be the ultimate CISC architecture); some, like the Intel x86 architecture (also used in AMD processors), have survived well into the new millennium.

CISCs tended to have machine language instruction sets reminiscent of high-level languages. They generally had many different machine instructions, and individual instructions often carried out relatively complex tasks. CISC machines usually had a wide variety of addressing modes to specify operands in memory or registers. To support many different types of instructions while keeping code size small (an important consideration given the price of memory 45, 50, or 60 years ago), CISC architectures often had variable-length instructions. This complicated control unit design, but by using microcode, designers were able to implement complex control logic without overly extending the time and effort required to design a processor. The philosophy underlying CISC architecture was to support high-level language programming by "bridging the semantic gap"—in other words, by making the assembly (and machine) language look as much like a high-level language as possible. Many computer scientists saw this trend toward higher-level machine interfaces as a way to simplify compilers and software development in general.

RISC architectures gained favor in the 1980s and 1990s as it became more and more difficult to extract increased performance from complicated CISC processors. The big, slow microprogrammed control units characteristic of CISC machines limited CPU clock speeds and thus performance. CISCs had instructions that could do a lot, but it took a long time to execute them. The RISC approach was not only to have fewer distinct machine language instructions (as the name implies), but for each instruction to do less work. Instruction sets were designed around the load–store philosophy, with all instruction formats kept to the same length to simplify the fetching and decoding process. Only instructions and addressing modes that contributed to improved performance were included; all extraneous logic was stripped away to make the control unit, and thus the CPU as a whole, "lighter" and faster.

The reader may recall that, in Chapter 1, we stated that the most meaningful performance metric was the overall time taken to run one's specific program of interest, whatever that program might happen to be. The less time it takes a system to execute the code to perform that task, the better. We can break down the time it takes to run a given program (in the absence of external events such as I/O, interrupts, etc.) as follows:

$$\frac{\text{Time}}{\text{Program}} = \frac{\text{Instructions}}{\text{Program}} \times \frac{\text{Clock cycles}}{\text{Instruction}} \times \frac{\text{Seconds}}{\text{Clock cycle}}$$

In order to reduce the overall time taken to execute a given piece of code, we need to make at least one of the three terms on the right side of the above equation smaller, and do so without causing a greater increase in any of the others. CISC machines address this performance equation by trying to minimize the first term in the equation: the number of machine instructions required to carry out a given task. By providing more functionality per instruction, programs are kept smaller and, all else being equal, they should thus execute faster.

The problem with this approach—and one of the chief reasons for the rise of RISC architectures—was that the very techniques used to cram more functionality into each machine instruction (e.g., microprogramming, which we will study in Section 3.3.3) hindered designers' ability to do anything about the remaining terms in the equation. Complex instructions generally took several clock cycles each to execute, and because of the long propagation delays through the control unit and datapath, it proved difficult to increase CPU clock frequencies (note that the rightmost term in the equation is the reciprocal of clock frequency).

With CISC performance stagnating, RISC architects adopted a very different approach. "Let's stop worrying about that first term in the equation," they said, "and focus on the other two." Yes, by making individual instructions simpler, it would take more of them to do the same amount of work, which meant that the size of programs would increase. If, however, this allows the resulting implementation to be streamlined (and, preferably, *pipelined* using the approach we will examine in Section 4.3), then it may be possible for us to minimize both the number of clock cycles per instruction and the clock cycle time. If the decrease in the second and third terms of the equation outweighed the increase in the first, then RISC architectures would win out. This was the wager placed by the creators of the early RISC chips, and it turned out to be a winning bet. Gradually (change takes time), RISC processors showed they could outperform their CISC counterparts, won acceptance, and increased their market share.

The main impetus for RISC architectures came from studies of high-level language code compiled for CISC machines. Researchers found that the time-honored 80/20 rule was in effect: Roughly 20% of the instructions were doing 80% (or more) of the work. Many machine instructions were too complex for compilers to find a way to use them, and thus, they never appeared in compiled programs at all. Expert assembly language programmers loved to use such instructions to optimize their code, but the RISC pioneers recognized that, in the future, assembly language would be used less and less for serious program development. Their real concern was optimizing system performance while running applications compiled from high-level code. Compilers were not taking full advantage of overly feature-rich CISC instruction sets, yet having the capability to execute these (mostly unused) instructions slowed down everything else. (The CPU clock is limited by the slowest logical path that must be traversed in a cycle.) Thus, RISC designers opted to keep only the 20% or so of instructions that were frequently used by the compilers. The few remaining needed operations could be synthesized by combining multiple RISC instructions to do the work of one seldom-used CISC instruction. RISC architectures required more effort to develop good, efficient, optimizing compilers—and the compilation process itself generally took longer—but a compiler must only be written once, and a given program must only be compiled one time once it is written and debugged, although it will likely be run on a given CPU many times. Trading off increased compiler complexity for reduced CPU complexity thus seemed like a good idea.

Two other motivating, or at least enabling, factors in the rise of RISC architecture were the advent of better very large-scale integration (VLSI) techniques and the development of programming languages for hardware design. With the availability of more standardized, straightforward VLSI logic design elements, such as programmable logic arrays, computer engineers could avoid the haphazard, time-consuming, and error-prone process of designing chips with random logic. This made hardwired control much more feasible than it had been previously. Of perhaps even more significance, *hardware description languages* (HDLs) were devised to allow a software development methodology to be used to design hardware. This ability to design hardware as though it were software negated one of the major advantages previously enjoyed by microprogrammed control unit design.

The reduced number and function of available machine instructions, along with the absence of variable-length instructions, meant RISC programs required more instructions and took up more memory for program code than would a similar program for a CISC. However, if the simplicity of the control unit and arithmetic hardware were such that the processor's clock speed could be significantly higher and if instructions could be completed in fewer clock cycles (ideally, one), the end result could be—and often was—a net increase in performance. This performance benefit, coupled with falling prices for memory over the years, has made RISC architectures more and more attractive, especially for high-performance machines. CISC architectures have not all gone the way of the dinosaurs, however, and many of the latest architectures borrow features from both RISC and CISC machines. This is why a course in computer architecture involves the study of the characteristics of both design philosophies.

## 3.2   The datapath

Having looked at factors that are important to the design of a machine's instruction set, we now turn our attention to the *datapath*, the hardware that is used to carry out the instructions. Regardless of the number or type of instructions provided or the philosophy underlying the instruction set design, any CPU needs circuitry to store binary numbers and perform arithmetic and logical operations on them. In this section, we examine the register set and computational hardware that make up a typical computer's datapath.

### 3.2.1   The register set

A machine's programmer-visible *register set* is the most obvious feature of its datapath because the register set is intimately intertwined with the ISA. Another way of saying this is that the visible register set is an architectural feature, and the particulars of the rest of the datapath are merely implementation details. (This is not meant to imply that implementation is trivial.) One machine may have two 64-bit ALUs and another may have a single 32-bit ALU that requires two passes to perform 64-bit operations, but if their register sets (and instruction sets) are the same, they will appear architecturally identical; the only difference will be in observed performance.

Registers do very little other than serve as temporary storage for data and addresses. (It is possible to implement the programmer's working registers as shift registers rather than as basic storage registers, but for reasons of speed and simplicity, the shifting capability is usually located further down the datapath, either as part of or following the ALU.) However, the fact that registers are explicitly referred to by program instructions that operate on data makes the register set one of the most important architectural features of a given machine.

The most obvious and significant attributes of a machine's register set are its size (the number of registers provided and the width in bits of each) and its logical organization. Assembly programmers like to have many registers—the more, the better. Compilers were once woefully inadequate at efficiently using a large register set but have become considerably better at this over the years. However, for both compilers and human programmers, there is a diminishing returns effect beyond some optimal number of registers. The number and size of the CPU registers was once limited by such simple but obvious factors as power consumption and the amount of available silicon "real estate." Now, the determining factors of the number of registers provided are more likely to be the desired instruction length (more registers means bigger operand fields), the overhead required to save and restore registers on an interrupt or context switch, the desire to maintain compatibility with previous machines, and the compiler's ability to make efficient use of the registers.

The size of individual registers depends to a considerable extent on the numerical precision required and the word width of the computational hardware. (It makes little sense to store 128-bit numbers if the ALU is only 32 bits wide, or if the machine's intended applications do not require 128-bit precision.) Because most architectures allow (or even require) some form of register-indirect or indexed addressing, register size also depends on the amount of memory the architects want to make addressable by a program. For example, 16-bit registers allow only 64 KB of address space (without resorting to some form of trickery, such as the segment registers in the Intel 8086, which allowed 20-bit addresses to be formed from two 16-bit values); 32-bit registers allow 4 GB of addressable space, which used to be considered plenty for almost any application but has proven insufficient for some of today's extremely large data sets. The next logical step, addressing using 64-bit registers, allows $2^{64}$ bytes (16,384 petabytes or 16 exabytes) of virtual (and perhaps one day physical) space to be addressed.

Given the range of integers (and double-precision floating-point values; see Section 3.2.3) that can be represented in 64 bits, it is unlikely that 128-bit registers will become common in the near future. (Of course, past predictions, such as "the 640K ought to be enough for anybody" attributed to Bill Gates, have shown the folly of saying "never.") In any case, most contemporary machines could have more (and larger) registers than they do, but designers have found other, more profitable ways to use the available chip area.

Whatever the size of the register set, its logical organization is very significant. Designers have taken widely different approaches to allocating (or, in some cases, not allocating) registers for different uses. Some architectures, such as the Intel 8086, had special-purpose register sets in which specific registers were used for specific functions. Multiplication results were always left in the AX and DX registers, for example, and CX was always used to count loop iterations. Others, for example, Motorola's 68000 and its descendants, divide the working registers only into the two general categories of data registers and address registers. All data registers are created equal, with the ability to provide operands for and receive results from any of the arithmetic and logical instructions. Address registers only have the capability for simple pointer arithmetic, but any of them can be used with any of the available addressing modes. Some architectures (such as VAX and MIPS) take the principle of equality to its fullest extent, making few, if any, distinctions in terms of functionality and allowing any CPU register to be used equally in any operation. Perhaps the most interesting approach is the one used in SPARC processors. SPARC's general-purpose registers may be used interchangeably for data or addresses, but are logically grouped by scope—in other words, based on whether they represent global variables, local variables within a given procedure, inputs passed to it from a caller, or outputs from it to a called subprogram.

Whatever the design of a machine's register set, compilers must be designed to be aware of its features and limitations in order to exploit the full performance potential of the machine. This is particularly true of RISCs (or any machine with a load–store ISA). A compiler that can effectively optimize code to take advantage of the features of a particular architecture is worth more than its figurative weight in megahertz.

## 3.2.2 Integer arithmetic hardware

Performing arithmetic on binary integers is perhaps the most fundamental and important capability possessed by digital computers. Numbers, after all, are composed of *digits* (binary digits in the case of essentially all modern computers), and to *compute* is to perform arithmetic operations on data. So, by definition, *digital computers* are machines that do arithmetic calculations. In this section, we examine some of the ways in which computer hardware can efficiently perform arithmetic functions on binary numbers.

The reader who has had a previous introductory course in digital logic or computer organization has probably encountered a basic datapath design that includes a simple arithmetic/logic unit. An elementary ALU can perform addition and subtraction on binary operands as well as some subset of standard, bitwise logic functions, such as AND, OR, NOT, NAND, NOR, XOR, etc. (A more advanced circuit might also be able to perform multiplication, division, and possibly other functions.) The datapath may also provide the capability to do bit shifting operations, either in the ALU itself or as a separate shifter block connected to the ALU outputs. A typical ALU is shown in block diagram form in Figure 3.11.

Notice that control inputs (which are output from the processor's control unit) are used to select which arithmetic or logical function the ALU and shifter perform at any given time. One of the ALU functions may be simply to transfer one of the operand inputs to the output without modification; this allows the shifter to operate directly on register contents and provides for simple register-to-register transfers via the datapath. The shifter has the ability to pass bits through unchanged, or move them to the left or

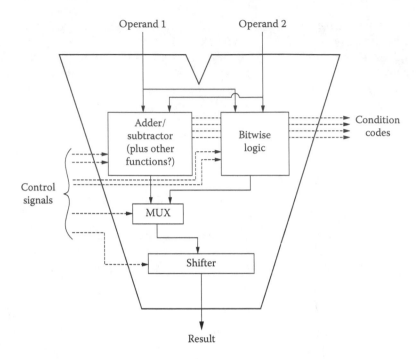

*Figure 3.11* Basic ALU.

right, either by one position or many (as in the case of a barrel shifter) at a time. Also notice that the ALU typically develops "condition code" or status bits that indicate the nature of the result produced (positive/negative, zero/nonzero, the presence or absence of an overflow or carry, etc.). The bitwise logical functions of the ALU are trivial to implement, requiring only a single gate of each desired type per operand bit; shifting is not much more complicated than that. The real complexity of the datapath is in the circuitry that performs arithmetic calculations. We devote the rest of this section to a closer look at integer arithmetic hardware.

### 3.2.2.1 *Addition and subtraction*
Addition and subtraction are the simplest arithmetic operations and the ones most frequently performed in most applications. Because these operations are performed so frequently, it is often worthwhile to consider implementations that improve performance even if they add somewhat to the complexity of the design, the amount of chip area required, or other costs.

The reader should be familiar with the basic *half adder* and *full adder* circuits (see Figures 3.12 and 3.13), which are the building blocks for performing addition of binary numbers in computers. The half adder is

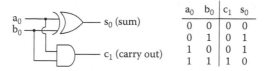

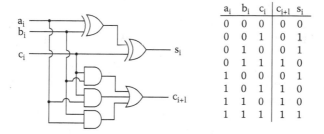

**Figure 3.12** Binary half adder circuit.

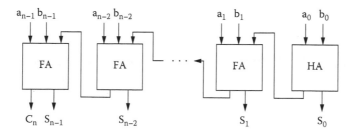

**Figure 3.13** Binary full adder circuit.

normally useful only for adding the least significant bits $a_0$ and $b_0$ of two binary numbers as it has no provision for a carry in. It simply adds two one-bit numbers and produces a sum bit $s_0$ and a carry out bit $c_1$.

The full adder circuit is useful for adding the bits (in any position) of two binary numbers. The corresponding bits of the two numbers, $a_i$ and $b_i$, are added to an input carry $c_i$ to form a sum bit $s_i$ and a carry out bit $c_{i+1}$. Any number of full adders can be cascaded together by connecting the carry out of each less significant position to the carry in of the next more significant position to create the classic *ripple carry adder* (see Figure 3.14). The operation of this circuit is analogous to the method used to add binary numbers by hand using a pencil and paper. Like hand addition, a ripple carry adder can be used to add two binary numbers of arbitrary size. However, because carries must propagate through the entire chain of

**Figure 3.14** Ripple carry adder.

adders before the final result is obtained, this structure may be intolerably slow for adding large binary numbers.

It is possible to design half and full subtractor circuits that subtract one bit from another to generate a difference bit and a borrow out bit. A full subtractor is able to handle a borrow in, and a half subtractor is not. These blocks can then be cascaded together in the same fashion as adder blocks to form a *ripple borrow* subtractor. However, in practice, separate circuits are rarely built to implement subtraction. Instead, signed arithmetic (most frequently, two's complement arithmetic) is used, with subtraction being replaced by the addition of the complement of the subtrahend to the minuend. Thus, one circuit can perform double duty as an adder and subtractor.

A combination adder–subtractor circuit may be built as shown in Figure 3.15. When input $S = 0$, the carry in is 0, and each XOR gate outputs the corresponding bit of the second operand B, producing the sum of the two input numbers $(A + B)$. When $S = 1$, each XOR gate produces the complement of the corresponding bit of B, and the carry in is 1. Complementing all bits of B gives its one's complement, and adding one to that value yields the two's complement, so the circuit performs the operation $[A + (-B)]$ or simply $(A - B)$. Because this technique can be used to perform subtraction with a ripple carry adder or any other circuit that will add two numbers, there is no need for us to study subtractor circuits separately.

The question then remains, "What other alternatives are available that will allow us to add and subtract numbers more quickly than a ripple carry adder?" A superior solution, in terms of computation speed, is provided by a *carry lookahead adder* (see Figure 3.16). The basic principle behind this circuit's operation is that the bits of the operands, which are available

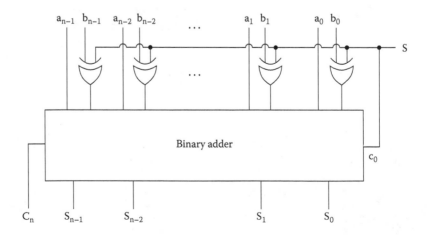

*Figure 3.15* Adder/subtractor circuit.

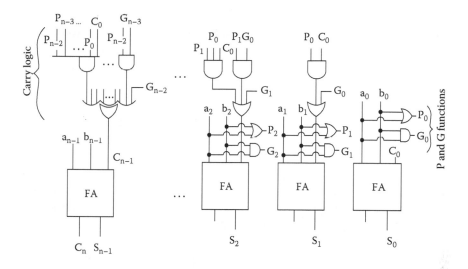

*Figure 3.16* Carry lookahead adder.

at the start of the computation, contain all the information required to determine all the carries at once. There is no need to wait for carries to ripple through the less significant stages; they can be generated directly from the inputs if one has room for the required logic.

The carry logic of this circuit is based on the fact that when we add two binary numbers, there are two ways that a carry from one position to the next can be caused. First, if either of the operand bits in that position ($a_i$ or $b_i$) are 1 and its carry in bit $c_i$ is also 1, the carry out bit $c_{i+1}$ will be 1. That is, if the logical OR of the input bits is true, this adder stage will *propagate* an input carry (if it occurs) to the next more significant stage. Second, if both input bits $a_i$ and $b_i$ are 1, the stage will produce a carry out $c_{i+1}$ of 1 regardless of the value of $c_i$. In other words, if the logical AND of $a_i$ and $b_i$ is true, this stage will *generate* a carry to the next stage. Each carry propagate function $P_i$ requires a single OR gate ($P_i = a_i + b_i$), and each carry generate function $G_i$ requires a single AND gate ($G_i = a_i b_i$). Each carry bit can be generated from these functions and the previous function as follows:

$$c_{i+1} = G_i + P_i c_i$$

If we assume that a carry into the least significant bit position ($c_0$) is possible (this would be convenient if we want to be able to cascade additions to handle large numbers), then the logical expression for the first carry out would be

$$c_1 = G_0 + P_0 c_0$$

By substituting this expression into the evaluation of $c_2$ and then using the expression for $c_2$ to develop $c_3$ and so on, we can obtain logical expressions for all the carries as follows:

$$c_2 = G_1 + P_1G_0 + P_1P_0c_0$$

$$c_3 = G_2 + P_2G_1 + P_2P_1G_0 + P_2P_1P_0c_0$$

$$c_4 = G_3 + P_3G_2 + P_3P_2G_1 + P_3P_2P_1G_0 + P_3P_2P_1P_0c_0$$

and so on. Each P function is a single OR function of input bits available at the start of the computation. Each G function is a single AND function of the same bits. Given a sufficient number of AND and OR gates, we can generate all the P and G functions simultaneously, within one gate propagation delay time. Using additional AND gates, a second gate delay will be sufficient to simultaneously generate all the product terms in the above expressions. These product terms can then be ORed, producing all the carries simultaneously in only three gate delays. Full adders can be built with two-level logic, so the complete sum is available in only five gate delays. (If three-input XOR gates are available, each of the sum bits could be computed by one such gate instead of two-level AND/OR logic, potentially further reducing delay.) This is true regardless of the "width" of the addition (and, thus, the number of carries required). For large binary numbers, this technique is much faster than the ripple carry approach.

Of course, as the old truism goes, there is no such thing as a free lunch. The carry lookahead adder provides superior speed, but the trade-off is greatly increased circuit size and complexity. Notice the trend in the carry equations above. The first one, for $c_1$, has two product terms with the largest term containing two literals. The second one has three product terms with the largest containing three literals, and so on. If we were designing a 32-bit adder, the final carry equation would have 33 terms, with 33 literals in the largest term. We would need an OR gate with a *fan-in* (number of inputs) of 33, plus 32 AND gates (the largest also having a fan-in of 33), just to produce that one carry bit. Similar logic would be needed for all 32 carries, not to mention the 32 AND gates and 32 OR gates required to produce the carry generate and propagate functions. Logic gates with high fan-in are difficult to construct in silicon, and large numbers of gates take up a great deal of chip space and increase power dissipation. It is possible to piece together an AND gate with many inputs by using two-level AND/AND logic (or to build a larger OR gate using two-level OR/OR logic), but this increases propagation delay and does nothing to reduce the size or power dissipation of the circuit. Now imagine using the carry lookahead logic required to build a 64-bit adder. It would be fast but huge.

It is possible to compromise between the higher speed of the carry lookahead adder and the simplicity and reduced size of the ripple carry adder by using both designs in a single circuit as shown in Figure 3.17. In this case, the designer has chosen to implement a 64-bit binary adder using four 16-bit carry lookahead blocks connected in a ripple carry arrangement. Because each block is much faster than a 16-bit ripple carry adder, the overall circuit will be considerably faster than a 64-bit ripple carry adder. Cascading the carries from one block to the next makes their delays additive rather than simultaneous, so the circuit is roughly one fourth the speed of a full 64-bit carry lookahead adder. However, it requires much less power and chip area and, thus, may represent a reasonable compromise between cost and speed for some applications.

Alternatively, we could also build a second level of carry lookahead logic to compute the carries between the blocks. Such a circuit would be faster (and more complex) than the circuit of Figure 3.17, but slower (and less complex) than a full 64-bit CLA. Yet another compromise approach would be the *carry select adder* shown in Figure 3.18. As in the previous design, we consider the implementation of a larger adder (in this case, 32 bits) by cascading smaller blocks (for example, 16 bits each) together. The output carry from the lower-order block would normally be fed to the carry input of a single high-order block; however, this would mean that

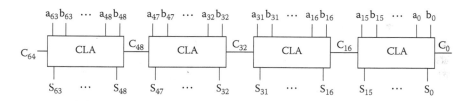

*Figure 3.17* 64-bit adder design using 16-bit carry lookahead blocks.

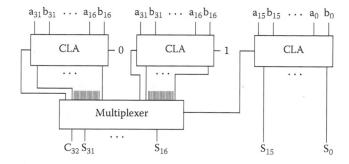

*Figure 3.18* 32-bit carry select adder.

after the first half of the addition is done, the full propagation delay of the second block must elapse before the answer is known to be correct.

In the carry select adder design, we completely replicate the logic of the high-order block. This requires 50% more hardware (three blocks instead of two) but gives an advantage in performance. The two redundant high-order blocks both compute the sum of the upper halves of the operands, but one assumes the carry in will be 0, and the other assumes it will be 1. Because we have covered both possibilities, one of the two high-order answers is bound to be correct. The two possible high-order sums are computed simultaneously with each other and with the low-order bits of the sum. Once the carry out of the low-order portion is determined, it is used to choose the correct high-order portion to be output; the other incorrect result is simply ignored. As long as the propagation delay of a multiplexer is less than that of an adder block (which is normally the case), the addition is performed more quickly than if two blocks were cascaded together in ripple carry fashion.

Which of these adder designs is the best? There is no universal answer; it depends on the performance required for a particular application versus the cost factors involved. This is just one example of a common theme in computer (or any other discipline of) engineering: trading off one aspect of a design to gain an improvement in another. Short of a major technological revolution, such as practical quantum computing—and likely even then—it will always be so.

Another good example of a design trade-off, in this case a decision between implementing functionality in hardware versus software, is the addition of binary coded decimal (BCD) numbers. BCD is often a convenient representation because most human users prefer to work in base 10 and because character codes used for I/O, such as ASCII, are designed to make conversion to and from BCD easy. However, it is difficult to perform arithmetic operations on numbers in BCD form because of the logic required to compensate for the six unused binary codes. Figure 3.19 shows a circuit that could be used to add two BCD digits together to produce a correct BCD result and a carry, if any.

Notice that the circuit implements the correction algorithm of adding 6 ($0110_2$) to the result if a binary carry out occurs (meaning the sum is 16 or greater) or if the most significant sum bit ($s_3$) and either of the next two bits ($s_2$ or $s_1$) are 1 (in other words, if the sum is between 10 and 15, inclusive). If the sum is in the range 0 to 9, then it is left unchanged by adding $0000_2$, and the carry out is 0.

We could cascade any number of these adders to process BCD values of arbitrary size. This takes up space on the chip and increases power consumption, but the designers of some systems (particularly those intended for business applications) have found BCD adders and machine instructions that use them to be worthy of inclusion in the CPU. The alternative

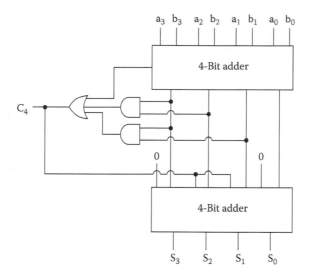

*Figure 3.19* Binary coded decimal adder.

is to provide only normal, binary addition circuits in the datapath. In this case, the designer could still provide a BCD correction instruction in microcode (see Section 3.3.3) or could choose not to support BCD addition in hardware at all. Absent a machine instruction to implement BCD addition, one could choose not to use BCD numbers (perhaps feasible if the machine is intended only for scientific or other very specific applications) or implement BCD addition as a software routine. All of these approaches have been tried and found worthwhile in one scenario or another, representing different hardware–software trade-offs.

One more type of binary addition circuit is worth mentioning at this time: the *carry save adder*. A carry save adder is just a set of full adders, one per bit position; it is similar to a ripple carry adder with the important exception that the carry out of each adder is not connected to the carry in of the next more significant adder. Instead, the carry in is treated as another input so that three numbers are added at once instead of two. The output carries are not propagated and added in at the next bit position; instead, they are treated as outputs from the circuit. The carry save adder thus produces a set of sum bits and a set of carry bits (see Figure 3.20).

The separate sets of sum and carry bits must eventually be recombined, with the carry bits added in one place to the left of where they were generated, before we can obtain a final result. The advantage of using carry save adders is only apparent when we want to add several numbers together rather than just two. We can do this using a tree structure of carry save adders (see Figure 3.21 for a carry save adder tree that adds

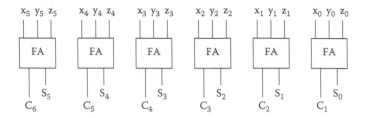

**Figure 3.20** Six-bit carry save adder.

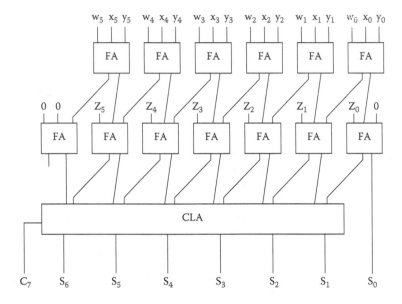

**Figure 3.21** Using a tree of carry save adders to add four 6-bit numbers.

four 6-bit numbers). If we are only interested in the overall sum of the four numbers and not the partial sums, this technique is faster and more efficient than adding the numbers two at a time. When we add two numbers at a time, we must propagate the carries to the next position (either via ripple carry or by carry lookahead) in order to produce the correct result, but when we are adding multiple numbers and do not care about the intermediate results, we can save the carries to be added in at a later stage. This means that every stage of our adder tree except the last one can be made of simple, fast carry save adders; only the last one must be a regular two-operand binary adder with carry propagation. This saves on logic, and because (compared to carry save adders) carry lookahead adders require at least three additional gate delays for the carry logic (and ripple carry adders are even slower), we compute the sum of all the numbers in less time.

Where would such a circuit be useful? If you have done any assembly language programming, you probably know that the ADD instruction in most computers takes only two operands and produces the sum of those two values. Thus, the ALU that performs that operation probably uses a carry lookahead adder; the normal ADD instruction would gain nothing from using a carry save adder. In some applications, it would be helpful to be able to add a large column of numbers quickly. Calculations of averages, as might be required for large spreadsheets, could benefit from this technique. It might also be useful for scientific applications that require taking the "dot" or inner product of vectors, which involves a summation of partial products. Digital filtering of sampled analog signals also requires summation; thus, one type of special-purpose architecture that definitely benefits from the capability of quickly adding multiple values is a digital signal processor (DSP). None of these, however, represent the most common use of carry save adders in computers. Rather, general-purpose machines that have no need for a multioperand addition instruction can and do use this idea to speed up a different arithmetic operation, which we will discuss next.

### 3.2.2.2 Multiplication and division

Multiplication and division are considerably more complex than addition and subtraction. Fortunately, most applications use these operations less frequently than the others, but in many cases, they occur sufficiently often to affect performance. One alternative, as in the case of BCD addition, is not to implement binary multiplication and division in hardware at all. In this case, if multiplication or division is required, we would have to write software routines (in a high-level language, in assembly, or even in microcode) to synthesize them using the more basic addition, subtraction, and shift operations. This is workable but slow, and thus not desirable if we need to multiply and divide numbers frequently.

Let us suppose, then, that we want to build a hardware circuit to multiply binary numbers. How would we go about it, and what would be required? Because the most basic addition circuit (the ripple carry adder) works analogously to pencil-and-paper addition, let's take a similar approach to the development of a multiplier circuit. Figure 3.22 shows, both in general and using specific data (11 × 13 = 143), how we could multiply a pair of 4-bit numbers to produce their 8-bit product. (It is no coincidence that 8 bits must be allowed for the product. In base 2, base 10, or any other base, multiplying a number of $n$ digits by a number with $m$ digits yields a product that can contain up to $n + m$ digits. Here $n = m = 4$, so we must allow for $2n = 8$ product bits.)

The binary operands A (the multiplicand, composed of bits $a_3a_2a_1a_0$) and B (the multiplier with bits $b_3b_2b_1b_0$) are used to generate four partial

|        |        |        | $a_3$    | $a_2$    | $a_1$    | $a_0$    |       |       |
|--------|--------|--------|----------|----------|----------|----------|-------|-------|
|        |        | *      | $b_3$    | $b_2$    | $b_1$    | $b_0$    |       |       |
|        |        |        | $a_3b_0$ | $a_2b_0$ | $a_1b_0$ | $a_0b_0$ | (PP0) |       |
|        |        | $a_3b_1$ | $a_2b_1$ | $a_1b_1$ | $a_0b_1$ | 0        |       | (PP1) |
|        | $a_3b_2$ | $a_2b_2$ | $a_1b_2$ | $a_0b_2$ | 0        | 0        |       | (PP2) |
| $a_3b_3$ | $a_2b_3$ | $a_1b_3$ | $a_0b_3$ | 0        | 0        | 0        |       | (PP3) |
| $P_7$  | $P_6$  | $P_5$  | $P_4$    | $P_3$    | $P_2$    | $P_1$    | $P_0$ |       |

```
      1 0 1 1
    * 1 1 0 1
      1 0 1 1
      0 0 0 0
    1 0 1 1
  1 0 1 1
  1 0 0 0 1 1 1 1
```

**Figure 3.22** Multiplication of two 4-bit numbers.

products PP0 through PP3. (There will always be as many partial products as there are multiplier bits.) Each successive partial product is shifted one place to the left as we do when multiplying numbers manually in any base; then the partial products are summed to get the final product P = A × B.

The convenient thing about designing a multiplier circuit this way is that it is very easy to generate the partial products. Each partial product bit is the product of one of the multiplicand bits $a_i$ times one of the multiplier bits $b_i$. The binary multiplication tables are trivial: $0 \times 0 = 0$; $0 \times 1 = 0$; $1 \times 0 = 0$; and $1 \times 1 = 1$. Notice that if you consider the two bits as Boolean variables (which they are), then this is exactly the same as the logical AND function. (Now you know why AND is sometimes called logical multiplication.) It so happens that in binary, logical multiplication and multiplication of 1-bit numbers are the same thing, so each partial product bit can be produced by a single AND gate with the appropriate inputs.

Once the partial products are generated, the remaining task is to add them all together to produce the final result. How can we best do this? If the previous mention of carry save adders rings a bell, congratulations! Multiplication is the number one application for a tree of carry save adders. (The most efficient arrangement of carry save adders is known as a Wallace tree in honor of its inventor.) Figure 3.23 shows how we could use this approach to compute the product of two 4-bit numbers.

This approach is not restricted to small numbers, although larger numbers require more hardware. A bigger multiplicand would require more AND gates to generate each partial product, and each carry save adder in the tree (plus the final carry lookahead adder) would need to be larger. A bigger multiplier would result in more partial products and thus require more carry save adder stages in the tree. Multiplying large numbers in this way can take up a good deal of chip space, but it is so efficient that it is a favorite technique of computer engineers.

The astute reader will recall that computers are called on to perform both signed and unsigned arithmetic. One of the reasons that two's

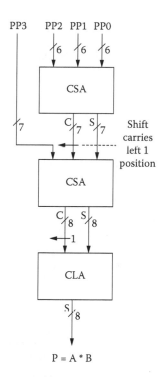

**Figure 3.23** Block diagram of Wallace tree multiplier for 4-bit numbers.

complement notation is widely preferred for representing signed numbers is that the same circuit can be used to add and subtract both signed and unsigned numbers. It is only the interpretation of the results that is different. (In the unsigned case, the result is always positive, and it is possible to generate a carry out, which represents an unsigned overflow. In the signed case, the most significant bit of the result tells us its sign; the carry out of the Most Significant Bit (MSB) is insignificant, but we must check for two's complement overflow.) With multiplication, we are not so lucky. The sample calculation shown is the unsigned multiplication of 11 ($1011_2$) times 13 ($1101_2$), producing the product 143 ($10001111_2$). Suppose we interpret these same numbers as signed, two's complement values. Binary 1011 now represents −5, and binary 1101 represents −3. The product of these two numbers is +15 or 00001111 in binary, but this is not the output from the circuit. Instead, we obtain −113 as the "product." Clearly this circuit is only good for unsigned multiplication.

How, then, can we handle the multiplication of signed values? If designers only anticipate an occasional need for signed multiplication, they might choose to implement only unsigned multiplication in hardware, leaving software to handle signed numbers. We could do this by checking the signs of both numbers before multiplying. If both are positive, the unsigned result will be correct. If both are negative, we know the product will be positive, so we simply complement both numbers first and then perform an unsigned multiplication. If one is positive and the other is negative, we would have to first complement the negative operand, then multiply, then complement the result to make it negative. It is easy to see that this approach would work, but would be unwieldy if we have to do many signed multiplications.

Another way to address the issue would be to build a circuit that directly multiplies signed numbers in two's complement format. Consider again the circuit of Figure 3.22, but with the understanding that either the multiplicand A (the 4-bit value $a_3a_2a_1a_0$), the multiplier B (the 4-bit value $b_3b_2b_1b_0$), or both, are negative if their leftmost bit is equal to 1. It is now possible for the partial products to be negative. Thus, to make the addition of partial products work out correctly, we would need to sign-extend each partial product all the way across to the leftmost bit position as shown in Figure 3.24.

Note that in this case, the last partial product needs to be subtracted rather than added. This is because in two's complement notation, the leftmost bit is the sign bit; in an $n$-bit number, its positional weight is $-(2^{n-1})$ rather than $2^{n-1}$. Because the last partial product is formed by multiplying the multiplicand by the multiplier's leftmost bit, its sign needs to be reversed by subtracting it instead of adding (or by taking its twos complement before adding). Doing so gives us the correct result, which, in the case of our example, is 00001111 in binary or +15 in decimal.

The approach used in Figure 3.24 works fine from a mathematical standpoint, giving the correct answer as a signed value when presented with signed inputs. However, it has the disadvantage of requiring a subtraction (or complementation) operation instead of only additions. More hardware is also required to add all the sign-extended partial products, which, in this example, are 8, 7, 6, and 5 bits wide, respectively, as opposed to the unsigned case in which all the partial products were only 4 bits wide. Fortunately, by doing some mathematical manipulation of the

| | | | | | | | | |
|---|---|---|---|---|---|---|---|---|
| | | | | | $a_3$ | $a_2$ | $a_1$ | $a_0$ |
| | | | | * | $b_3$ | $b_2$ | $b_1$ | $b_0$ |
| | $a_3b_0$ | $a_3b_0$ | $a_3b_0$ | $a_3b_0$ | $a_3b_0$ | $a_2b_0$ | $a_1b_0$ | $a_0b_0$ |
| + | $a_3b_1$ | $a_3b_1$ | $a_3b_1$ | $a_3b_1$ | $a_2b_1$ | $a_1b_1$ | $a_0b_1$ | |
| + | $a_3b_2$ | $a_3b_2$ | $a_3b_2$ | $a_2b_2$ | $a_1b_2$ | $a_0b_2$ | | |
| – | $a_3b_3$ | $a_3b_3$ | $a_2b_3$ | $a_1b_3$ | $a_0b_3$ | | | |
| | $P_7$ | $P_6$ | $P_5$ | $P_4$ | $P_3$ | $P_2$ | $P_1$ | $P_0$ |

| | | | | | | | | | |
|---|---|---|---|---|---|---|---|---|---|
| | | | | | | 1 | 0 | 1 | 1 |
| | | | | | * | 1 | 1 | 0 | 1 |
| | | 1 | 1 | 1 | 1 | 1 | 0 | 1 | 1 |
| | + | 0 | 0 | 0 | 0 | 0 | 0 | 0 | |
| | + | 1 | 1 | 1 | 0 | 1 | 1 | | |
| ~~1~~ | | 1 | 1 | 1 | 0 | 0 | 1 | 1 | 1 |
| | – | 1 | 1 | 0 | 1 | 1 | | | |
| | | 0 | 0 | 0 | 0 | 1 | 1 | 1 | 1 |

***Figure 3.24*** Multiplication of two signed 4-bit numbers.

partial products, it is possible to come up with an alternative solution (the Baugh-Wooley algorithm) that produces the same result without the need for extended partial products or subtraction (while replacing some of the AND gates in the previous circuits with NAND gates to complement specific terms). This alternative multiplication method is illustrated in Figure 3.25.

Another way to multiply signed numbers efficiently is to use Booth's algorithm. This approach handles two's complement signed numbers directly, without any need to convert the operands to positive numbers or complement results to make them negative. The basic idea behind Booth's algorithm is that the multiplier, being a binary number, is composed of 0s and 1s. Sometimes an individual 0 is followed by a 1 (or vice versa); sometimes there are strings of several 0s or 1s consecutively. Depending on whether we are at the beginning, the end, or the middle of one of these strings, the steps involved in accumulating the partial products are different.

When we are multiplying and the multiplier contains a string of consecutive zeroes, we know that each partial product will be zero; all that is necessary is to shift the previously accumulated product by one bit position for each zero in the string so that the next partial product will be added in that many places to the left. When the multiplier contains a string of ones, we can treat this as a multiplication by the number $(L - R)$, where R is the weight of the 1 at the right (least significant) end of the string, and L is the weight of the zero to the left of the 1 at the left (most significant) end of the string.

For example, say we are multiplying some number times +12 (01100 binary, as a 5-bit value). The two 0s at the right end of the multiplier and the 0 at the left end contribute nothing to the result because the partial products corresponding to them are zero. The string of two 1s in the middle of the number will determine the result. The rightmost 1 has a weight of $2^2 = 4$, and the zero to the left of the leftmost 1 has a weight of $2^4 = 16$. The value of the multiplier is $16 - 4 = 12$; we can achieve the effect of

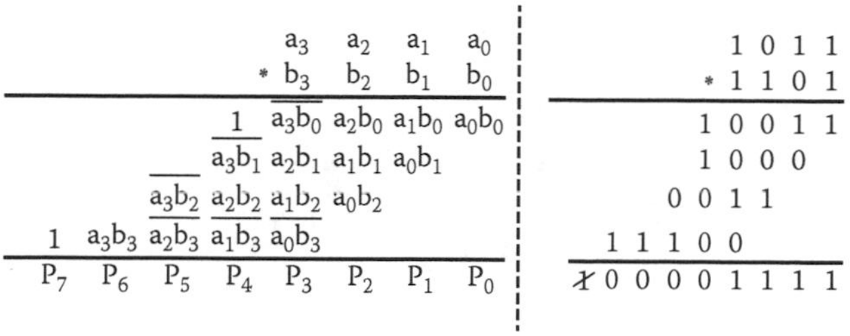

$$
\begin{array}{ccccc}
 & a_3 & a_2 & a_1 & a_0 \\
* & b_3 & b_2 & b_1 & b_0 \\
\hline
1 & a_3b_0 & a_2b_0 & a_1b_0 & a_0b_0 \\
 & a_3b_1 & a_2b_1 & a_1b_1 & a_0b_1 \\
 & a_3b_2 & a_2b_2 & a_1b_2 & a_0b_2 \\
1 & a_3b_3 & a_2b_3 & a_1b_3 & a_0b_3 \\
\hline
P_7 & P_6 & P_5 & P_4 & P_3 & P_2 & P_1 & P_0
\end{array}
$$

$$
\begin{array}{cccccccc}
 & & & & 1 & 0 & 1 & 1 \\
 & & & * & 1 & 1 & 0 & 1 \\
\hline
 & & & 1 & 0 & 0 & 1 & 1 \\
 & & & 1 & 0 & 0 & 0 & \\
 & & & 0 & 0 & 1 & 1 & \\
 & & 1 & 1 & 1 & 0 & 0 & \\
\hline
\cancel{1} & 0 & 0 & 0 & 0 & 1 & 1 & 1
\end{array}
$$

*Figure 3.25* Simplified circuit (Baugh-Wooley) for multiplication of two signed 4-bit numbers.

multiplying by +12 by taking the product of the multiplicand with 16 and subtracting the product of the multiplicand with 4. Algebraically, $12x = (16 - 4)x = 16x - 4x$.

Suppose instead that we want to multiply a number times –13 (10011 as a two's complement binary number). Once again, the string of two 0s in the middle of the number contributes nothing to the final result; we only need to concern ourselves with the two 1s at the right end of the multiplier and the 1 at the left end. The two 1s at the right end may be treated as $2^2 - 2^0 = (4 - 1) = 3$. The leftmost 1 is in the $2^4 = 16$ position. There are no bits to the left of this bit (it is the sign bit for the multiplier), so it simply has a weight of $-(2^4) = -16$. Thus we can treat multiplication by –13 as a combination of multiplying by –16, +4, and –1.

To take one more example, assume the multiplier is binary 10101, which is –11 in two's complement form. There are three strings of 1s, each just one bit long. The rightmost is treated as $2^1 - 2^0 = 2 - 1 = 1$; the middle string is treated as $2^3 - 2^2 = 8 - 4 = 4$; the leftmost is once again treated as $-(2^4) = -16$. Multiplying by –11 can thus be treated as multiplying by –16, +8, –4, +2, and –1 with the signed partial products added together to produce the result. All partial products can be generated as positive numbers, but each is either added or subtracted depending on whether its contribution is positive or negative. Alternatively, the partial products may be generated in two's complement form and simply added together.

This last numerical example illustrates the worst case for Booth's algorithm, in which there are no strings of more than one bit equal to 1. In this case, there is a positive or negative partial product generated for every bit of the multiplier. However, when there are longer strings of 1s, Booth's algorithm may significantly reduce the number of partial products to be summed. This is because a string of 1s of any length generates only one positive and one negative partial product.

Given the basic idea outlined above, how could we implement Booth's multiplication technique in hardware? The key is that as we move through the multiplier from right to left, we consider every two adjacent bits (the current bit and the bit to its right) to determine whether we are in the beginning, middle, or end of a bit string as shown in Table 3.1. Note that

*Table 3.1* Using bit pairs to identify bit string position in Booth's algorithm

| Current bit (bit $i$) | Bit to the right (bit $i - 1$) | Explanation | Example |
|---|---|---|---|
| 0 | 0 | Middle of a string of 0s | 01111100 |
| 0 | 1 | End of a string of 1s | 01111100 |
| 1 | 0 | Beginning of a string of 1s | 01111100 |
| 1 | 1 | Middle of a string of 1s | 01111100 |

an imaginary 0 is inserted to the right of the rightmost (least significant) bit of the multiplier to get the process started.

We could build a hardware circuit to implement signed multiplication in sequential fashion as shown in Figure 3.26. To perform an $n$-bit by $n$-bit multiplication, an $n$-bit adder is needed to add the partial products. The multiplicand is stored in an $n$-bit register M; a $2n$-bit register P is used to accumulate the product. (P must be capable of implementing the arithmetic shift right operation, which is a shift to the right while maintaining the leftmost or sign bit.) Initially, the lower half of P contains the multiplier and the upper half contains zeroes.

This circuit computes the product in $n$ steps with the action at each step being determined based on which of the four possible scenarios from Table 3.1 applies. If the current pair of bits is 00 or 11 (the middle of a string), nothing is added to or subtracted from P. If the current pair of bits is 01 (the end of a string of 1s), the contents of M (the multiplicand) are added to the upper half of P; if the current pair of bits is 10 (the beginning of a string of 1s), the multiplicand in M is subtracted from the upper half of P (or, equivalently, the two's complement of M is added to the upper half of P). Following the addition or subtraction, if any, the contents of P are arithmetically shifted one place to the right. This process continues until all $n$-bit pairs of the multiplier have been considered. (Notice that the multiplier bits, which originally occupy the right half of the product register P, are shifted out and lost one at a time until the product occupies the entire width of P. One extra flip-flop C is needed at the right end of P to

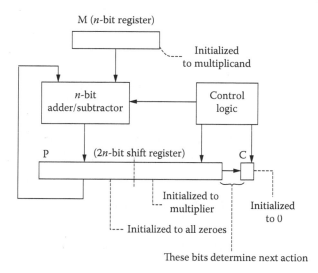

*Figure 3.26* Sequential circuit for multiplying signed numbers using Booth's algorithm.

hold the previous rightmost multiplier bit until it can safely be discarded after the next iteration.) At the end, C can be ignored, and the $2n$-bit signed product resides in P.

Figures 3.27 and 3.28 show examples of multiplying 4-bit numbers using Booth's algorithm. In the first example, +2 (0010 binary) is multiplied by –3 (1101) to produce a product of –6 (11111010). In the second, –4 (1100) times –6 (1010) yields a product of +24 (00011000).

The circuit shown in Figure 3.26 is sequential, requiring $n$ clock cycles to multiply two $n$-bit numbers to obtain a $2n$-bit product. Such a sequential implementation may be intolerably slow if multiplications are to be performed frequently. To speed the process, one can implement the logic

| P | C | |
|---|---|---|
| 0 0 0 0    1 1 0 1 | 0 | |
| + 1 1 1 0 | | Begin string of 1s |
| 1 1 1 0    1 1 0 1    0 | | add –M to upper bits of P |
| 1 1 1 1    0 1 1 0    1 | | and shift right |
| + 0 0 1 0 | | End string of 1s; add |
| 0 0 0 1    0 1 1 0    1 | | +M to upper bits of P |
| 0 0 0 0    1 0 1 1    0 | | and shift right |
| + 1 1 1 0 | | Begin string of 1s |
| 1 1 1 0    1 0 1 1    0 | | add –M to upper bits of P |
| 1 1 1 1    0 1 0 1    1 | | and shift right |
| 1 1 1 1    1 0 1 0    1 | | Middle string of 1s; just shift right |
| Product | (Disregard) | |

*Figure 3.27* Multiplication of (+2) * (–3) using Booth's algorithm.

| P | C | |
|---|---|---|
| 0 0 0 0    1 0 1 0 | 0 | Middle string of 0s; just shift right |
| 0 0 0 0    0 1 0 1 | 0 | |
| + 0 1 0 0 | | Begin string of 1s add –M to upper bits of P |
| 0 1 0 0    0 1 0 1    0 | | and shift right |
| 0 0 1 0    0 0 1 0    1 | | |
| + 1 1 0 0 | | End string of 1s |
| 1 1 1 0    0 0 1 0    1 | | add +M to upper bits of P |
| 1 1 1 1    0 0 0 1    0 | | and shift right |
| + 0 1 0 0 | | Begin string of 1s |
| 0 0 1 1    0 0 0 1    0 | | add –M to upper bits of P |
| 0 0 0 1    1 0 0 0    1 | | and shift right |
| Product | (Disregard) | |

*Figure 3.28* Multiplication of (–4) * (–6) using Booth's algorithm.

of Booth's algorithm in purely combinational form using a tree of carry save adders similar to that used for unsigned multiplication, rather than a single adder used repetitively as in Figure 3.26. The shift register, which was needed to move the previously accumulated product one place to the right before the next addition, is eliminated in the combinational circuit by adding each more significant partial product one place to the left. The result is a Wallace tree multiplier much like the one shown in Figure 3.23 except that the partial products may be positive or negative.

It is also possible to refine the basic Booth's algorithm by having the logic examine groups of three or more multiplier bits rather than two. This increases circuit complexity for generating the partial products but further reduces the number of partial products that must be summed, thus decreasing overall propagation delay. Once again, as in the case of the carry lookahead adder, we note that circuit complexity can be traded off for speed.

Division of binary integers is even more complex than multiplication. A number of algorithms for dividing binary numbers exist, but none are analogous to Booth's algorithm for multiplication; in other words, they work only for unsigned numbers. To perform division of signed integers, the dividend and divisor must be preprocessed to ensure that they are positive; then, after the operation is complete, the quotient and remainder must be adjusted to account for either or both operands being negative.

The most basic algorithm for dividing binary numbers, known as *restoring division*, operates analogously to long division by hand; that is, it uses the "shift and subtract" approach. The bits of the dividend are considered from left to right (in other words, starting with the most significant bit) until the subset of dividend bits forms a number greater than the divisor. At this point, we know that the divisor will "go into" this *partial dividend* (that is, divide it with a quotient bit of 1). All previous quotient bits to the left of this first 1 are 0. The divisor is subtracted from the partial dividend, forming a *partial remainder*; additional bits are brought down from the rest of the dividend and appended to the partial remainder until the divisor can again divide the value (or until no more dividend bits remain, in which case the operation is complete and the partial remainder is the final remainder). Any comparison showing the divisor to be greater than the current partial dividend (or partial remainder) means the quotient bit in the current position is 0; any comparison showing the divisor to be smaller produces a quotient bit of 1. Figure 3.29 shows how we would perform the operation 29/5 (11101/101) using restoring division.

This algorithm is called *restoring* division because at each step the necessary comparison is performed by subtracting the divisor from the partial dividend or partial remainder. If the result is positive or zero (that is, if no borrow is required), then a 1 is entered in the quotient, and we move on. If the result is negative (meaning that a borrow would be needed

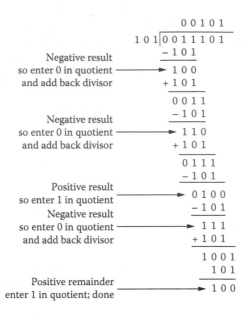

*Figure 3.29* Binary restoring division example.

for the difference to be positive), then we know the divisor was too large to go into the other number. In this case, we enter a 0 in the quotient and add back (restore) the divisor to undo the subtraction that should not have been done, then proceed to bring down the next bit from the dividend.

In the binary example of Figure 3.29, we first try 1 – 101 (1 – 5 decimal), get a borrow, and have to add the 101 back to restore the partial dividend. Then we try 11 – 101 (3 – 5 decimal) and generate another borrow; this tells us that the divisor is still greater than the partial dividend and results in another restoration. At the next bit position, we perform 111 – 101 = 010, giving us the first 1 in the quotient. The partial remainder is positive, and thus, no restoration is needed. We append the next dividend bit to the partial remainder, making it 0100. Subtracting 0100 – 0101 requires another restoration, so the next quotient bit is zero. Finally, we append the rightmost dividend bit to the partial remainder, making it 01001. Subtracting 1001 – 101 gives us the final remainder, 100 (four decimal), and makes the rightmost quotient bit 1 for a final quotient of 00101 (five decimal).

Restoring division is cumbersome and not very efficient because of the need to add the divisor back to the partial dividend/remainder for every bit in the quotient that is 0. Other widely used approaches are also based on the shift and subtract approach but with some refinements that increase speed at the expense of circuit complexity. These other division algorithms, in particular *nonrestoring* division (which generates quotient bits representing 1 and –1 instead of 1 and 0) and the faster Sweeney,

Robertson, and Tocher (SRT) method (which generates two quotient bits per iteration instead of one), eliminate the need to add back the divisor but complicate the design in other ways, such as having to handle negative remainders and correct quotient bits after the fact. (To answer an unasked trivia question, an error in the implementation of the SRT algorithm—not in the algorithm itself—was the cause of the widely publicized division bug discovered in Intel's Pentium CPU in 1994.)

Although the details of these advanced division algorithms are beyond the scope of this text (they are typically covered in an advanced course on computer arithmetic), it is worth pointing out that even the best techniques for division are more expensive in terms of time and hardware required than those used to perform the other three arithmetic operations. It is therefore usually worthwhile for an assembly language programmer (or a compiler) to try to generate code containing as few divide instructions as possible, especially within loops or frequently called routines. Tricks such as replacing division by powers of two with right shifts, using lookup tables, and precomputing divisions by constants can significantly speed execution. In this as well as many other scenarios, well-written software cannot eliminate, but can certainly mitigate, the performance penalty incurred by operations done in slow hardware.

## 3.2.3   Arithmetic with real numbers

All of the arithmetic circuits discussed in the previous section operate on binary integers. Many quantities in the real world may have both integer and fractional parts. This is particularly true of measurements of physical attributes, such as distance, mass, time, etc., which are frequently encountered in scientific and engineering calculations. When we want to use a computer to automate such calculations, we must have some way of representing and operating on these real values. The usual approach is to employ a floating-point representation for real numbers.

### 3.2.3.1   Why use floating-point numbers?

Sometimes integers are just not practical; many real-world applications deal with quantities that take on real values: numbers with both an integer and a fractional part. It is possible to employ a *fixed-point* convention and use integer arithmetic hardware to perform computations on numbers with both integer and fractional parts, but this approach has its limitations, especially when the range of values to be expressed is wide. Scientific calculations, for example, must deal with both very large and very small numbers; it is hard to handle both in a fixed-point system. For a given number of bits in a word, the range of integer (or fixed-point) values is limited. Signed, 32-bit integers cover a range of only +2,147,483,647 to –2,147,483,648, clearly insufficient to calculate the national debt or the number of stars in a galaxy. If some of the bits are allocated for the fraction, the integer range is even smaller. Using 64 bits for integers allows a range of approximately $\pm 9.223 \times 10^{18}$ without fractions, which is still not adequate for many purposes. Even going to 128-bit integer representation gives us only a range of about $\pm 1.701 \times 10^{38}$, which again would have to be reduced in a fixed-point format in which some integer bits are sacrificed to store the fractional part of numbers. A word size of 128 bits is double that of today's most advanced microprocessors, and even larger word sizes would be required to make fixed-point representation of very large and very small numbers feasible. Larger word sizes also waste memory for most ordinary numbers and make computational hardware larger, slower, and more expensive. "There must be a better way to handle a wide range of numeric values," the reader may be thinking, and there is.

Consider the way we handle decimal numbers in a science class. In order to be able to conveniently handle very large and very small numbers without writing many zeroes, we write numbers in *scientific notation* with a sign, a *mantissa* (or set of significant digits), a *base* or *radix* (normally 10), and an *exponent* or *power* of the base. A given number might be expressed in scientific notation as the following:

$$-4.127 \times 10^{+15}$$

where the sign of the number is negative, the significant digits that give the number its value are 4.127, the base is 10, and the base is raised to the positive 15th power. Some scientific calculators would display the number as −4.127 E+15, where the E indicates the exponent, and the base is understood to be 10. Written in normal integer format, the number is −4,127,000,000,000,000. In this case, by using scientific notation, we save ourselves from manipulating 12 zeroes that serve only as place holders. The reader can easily see that for very large or very small numbers, scientific notation is a wonderful convenience.

### 3.2.3.2   Floating-point representation

Floating-point representation in computers is based on exactly the same idea as the scientific notation we use for hand calculations. Rather than try to store and manipulate an entire number in a monolithic, fixed-point format, we divide it into separate bit fields for the sign, mantissa, and exponent. Because we are dealing with binary rather than decimal representation, the base is usually 2 rather than 10 (although other bases such as 4, 8, or 16 have also been used). Other than the use of a different base, floating-point representation of real numbers works exactly the same way as decimal scientific notation. It offers essentially the same advantage: the ability to store (and process) very large and very small numbers with integer and fractional parts without having to store (and process) many bits that are either zero or insignificant.

Over the years, computer manufacturers have used many different floating-point formats. (The first floating-point representation in hardware dates back to the IBM 704, produced in 1954.) They have differed in the base chosen, the total number of bits used (from as few as 32 up to 48, 60, 64, or even more bits), the size of the mantissa and the exponent, the format in which the mantissa and exponent are represented (one's complement, two's complement, sign-magnitude, and "excess" or *biased* notation have all been used at one time or another), specifications for rounding and handling overflows and underflows, and so on.

As one can imagine, the wide variety of floating-point formats caused many problems. Some manufacturers' standards left quite a bit to be desired, mathematically speaking. Many arcane numerical problems arose, often causing loss of precision or outright errors in computations. It was not uncommon for the same program to give different results on different systems (or even to crash on some while running just fine on others). The most obvious problem was that binary data files containing floating-point values were not portable between systems. In order to transfer floating-point data from one architecture to another, users had to convert the values and output them to a text file, then copy the text file to the target system and reconvert the data to that machine's floating-point format. Much time, not to mention precision, was lost in doing these conversions.

After many years of programmers fighting new problems every time they had to port a program with floating-point calculations over to a new system, an industry consortium decided to create a single standard for binary floating-point arithmetic. This standard was endorsed by the Institute of Electrical and Electronics Engineers (IEEE), a professional organization to which many computer engineers belong; it became officially known as IEEE 754-1985 (the year it was adopted) or simply IEEE 754. (A later standard, IEEE 854-1987, generalized 754 to cover decimal as well as binary numbers.) Most computers manufactured in the past 25 to 30 years have adopted the IEEE standard for floating-point arithmetic. This change was inconvenient at first, but has allowed for more consistent operation of programs and easier transfer of data, resulting in far fewer headaches all around.

The IEEE 754 floating-point standard is not just a specification for a single floating-point format to be used by all systems. Rather, its designers recognized the need for different formats for different applications; they specified both single and double precision floating-point data formats along with rules for performing arithmetic operations (compliant systems must obtain the same results, bit for bit, as any other system implementing the standard) and several rounding modes. The standard also defines representations for infinite values and methods for handling exceptional cases such as overflows, underflows, division by zero, and so on. IEEE 754 is not merely a floating-point number format, but a comprehensive standard for representing real numbers in binary form and performing arithmetic operations on them.

The two basic floating-point data formats provided in IEEE 754 are *single precision* (32 bits) and *double precision* (64 bits). The standard also provides for *single extended* and *double extended* precision formats; these are intended to be used for intermediate calculations so that final values expressed in single or double precision will exhibit less rounding error. The standard strongly recommends that "implementations should support the extended format corresponding to the widest basic format supported." Because most hardware floating-point units support both single and double precision calculations, the single extended precision format is rarely seen; most frequently, machines support the single, double, and double extended (80-bit) precision formats. All IEEE numeric formats are set up according to the same basic plan; only the sizes of the bit fields (and details directly related to those sizes) are different. Each format consists of a sign bit, a *significand* field, and a biased exponent field. The two basic formats are depicted in Figure 3.30.

The leftmost bit in each format is the sign bit for the number. As in most integer representations, a sign of 0 indicates a positive number, and a sign of 1 means the number is negative. By placing the sign bit in the leftmost position and defining it in the same way it is defined for integers, the designers of the IEEE standard ensured that the same machine instructions that test the sign of an integer can be used to determine the sign of a floating-point number as well.

The next field, the exponent (or power of two), is 8 bits long in the single precision format and 11 bits long in the double precision format. In each case, exponents use a *biased* notation; that is, they are stored and treated as unsigned values even though they represent a signed value. Single precision exponents are expressed in excess-127 notation. This means that an exponent of 0 would be stored as $01111111_2$ ($127_{10}$). The smallest allowable exponent, –126, would be stored as $00000001_2$ ($1_{10}$); the largest, +127, would be stored as $11111110_2$ ($254_{10}$). The two remaining exponent patterns are all

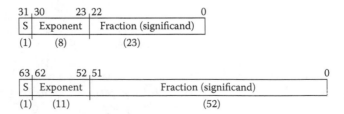

*Figure 3.30* IEEE standard floating-point formats.

zeroes ($00000000_2$) and all ones ($11111111_2^.$); these are used to handle special cases including zero and infinity. Double precision exponents work the same way except that they are stored in excess-1023 format, ranging from $-1022$ ($00000000001_2$) to $+1023$ ($11111111110_2$).

The remaining bits (23 in single precision format, 52 in double precision) make up the significand (or fraction). The significand represents the fractional portion of the normalized mantissa. What is *normalization*? It refers to the process of expressing the mantissa so that the first digit is nonzero and the radix point (in this case, the binary point) is in a known location. Consider the following representations of the decimal number 34,720:

$$34.720 \times 10^3$$

$$3.4720 \times 10^4$$

$$0.3472 \times 10^5$$

$$0.03472 \times 10^6$$

Although they are written differently, each of the above values in scientific notation represent the same number. With a little practice, a person performing hand calculations can learn to recognize that they represent the same thing and easily convert from one representation to another. However, when building computer hardware to process floating-point numbers, it greatly simplifies the design to adopt a single format for the mantissa. Just as with most calculators that display decimal numbers in scientific notation, the usual choice in floating-point systems (including IEEE 754) is to keep one digit to the left of the radix point. The remaining digits to the right of the radix point represent the fraction. Thus, the decimal number 34,720 would be written in normalized form as $3.4720 \times 10^4$. To take a binary example, we could write the number 13 ($1101_2$) as $11.01 \times 2^2$ or $0.1101 \times 2^4$, but its normalized form would be $1.101 \times 2^3$.

Normalized form is the only form used to store values, and it is the form in which the hardware expects to receive all operands. If any intermediate calculation yields a result that is not normalized, it is immediately renormalized before further calculations are done. The exception to this rule in the IEEE 754 standard is for numbers that are nonzero but too small to be normalized (that is, less than 1.0 times 2 to the smallest exponent). Special procedures are defined for handling such *denormalized* numbers, which are used to provide a gradual, rather than abrupt, reduction of precision for very small but nonzero numbers.

Why is the IEEE 754 field containing significant digits called the significand rather than the mantissa? It is because the engineers who designed the standard used a little trick to gain one extra bit of precision for each format. Consider the mantissa of a decimal number. The first digit (the one to the left of the decimal point) can be any number from 1 to 9, inclusive. It cannot be zero because then the number would not be normalized. ($0.8351 \times 10^4$, for example, would be renormalized to $8.351 \times 10^3$.) The same idea applies to any base: Normalized numbers never begin with a zero. That means, in the case of binary numbers, that all normalized numbers begin with the digit 1. (There's no other choice.) All normalized mantissas are in the range $1.0000 \ldots 0 \leq m \leq 1.1111 \ldots 1$ (in decimal, $1 \leq m < 2$).

Because all normalized mantissas begin with "1.," there is no need to store the leading 1 in a memory location or register; the hardware can just assume it is there, inserting it into computations as required. Omitting or *eliding* the leading 1 allows for one additional bit of precision to be retained at the least significant bit position and stored in the same number of bits. This stored mantissa with the implied (or hidden) leading "1." is known as the *significand*. Because of this convention, normalized single precision numbers get 24 bits of precision for the price (in storage) of 23; the double precision format stores 53 significant bits in 52 bits of memory. Note that in most high-level languages, variables declared with type *float* are stored in the single precision format, and variables declared with type *double* use the double precision format.

Putting it all together, the three-bit fields (sign, exponent, and significand) of a normalized floating-point number stored in the IEEE 754 single precision format are mathematically combined as follows:

$$(-1)^{sign} \times (1.significand) \times 2^{(exponent - 127)}$$

The double precision format works the same way except that 1023 is subtracted from the stored exponent to form the actual exponent.

To illustrate the IEEE 754 floating-point format with an actual number, consider how we would represent the decimal number $-0.8125$ as a single precision floating-point value. Our first step would be to represent the number in scientific notation but using base 2 instead of base 10:

$$-0.8125_{10} = -0.1101_2 = -0.1101 \times 2^0 = -1.101 \times 2^{-1}$$

The number is negative, so the sign bit would be 1. The exponent is $-1$, but exponents are represented in the excess-127 format, so the stored exponent would be $(-1) + (+127) = +126$ decimal or 01111110 binary. The leading 1 of the mantissa is elided, leaving the significand to be stored as 10100 ... 0 (23 bits total). Figure 3.31 shows how these parts would be fit together

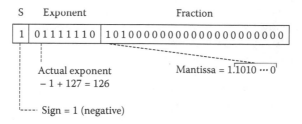

*Figure 3.31* Representing –0.8125 as a single precision floating-point number.

and stored in memory. The stored bit pattern would be 10111111010100000 000000000000000 binary or BF500000 hexadecimal.

Some special cases (see Table 3.2) in the IEEE 754 standard are defined differently from normalized numbers. Zero and positive and negative infinity are numbers that cannot be normalized. Positive and negative infinity have no definite values, and zero contains no bits equal to 1. (Because normalization is defined as adjusting the exponent until the most significant 1 is in the units position, a number with all bits equal to zero cannot be normalized.) Zero is represented in the IEEE format with the exponent and significand bits all equal to zero. If the sign bit is also 0 as is normally the case, then floating-point zero is represented exactly the same way as integer zero, making it simple to check whether any operand or result in any format is zero. (There is a distinct representation of –0.0 although it is, mathematically speaking, no different from +0.0.) Numbers that are too small to be normalized are stored in denormalized form with an exponent field of all zeroes but a nonzero significand. Infinity is represented by a significand with all bits equal to zero and an exponent with all bits set to one. The sign bit can be either 0 or 1, allowing for the representation of both +∞ and –∞.

There is also a special format for a computed result that is *not a number* (NaN). Examples of computations for which the result cannot be interpreted as a real number include taking the square root of a negative number, adding +∞ to –∞, multiplying 0 times ∞, and dividing 0 by 0 or ∞ by ∞. In each of these cases, an IEEE compliant system will return NaN. (Note

*Table 3.2* Special values in the IEEE 754 format

| Significand | Exponent | Meaning |
| --- | --- | --- |
| Any bit pattern | 00 ... 01 through 11 ... 10 | Normalized real number |
| 00 ... 00 (all bits zero) | 00 ... 00 (all bits zero) | Zero |
| Nonzero | 00 ... 00 (all bits zero) | Denormalized real number |
| 00 ... 00 (all bits zero) | 11 ... 11 (all bits one) | Infinity |
| Nonzero | 11 ... 11 (all bits one) | Not a number (NaN) |

that dividing a finite positive or negative number by 0 yields $+\infty$ or $-\infty$, respectively.) Having a special, defined representation for the results of undefined calculations avoids the need to ungracefully terminate a program when they occur. Instead, programs can check for special conditions as needed and handle them as appropriate.

It should be noted that a revised version of the IEEE 754 standard was published in 2008. Some of the changes put forth as part of IEEE 754-2008 included merging it with the (radix-independent) standard 854 and adding decimal formats; resolving certain ambiguities that had been discovered in 754 since it was first adopted, and clarifying and/or renaming some of the terminology (for example, denormalized numbers are now referred to as *subnormal*). Also of note is the creation of two new formats: *half precision* (16 bits) and *quadruple precision* (128 bits). The half precision format is not intended for performing computations, but rather to save storage space for numbers that do not need the range and precision of the 32-bit format. It consists of a sign bit, a 5-bit (excess-15) exponent, and a 10-bit fraction (yielding 11 bits of effective precision). The 128-bit format, usually referred to by the shortened name of "quad precision," has a sign bit, a 15-bit (excess-16383) exponent, and a 112-bit fraction. Guidelines were also put in place for adding new formats with even more bits. With these latest modifications, IEEE 754 should be a useful, widely accepted standard for floating-point arithmetic for a long time to come.

### 3.2.3.3    *Floating-point arithmetic hardware*

Floating-point arithmetic hardware is somewhat more complex than (and thus generally not as fast as) the circuitry used to process integers. Because both the exponents and mantissas must be processed individually as fixed-point values, floating-point arithmetic circuits make use of integer addition, subtraction, multiplication, division, and shifting circuits as building blocks. (Floating-point arithmetic can be "emulated" or implemented in software routines using only integer instructions, but such implementations tend to be very slow.) The main complicating factor is that operations on the exponents can sometimes affect what must be done with the mantissas (and vice versa).

Floating-point multiplication and division are fairly straightforward to understand and implement once one has the capability to multiply and divide integer or fixed-point numbers. To multiply two floating-point numbers (see Figure 3.32), all that is required is to add the exponents (checking, of course, for overflows and underflows, and—assuming IEEE format is being used—subtracting out the bias to avoid double-biasing the result) and then multiply the mantissas. If both mantissas are normalized before the computation, we know that each satisfies the inequality $1 \leq m < 2$. Thus, their product must satisfy the inequality $1 \leq m < 4$. The last stage in a floating-point multiplication, then, involves checking to see if the

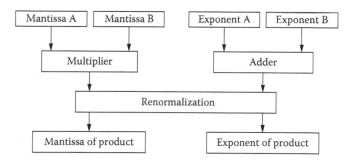

**Figure 3.32** Block diagram of a floating-point multiplier.

mantissa is greater than or equal to 2 and, if so, renormalizing it by moving the binary point one place to the left while adding one to the exponent.

Floating-point division is similar to floating-point multiplication. Exponents are subtracted rather than being added, and (the only difficult part) the mantissa of the dividend is divided by the mantissa of the divisor. Renormalization, if it is required because of a quotient mantissa less than one, involves moving the binary point to the right and subtracting one from the exponent.

Floating-point addition and subtraction are actually somewhat more complex than multiplication and division. This is because the mantissas must be *aligned* such that the exponents are the same before addition or subtraction are possible. To take a decimal example, we can write the numbers 3804 and 11.25 as $3.804 \times 10^3$ and $1.125 \times 10^1$, respectively, but it makes no mathematical sense to begin the process of adding these two values by performing the operation $3.804 + 1.125$. The result, 4.929, is meaningless. Rather, we must express the second number as $0.01125 \times 10^3$ (or the first number as $380.4 \times 10^1$) such that they have the same exponent before adding the mantissas. The correct result, $3.81525 \times 10^3$, is then obtained by adding the aligned mantissas and leaving the exponent unchanged. Subtraction is done exactly the same way except that we subtract the aligned mantissas instead of adding.

As with multiplication and division, floating-point addition and subtraction (see Figure 3.33) may require renormalization after the initial computation. For example, if we add $1.1101 \times 2^3 + 1.0110 \times 2^3$, we obtain the result $11.0011 \times 2^3$, which must be renormalized to $1.10011 \times 2^4$ before we store it for later use. Although adding two numbers of like sign cannot produce a result that is much greater than the larger of the operands, subtracting (or adding numbers of opposite signs) can produce a result that is much smaller than either operand. For example, $1.110011 \times 2^3 - 1.110010 \times 2^3 = 0.000001 \times 2^3 = 1.000000 \times 2^{-3}$; in this case, a shift by six bit positions is necessary to renormalize the result. In the worst case, the operation could

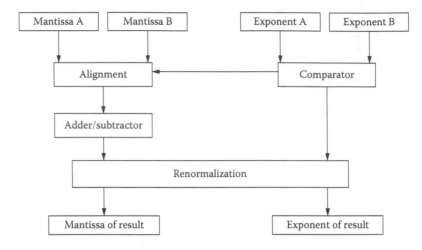

*Figure 3.33* Block diagram of a floating-point adder/subtractor.

yield a result mantissa that is all zeroes except for a one in the rightmost position, requiring a shift by the full width of the mantissa to renormalize the number. This is one more way in which a floating-point adder/subtractor is more complex than a multiplier or divider.

Although incremental improvements and refinements have been and continue to be made, the basics of integer and floating-point arithmetic hardware described in this section have not changed much in quite a few years. The changes we have seen, like those in memory systems, have been primarily the results of reductions in the size and propagation delay of individual circuit components. As transistors and the gates built from them have become smaller and faster, manufacturers have been able to construct arithmetic hardware with larger word widths (64-bit integers are now common, and 128-bit floating-point numbers are coming into use as well). At the same time, circuit speeds have increased considerably. Thus, even commodity computers can now achieve integer and floating-point computational performance that a few years ago was limited to supercomputers with multimillion-dollar price tags.

## 3.3   The control unit

Now that we have discussed computer instruction sets, the register sets used to hold operands and addresses for the instructions, and the arithmetic hardware used to carry out the operations specified by the instructions, it is appropriate to address the part of the CPU that puts everything else in motion: the *control unit*. It is the control unit that determines which machine instruction is to be executed next, fetches it from memory, decodes the instruction to find out which one it is, and activates the proper signals to tell the datapath (and external hardware, such as I/O and memory devices) what to do and when to do it in order to carry out that instruction and save the result. Because the control unit is the brains of the outfit, the prime mover behind everything that happens in the processor and the rest of the system, its design is a very important aspect of the overall system design and a key factor in its performance. In this section, we examine the functions of the control unit in a typical CPU and discuss different design methodologies that have historically been used to realize those functions.

## 3.3.1 A simple example machine

In order to illustrate the operation of a processor's control unit, it helps to have an example to refer to. Contemporary CPUs are much too complex to allow a thorough explanation of their control units in a reasonable space; there are far too many "trees" for us to be able to see the "forest" very well. For illustrative purposes, we will consider a hypothetical machine that is simple enough to illustrate the important points without miring the reader in myriad details.

Consider the simple but functional CPU depicted in Figure 3.34. It contains a set of eight general-purpose registers and an ALU capable of

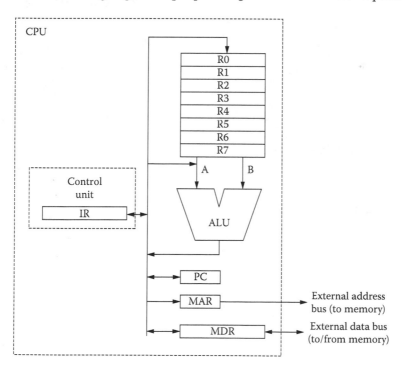

*Figure 3.34* Simple example CPU.

performing basic integer arithmetic and logical operations. Other registers include a *program counter* (PC) to keep track of instruction execution, a *memory address register* (MAR) used to place an address on the external address bus, a *memory data register* (MDR) used to send or receive data to or from the external data bus, and an *instruction register* (IR), which holds the currently executing machine instruction so its bits can be decoded and interpreted by the control unit. To keep things as simple as possible, we will assume the system can address only 256 ($2^8$) memory locations. This means the PC and MAR are 8 bits wide. The ALU, each memory location, all the general-purpose registers, and the MDR are 16 bits wide. Machine instructions occupy one word (16 bits) of memory, so the IR is 16 bits wide also.

Our example machine has two instruction formats: one for instructions that operate on a memory location and a register, and another for instructions that operate only on registers. These instruction formats are laid out as shown in Figure 3.35.

Notice that the leftmost bit (bit 15) of each instruction is used to distinguish the two formats. If it is 0, the remaining 15 bits are interpreted in the register–memory format; if it is 1, the remaining bits are interpreted in the register–register format. The first format uses the eight low-order bits (bits 0 through 7) as a memory address and the next three (bits 8, 9, and 10) to specify the register involved. This leaves four bits (11 to 14) for the op code, meaning there can be no more than $2^4 = 16$ machine instructions of this type. The second format uses the nine rightmost bits as three 3-bit fields (bits 0 to 2, 3 to 5, and 6 to 8) to specify a destination register and up to two source registers. Six bits (9 to 14) are available for op codes, implying that the machine can have up to $2^6 = 64$ register-only instructions. The total possible number of machine instructions is $16 + 64 = 80$, which is fewer than most actual architectures have, but many more than we need for illustrative purposes.

The job of the control unit is, for every instruction the machine is capable of executing, to carry out the steps of the von Neumann machine

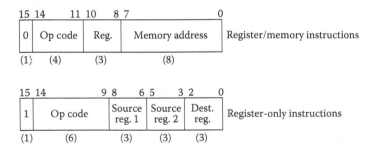

*Figure 3.35* Instruction formats for example CPU.

cycle shown in Figure 1.2. That is, it must fetch the next instruction from memory, decode it to determine what is to be done, determine where the operand(s) for the instruction are and route them to the ALU (unless the operation is a simple data or control transfer requiring no computation), perform the specified operation (if any), and copy the result of the operation to the appropriate destination. To accomplish all this, the control unit must develop and send the necessary control signals at the proper times so that all these steps are carried out in sequence.

Let us examine the specific steps that would be required to execute certain machine instructions. First, consider that the architecture specifies a load instruction using direct addressing that performs a data transfer from a specified memory location to a specified register. In assembly language, loading a value from memory location 64 ($01000000_2$) into register 3 ($011_2$) might be written like this:

LOAD [64], R3

Assuming that the op code for the load direct instruction is 0000, the machine language instruction would appear as shown in Figure 3.36.

What would be the sequence of steps required to carry out this instruction? Figure 3.37 illustrates the flow of information to, from, and through the CPU as the instruction is fetched and executed. Execution always starts with the address of the instruction to be executed in the PC. The sequence of required steps in register transfer language (RTL) and in words would be as follows:

1. MAR←PC: Copy the contents of the PC (the address of the instruction) to the MAR so that they can be output on the address bus.
2. Read; PC←PC + 1: Activate the read control signal to the memory system to initiate the memory access. While the memory read is taking place, increment the PC so that it points to the next sequential instruction in the program. (We are assuming that the PC is built as a counter rather than just a storage register, so it can be incremented without having to pass its contents through the ALU.)
3. MDR←[MAR]: When the memory read is complete, transfer the contents of the memory location over the data bus and latch them into the MDR.

| 15 14 | | 11 10 | 8 7 | | 0 |
|---|---|---|---|---|---|
| 0 | 0000 | 011 | 01000000 | | |

Op code      Reg.      Address

*Figure 3.36* Example machine language LOAD instruction.

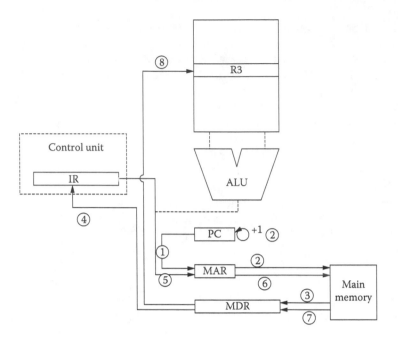

***Figure 3.37*** Execution of the LOAD instruction.

4. IR←MDR: Transfer the contents of the MDR (the machine language instruction) to the IR and decode the instruction. At this point, the control unit discovers that this is a load direct instruction.
5. MAR←IR$_{low}$: Transfer the lower 8 bits from the IR (the operand address) to the MAR to prepare to read the operand.
6. Read: Activate the read control signal to the memory system to initiate the memory access for the operand.
7. MDR←[MAR]: When the memory read is complete, transfer the contents of the memory location over the data bus and latch them into the MDR.
8. R3←MDR: Transfer the contents of the MDR (the operand) to the destination register. Execution of the current instruction is now complete and the control unit is ready to fetch the next instruction.

As another example, let's consider the steps that would be required to execute the instruction ADD R2, R5, R7 (add the contents of register 2 to those of register 5, storing the sum in register 7):

1. MAR←PC: Copy the contents of the PC (the address of the instruction) to the MAR so that they can be output on the address bus.
2. Read; PC←PC + 1: Activate the read control signal to the memory system to initiate the memory access. While the memory read is

taking place, increment the PC so that it points to the next sequential instruction in the program.

3. MDR←[MAR]: When the memory read is complete, transfer the contents of the memory location over the data bus and latch them into the MDR.

4. IR←MDR: Transfer the contents of the MDR (the machine language instruction) to the IR and decode the instruction. (Notice that the first four steps are the same for this instruction as they would be for LOAD or any other instruction.) After decoding is complete, the control unit discovers that this is a register add instruction.

5. $R2_{outA}$; $R5_{outB}$; Add: Transfer the contents of registers 2 and 5 to the ALU inputs and activate the control signal telling the ALU to perform addition. Note that if there were only one bus (instead of two, marked A and B in Figure 3.34) between the register file and the ALU, these operations would have to be done sequentially.

6. R7←ALU: Transfer the output of the ALU (the sum of the operands) to the destination register. Execution of the current instruction is now complete, and the control unit is ready to fetch the next instruction.

For each of these operations within a machine operation, or *micro-operations*, the control unit must activate one or more control signals required to cause that action to take place, while deactivating others that would cause conflicting actions to occur. Even a simple machine, such as the one in our example, may require dozens of control signals to allow all required micro-operations to be performed. For example, each of the machine's registers must have one control signal to enable loading a new value and another to enable it to output its current contents (so that they can be sent to another register or the ALU). We might refer to these signals as $R0_{in}$, $IR_{in}$, $PC_{out}$, etc. Registers that can send data to the ALU will need two output control signals (e.g., $R4_{outA}$, $R4_{outB}$) so that they can transfer data to either of its inputs as needed. Some or all registers may have built-in counting (incrementing and decrementing) capabilities; if so, the control unit will need to generate signals, such as $PC_{increment}$, $R5_{decrement}$, etc. (If this capability is not in place, registers will have to be incremented or decremented by using the ALU to add or subtract one and then writing the result back to the same register.) Several more signals may be needed to select an operation for the ALU to perform. These may be in the form of separate, decoded signals for *Add*, *Subtract*, *Nand*, *Xor*, and so on, or an encoded value that is interpreted by the ALU's internal logic to select a function. Memory control signals, such as *Read* and *Write*, must also be activated at the appropriate times (but never at the same time) to allow the CPU to interact with memory.

To execute a single micro-operation (or step in the execution of a machine language instruction), some of these control signals must be made active while the rest are inactive. (Signals may be active high, active low, or some combination thereof; in our examples in this section, we assume that the active state is a logic 1 unless otherwise specified.) For example, to perform step 1 for any instruction (the micro-operation MAR←PC), the control unit would make $PC_{out} = 1$, $MAR_{in} = 1$, and all other signals 0 (inactive). This set of outputs would be maintained for as long as it takes to perform the given micro-operation (typically one clock cycle) and then would be replaced by the set of control signals needed to perform step 2. In this way the control unit would cause the hardware to sequence through the micro-operations necessary to fetch and execute any machine instruction.

How exactly does the control unit develop and sequence the dozens (likely hundreds in a more complex CPU) of control signals needed to execute machine instructions and thus run programs? Two principal design approaches have been used over the modern history of computing. We examine them in some detail in the following sections.

### 3.3.2   Hardwired control unit

The original method used to design control units was simply to use standard combinational and sequential logic design techniques. The control unit, after all, is nothing more or less than a synchronous, sequential state machine with many inputs and outputs. Because of the large number of inputs and outputs, it is considerably more complex than the simple state machine examples found in most introductory digital logic texts, but the same basic principles and design techniques apply. A control unit designed by using standard logic design techniques is called a *hardwired* control unit. (Because this was the original approach used to design control unit logic in the 1940s and 1950s, a hardwired control unit is also referred to as a *conventional* control unit.)

Figure 3.38 shows the basic layout of a simple hardwired control unit. The instruction register (IR) is used to hold the bits of the currently executing machine instruction. Its outputs, particularly those that correspond to the op code bits, are connected to a decoder that generates a unique output corresponding to that instruction. This is how the machine knows what it is supposed to be doing. The CPU's clock signal is connected to a counter (referred to as the *control step counter*); by decoding the current counter state, the control unit is aware of which step in the execution of the current instruction it is now performing. These decoder outputs, possibly in addition to other machine state information, are input to a block of AND/OR combinational logic (in the early days, implemented as "random" logic; in modern designs, as a programmable logic array) that generates all the required control signals.

Consider the subset of control logic required to develop one of the machine's control signals, the memory Read control signal. This signal

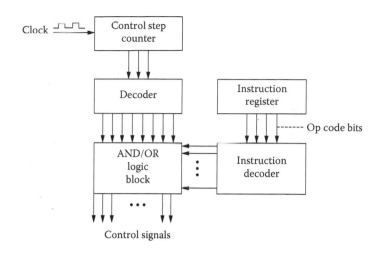

**Figure 3.38** Block diagram of a hardwired control unit.

must be activated (we will assume that means the logic 1 state) during any step of any instruction when the CPU must fetch an instruction or load an operand from memory. It should be deactivated (logic 0) at all other times. In our example machine, instruction fetch occurs during step 2 of every instruction. Certain specific instructions, such as LOAD, ADDM (add memory to register), etc., also require subsequent reads to obtain a data value from memory. In the case of LOAD and ADDM, this occurs during step 6; one would have to analyze the operation of all other machine instructions to see when memory reads would have to occur. Once the timing of memory read operations is defined for every machine instruction, it is a fairly straightforward process to generate the logic for the Read control signal (see Figure 3.39). All that is required is to logically AND the appropriate instruction decoder and control step decoder outputs identifying each case in which this signal needs to be active, and then OR all of these so that the signal is activated when any one of these conditions is true. In Boolean logic, one could write a sum of products expression for the Read control signal as the following:

$$\text{Read} = T_2 + T_6 \cdot \text{LOAD} + T_6 \cdot \text{ADDM} + \dots \text{ (additional terms}$$
$$\text{depending on other instructions that read memory locations)}$$

where LOAD, ADDM, etc., represent the instruction decoder outputs corresponding to those machine instructions.

The logic shown is only that required to produce one of the machine's many control signals. The same sort of design process would have to be employed for all the other signals in order to determine which

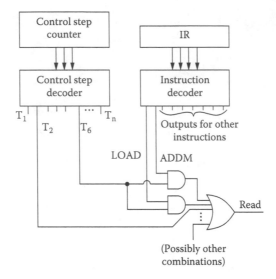

*Figure 3.39* Hardwired control logic for the example machine's memory read signal.

instructions, at which time steps, would need each signal to be activated. The reader can no doubt appreciate that although this process would not be particularly difficult for a simple machine, such as the one in our example, it could become quite complex, tedious, and error-prone (especially without modern design tools) for a processor with many machine instructions, addressing modes, and other architectural features. Thus, although hardwired control unit design was manageable and appropriate for the first- and second-generation machines of the 1940s and 1950s, the increasing complexity of the third- and fourth-generation architectures of the 1960s and 1970s demanded a more flexible, robust design technique (which we examine next). It would take another 20 years (until the RISC revolution of the 1980s and 1990s) for performance concerns to dictate, and for advances in hardware design to enable, the great comeback of hardwired control.

### 3.3.3   Microprogrammed control unit

*Microprogramming* as a control unit design technique was invented in the early 1950s by computer pioneer Maurice Wilkes, who used it in the design of the EDSAC 2. It gained widespread popularity among computer designers in the 1960s and 1970s, which was not coincidentally the era of the great CISC architectures. (The IBM 360 and 370 and the DEC PDP-11 and VAX series machines were all microprogrammed.) It was microprogramming that made CISC processor design practical. CISC architectures were characterized by large, complex, feature-rich instruction sets reminiscent of high-level languages. Instructions were often variable in length and made use of many addressing modes. As CPUs became more complex, the difficulty of implementing all the required control logic directly in hardware became (for the time, at least) insurmountable, and computer designers turned to microprogramming to design control units.

What is microprogramming? In a nutshell, it is the "computer within a computer" approach to developing control signals for a processor. Wilkes' basic idea was that if we can write a computer program as a sequence of machine language instructions and have a CPU execute those machine instructions to perform some task, why can we not treat each individual machine language instruction as a (more basic) task to implement using software techniques? Designers would thus program each machine instruction using very simple, hardware-level *microinstructions*, which are executed (a better term would be *issued*) by a very simple, low-level sequencing unit inside the processor. In keeping with the analogy to higher-level software, the collection of microinstructions containing the machine's entire control strategy is considered its *microprogram*, while the subsets of microinstructions that carry out individual tasks, such as fetching a machine instruction, handling an interrupt request, or executing a particular operation, are referred to as *microroutines*. Just as high- or machine-level computer instructions are generically dubbed "code," control information in a microprogrammed machine is often referred to simply as *microcode*.

Just as a CPU fetches machine instructions from main memory and executes them, in a microprogrammed control unit the microinstructions making up the microprogram must also be fetched from memory. Fetching them from main memory would be intolerably slow; instead, they are stored in a special memory inside the processor, known as the *control store* or *microprogram memory*. In microprocessors, the control store is fabricated onto the chip and is virtually always a read-only memory (often referred to as *microROM* or *μROM*). The microprogram is written and burned into the chip at the factory and cannot be altered by the end user, who may not even know (or care) that the CPU uses microprogrammed control. Some past systems, however (DEC's VAX minicomputers of the late 1970s and early 1980s are a notable example) actually implemented part or all of their control store as writable memory. This enabled field service technicians to load diagnostics and patches to the microcode as needed. It also provided the enterprising end user with the ability to custom-tailor the machine language instruction set to his or her specific needs. If the machine did not have an instruction to do something, or if the end user did not like the way a given operation was implemented, he or she could simply upload a new microroutine to the control store and change the behavior of the

CPU. Such field upgrades were impossible on processors designed with hardwired control units.

Regardless of the specific technology used to implement the microprogram memory, a microprogrammed control unit generates control signals by fetching microinstructions in the proper sequence from that memory. The bits of each microinstruction may represent individual control signals on a bit for bit basis, or they may be organized into bit fields that represent encoded values that are then decoded to produce groups of related control signals. These control signals obtained from the control store are then output to the controlled elements (registers, ALU, memory devices, etc.) just as they would be output from a hardwired control unit. The function of the control signals is the same as it would be in a machine with conventional control; only the method of generating them is different.

A block diagram of a typical microprogrammed control unit is shown in Figure 3.40. As with a hardwired control unit, the sequence of micro-operations to be carried out depends on the machine instruction stored in the instruction register. In this case, the instruction op code bits are used to locate the starting address in the control store of the microroutine required to carry out that machine operation. This address is loaded into the *microprogram counter* (µPC), which serves the same function within the control unit that the program counter does at the next higher level. Just as the PC is used to determine which machine language instruction to execute next, the µPC is used to locate the next microinstruction to issue.

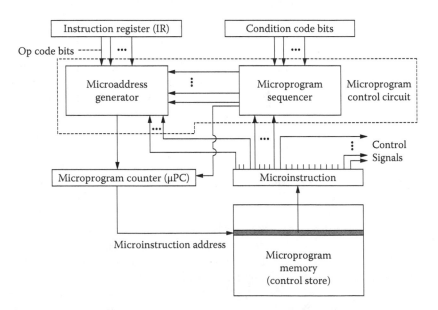

*Figure 3.40* Block diagram of a microprogrammed control unit.

Normally, within a given microroutine, the μPC is simply incremented to point to the next microinstruction; however, in order to get into and out of a given microroutine (and to implement some machine language operations that transfer control), it is necessary to be able to transfer control within the microprogram as well. All useful machine languages include a conditional branching capability; to implement this in a microprogrammed machine, conditional branching in microcode must be supported. This is why the microprogram control circuit (shown in Figure 3.40 as including a microaddress generator and a microprogram sequencer) is connected to the machine's condition code register and the microprogram memory. The condition code bits may be needed to determine whether or not a branch in the microprogram succeeds; if it does, the destination address must be specified in (or calculated from information encoded in) the microinstruction.

A microprogrammed control unit design may have some microinstructions that are only used for branching. (In other words, they contain a special bit pattern that identifies the remaining bits as branching information rather than control information.) This saves space but may slow execution because no useful work will be done by branching microinstructions. Alternatively, additional bits used to determine branch conditions and destination addresses may be appended to the *control word* (the bits that specify control signal values) to make wider microinstructions that can specify hardware operations and control transfers simultaneously. This increases the width of the microprogram memory but ensures that no clock cycles need be wasted merely to transfer control within the microprogram.

Except for information needed to implement control transfers within the microprogram, the bulk of the microprogram memory's contents is simply the information for each step in the execution of each machine operation regarding which control signals are to be active and which are to be inactive. Typically, there would be one shared, generic microroutine for fetching any machine instruction from main memory plus a separate, specific microroutine for the execution phase of each different machine instruction. The last microinstruction of the instruction fetch microroutine would be (or include) a microprogram branch based on the contents of the instruction register (the new instruction just fetched) such that the next microinstruction issued will be the first one in the correct execution microroutine. The last microinstruction of each execution microroutine would cause a branch back to the beginning of the instruction fetch microroutine. In this way, the machine will continue to fetch and execute instructions indefinitely.

*Horizontal microprogramming* is the simplest and fastest but least space-efficient approach to microprogrammed control unit design. A horizontal microprogram is one in which each control signal is represented as a distinct bit in the microinstruction format. If a given machine requires, for example, 118 control signals, each control word will be 118 bits wide. (The stored microinstructions will be somewhat wider if additional bits are included to handle branching within the microprogram.) If our simple example machine requires 43 control signals, its control words would be 43 bits wide (see Figure 3.41).

The obvious advantage of horizontal microprogramming is (relative) speed: The control signals are available to be issued (output) as soon as a given microinstruction is fetched from the control store. It is also the most flexible approach because any arbitrary set of control signals may be active at the same time. However, much microprogram memory is wasted because usually only a few control signals are active during each time step and many sets of signals may be mutually exclusive. A large control store takes up space that could be used for other CPU features and functions.

*Vertical microprogramming* saves microcode storage space by encoding parts of the microinstruction into smaller bit fields. This is often possible because, in theory, certain control signals are (or in practice turn out to be) mutually exclusive. One simple example is that the memory read and write signals will never both be asserted at the same time. An even better illustration is that if there are several registers in the CPU with outputs connected to a common bus, only one register may be designated to output data at a time. Rather than having separate bits in the microinstruction format for $R0_{outA}$, $R1_{outA}$, $R2_{outA}$, etc., a vertically microprogrammed control unit might have a bit field designated as $DataReg_{outA}$. Eight register outputs could be controlled with 3 bits, or 16 with 4 bits, and so on, thus saving several bits in every microinstruction. A vertical microprogram for our example machine (see Figure 3.42) could take advantage of this, reducing the number of bits per control word from 43 to 19.

Given that a control unit might have thousands of microinstructions in its microprogram, reducing the amount of memory required for each microinstruction could save quite a bit of chip area. The trade-off with vertical microprogramming is that decoders are required to interpret the

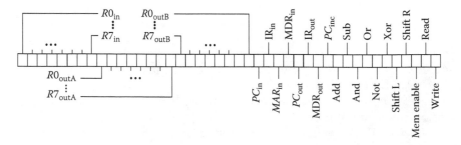

*Figure 3.41* Horizontal control word for example machine.

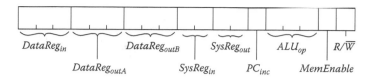

*Figure 3.42* Vertical control word for example machine.

encoded bit fields of a microinstruction and generate the required control signals. Because the decoders introduce propagation delay, all else being equal, a vertically microprogrammed processor will be slower than the same design implemented with a horizontal microprogram. (Given that most memory technologies are slower than hardwired logic, micropro-grammed control in general suffers from a performance disadvantage compared with conventional control; vertical microprogramming exacer-bates this problem.) Also, once encoded bit fields are determined, the vari-ous control signals assigned to them become mutually exclusive even if they were not inherently so. This may take a certain amount of flexibility away from the designer.

*Two-level microprogramming* (or *nanoprogramming*) is another tech-nique, similar to vertical microprogramming, that can be used to reduce the required size of microprogram memory. Instead of being a single, monolithic microprogram memory, the control store is divided into two hierarchical levels: *micromemory* and *nanomemory* (μROM and nROM, assuming that control store is not writable). Some (if not all) of the bits of each microinstruction represent an address in the second level control store (nROM). When the microinstruction is read from the micromemory, this address is used to retrieve the required control information from the nanomemory. Figure 3.43 illustrates the process.

The reason two-level microprogramming can often save a great deal of space is that many machine instructions are similar. For example, every step of the ADD and SUB instructions in most machines is identical except for the step in which the actual operation is done in the ALU. Logical and shift instructions may also be similar, as they generally obtain oper-ands from and store results in, the same places as arithmetic instructions. Memory reads for operands require the same control signals as memory reads for instructions, and so on. In a machine with a single-level micro-program, this means the control store may contain a number of similar, but slightly different, microroutines. If a number of machine instructions require, for example, seven of eight identical steps, two-level micropro-gramming allows the seven corresponding common control words to be stored only once in the nanomemory. Any number of microinstructions can point to a given stored control word and cause it to be retrieved when necessary. By storing commonly used control words only once and using

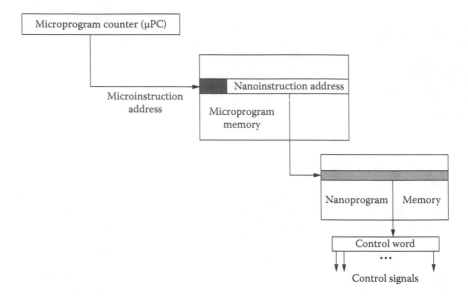

*Figure 3.43* Two-level microprogram for example machine.

microinstructions to point to them, a considerable amount of chip area can be saved. The savings in silicon real estate obtained by using two-level microprogramming was a key factor that allowed Motorola to implement its complex, ambitious 68000 CPU on a single chip using 1980 technology.

Of course, the reduction in on-chip control memory size realized with two-level microprogramming comes at a price. As in the case of vertical microprogramming, there is an additional delay after a given microinstruction is fetched before the control signals can be issued. This delay may be even longer in the case of a second level (nanomemory) lookup because ROM generally has an access time in excess of the decoder delays inherent to vertical microprogramming.

It is possible (and once was fairly common) to combine horizontal, vertical, and two-level microprogramming in the same control unit. Control signals that must be available early in a given time step (perhaps because of longer delays in the hardware they control or a need to send the results somewhere else after they are produced) may be represented bit for bit (horizontally) in the control word, and other signals that are less timing-critical may be encoded (vertically) or even kept in a second level lookup table. It is the longest total delay from the beginning of a clock cycle through the generation of a given control signal and the operation of the element it controls that places a lower bound on the cycle time of the machine and thus an upper bound on its clock frequency. A chain is only as strong as its weakest link, and a processor is only as fast as its slowest logic path.

Hardwired control and microprogrammed control are two ways of accomplishing the same task, generating all the control signals required for a computer to function. Each approach has its own set of advantages and disadvantages. The main advantage of hardwired control is simple: All else being equal, it almost always results in a faster implementation. Generating control signals in less time, as is possible with two-level logic versus a memory lookup, yields a shorter clock cycle time (and thus a faster processor) given the same datapath. The main disadvantage of conventional control is the complexity and difficulty of designing large logic circuits. This difficulty proved essentially insurmountable for the CISC machines of the 1960s and 1970s, but was later overcome by innovations in hardware design (programmable logic) by the development of hardware design languages and other software tools for chip design and by taking much of the complexity out of ISAs (in other words, the advent of RISC machines).

Microprogramming offers the advantage of relatively simple, systematic control unit design. Because microcode may be reused, it is easier to build a family of compatible machines; the microroutines for added features can simply be added to the previously written microprogram. The additional cost to add features incrementally is small, provided the additional microcode does not exceed the space available for control store. If it is necessary or profitable to emulate the behavior of previous systems (as it was with the IBM 360), it is easier to implement that emulation in microcode than directly in logic. If the control store is writable, it is possible to load diagnostic microroutines or microcode patches in the field—an advantage that may not be crucial, but is definitely nontrivial.

The main disadvantage of microprogramming (as seen by the end user) is that, due to the overhead of sequencing microinstructions and fetching them from control store (plus decoding them if vertical microcode is used), the resulting CPU is slower than a hardwired implementation built with similar technology. Because the control store takes up a great deal of space on a chip, microprogrammed control units are generally much larger than conventional control units with an equivalent amount of logic. Thus, less chip area remains for building registers, computational hardware, on-chip cache memory, or other features. From a designer's point of view, microprogramming (although ideal for complex designs) has a high startup cost and may not be cost-effective (in terms of design time and ultimately money) for simpler systems. In addition, because CPU designs are generally unique and proprietary, very few design support tools were or are available to support microprogram development. The microcode compilers developed for particular projects were generally too specific to be of much use for other designs, and the few that were more general tended to produce very inefficient microcode.

With the state of logic design as it was in the 1960s and 1970s, microprogramming was the only viable response to users' hunger for more capable, feature-rich architectures. CISC was the prevailing architectural philosophy of the day; given the available hardware and software technology, it made perfect sense. Compilers were primitive; memory was expensive; hardware design tools consisted mainly of pencils and paper. There was a real desire to bridge the semantic gap and make machine languages more like high-level languages. By making one machine instruction perform a complex task, executable programs (and thus main memory) could be kept relatively small. Microprogramming was a way—the only way at the time—of achieving these goals and of designing hardware using a software methodology. (Contrary to what many computer programmers think, software has always been easier to design than hardware.)

By the 1980s and 1990s, hardware and software technologies had changed quite a bit. Compilers were considerably more sophisticated. Memory, although not dirt cheap, was much less expensive than it was 20 years before. Disk drives and semiconductor memory devices had much larger capacities, and virtual memory was becoming ubiquitous even in microprocessor-based systems. The speed of CPUs with microprogrammed control units was gradually improving, but not rapidly enough to satisfy user demands. Gradually, the amount of memory taken up by programs became less important than the speed of execution. Thus, RISC was born. Designers realized that stripping extraneous, seldom-used features from the machine, coupled with the use of programmable logic, hardware description languages, and other modern design tools would enable them to return to the use of faster, hardwired control units.

Will microcode ever make a comeback? It certainly doesn't appear that way at present, but wise observers have learned never to say "never." If the main disadvantage of microprogrammed control, the relatively slow access to the control store, could be eliminated, its advantages might once again catapult it to prominence. In particular, if magnetic RAM (see Section 2.1.2) proves to be fast, economical, and easy to integrate onto a chip with other types of logic (admittedly an ambitious list of goals for a new and unproven technology), then all bets are off. We could see a day in the not-too-distant future when processors have at least some writable, magnetic control store as standard equipment.

## 3.4   Chapter wrap-up

Although all the subsystems of a modern computer are important, no part of the machine is as complex as its central processing unit (or units; parallel machines are discussed in Chapter 6). Under the direction of a highly specialized state machine known as the control unit, computational hardware must carry out arithmetic, logical, shifting, and other operations on data stored in memory locations or registers as specified by a machine language program written (or compiled) in accordance with the machine's native instruction set. This instruction set may be very simple, extremely complex, or anywhere in between.

In order to understand and keep up with technical developments in their field, it is important for all computer professionals to have a good understanding of the major architectural and implementation considerations that go into the design of a modern CPU. Any student of computer science or computer engineering should learn about the wide variety of philosophies embodied in various ISAs; the techniques for designing fast, efficient datapath hardware to execute those instructions on integer and real number values; and the different approaches to designing the control unit that oversees all activities of the system.

In Chapter 1, we discussed the differences between architectural design and implementation technology while noting that neither exists in a vacuum. Perhaps in no other part of a computer system do architecture and implementation influence each other so much as they do inside the CPU. Computer designers, who like to eat as much as anyone else (and who therefore want their companies to sell as many microprocessor chips as they can), want to make the processor as efficient as possible; they accomplish this by creatively implementing their architectural ideas using available implementation technologies.

In this chapter, we examined some of the basic CPU design concepts mentioned above in their historical context and in a fair amount of detail with regard to both architecture and implementation. At this point, the reader should have an appreciation of the history of CPU design and a good understanding of the general principles and key concepts involved in the process of constructing a working, general-purpose digital processor. By studying and understanding the basics of processor architecture and implementation covered in this chapter, we have created a firm foundation for the more advanced, performance-enhancing approaches covered in the next chapter.

## REVIEW QUESTIONS

1. Does an architecture that has fixed-length instructions necessarily have only one instruction format? If multiple formats are possible given a single instruction size in bits, explain how they could be implemented; if not, explain why this is not possible.
2. The instruction set architecture for a simple computer must support access to 64 KB of byte-addressable memory space and eight 16-bit general-purpose CPU registers.
   a. If the computer has three-operand machine language instructions that operate on the contents of two different CPU registers

to produce a result that is stored in a third register, how many bits are required in the instruction format for addressing registers?

b. If all instructions are to be 16 bits long, how many op codes are available for the three-operand, register operation instructions described above (neglecting, for the moment, any other types of instructions that might be required)?

c. Now assume (given the same 16-bit instruction size limitation) that, in addition to the instructions described in (a), there are a number of additional two-operand instructions to be implemented, for which one operand must be in a CPU register while the second operand may reside in a main memory location or a register. If possible, detail a scheme that allows for at least 50 register-only instructions of the type described in (a) plus at least 10 of these two-operand instructions. (Show how you would lay out the bit fields for each of the machine language instruction formats.) If this is not possible, explain in detail why not and describe what would have to be done to make it possible to implement the required number and types of machine language instructions.

3. What are the advantages and disadvantages of an instruction set architecture with variable-length instructions?

4. Name and describe the three most common general types (from the standpoint of functionality) of machine instructions found in executable programs for most computer architectures.

5. Given that we wish to specify the location of an operand in memory, how does indirect addressing differ from direct addressing? What are the advantages of indirect addressing, and in what circumstances is it clearly preferable to direct addressing? Are there any disadvantages of using indirect addressing? How is register-indirect addressing different from memory-indirect addressing, and what are the relative advantages and disadvantages of each?

6. Various computer architectures have featured machine instructions that allow the specification of three, two, one, or even zero operands. Explain the trade-offs inherent to the choice of the number of operands per machine instruction. Pick a current or historical computer architecture, find out how many operands it typically specifies per instruction, and explain why you think its architects implemented the instructions the way they did.

7. Why have load–store architectures increased in popularity in recent years? (How do their advantages go well with modern architectural design and implementation technologies?) What are some of their less desirable trade-offs compared with memory–register architectures, and why are these not as important as they once were?

8. Discuss the two historically dominant architectural philosophies of CPU design:
   a. Define the acronyms CISC and RISC and explain the fundamental differences between the two philosophies.
   b. Name one commercial computer architecture that exemplifies the CISC architectural approach and one other that exemplifies RISC characteristics.
   c. For each of the two architectures you named in (b), describe one distinguishing characteristic not present in the other architecture that clearly shows why one is considered a RISC and the other a CISC.
   d. Name and explain one significant advantage of RISC over CISC and one significant advantage of CISC over RISC.

9. Discuss the similarities and differences between the programmer-visible register sets of the 8086, 68000, MIPS, and SPARC architectures. In your opinion, which of these CPU register organizations has the most desirable qualities and which is least desirable? Give reasons to explain your choices.

10. A circuit is to be built to add two 10-bit numbers $x$ and $y$ plus a carry in. (Bit 9 of each number is the most significant bit [MSB], and bit 0 is the least significant bit [LSB]. $c_0$ is the carry in to the LSB position.) The propagation delay of any individual AND or OR gate is 0.4 ns, and the carry and sum functions of each full adder are implemented in sum of products form.
    a. If the circuit is implemented as a ripple carry adder, how much time will it take to produce a result?
    b. Given that the carry generate and propagate functions for bit position $i$ are given by $g_i = x_i y_i$ and $p_i = x_i + y_i$, and that each required carry bit ($c_1 \ldots c_{10}$) is developed from the least significant carry in $c_0$ and the appropriate $g_i$ and $p_i$ functions using AND-OR logic, how much time will a carry lookahead adder circuit take to produce a result? (Assume AND gates have a maximum fan-in of 8 and OR gates have a maximum fan-in of 12.)

11. Under what circumstances are carry save adders more efficient than normal binary adders that take two operands and produce one result? Where, in a typical general-purpose CPU, would one be most likely to find carry save adders?

12. Given two 5-bit, signed, twos complement numbers $x = -6 = 11010_2$ and $y = +5 = 00101_2$, show how their 10-bit product would be computed using Booth's algorithm (you may wish to refer to Figures 3.26, 3.27, and 3.28).

13. Discuss the similarities and differences between scientific notation (used for manual calculations in base 10) and floating-point representations for real numbers used in digital computers.

14. Why was IEEE 754-1985 a significant development in the history of computing, especially in the fields of scientific and engineering applications?
15. In 2008, the IEEE modified standard 754 to allow for "half-precision" 16-bit floating-point numbers. These numbers are stored in similar fashion to the single precision 32-bit numbers but with smaller bit fields. In this case, there is 1 bit for the sign, followed by 5 bits for the exponent (in excess-15 format), and the remaining 10 bits are used for the fractional part of the normalized mantissa. Show how the decimal value +17.875 would be represented in this format.
16. Show how the decimal value –267.5625 would be represented in IEEE 754 single and double precision formats.
17. Consider a simple von Neumann architecture computer like the one discussed in Section 3.3.1 and depicted in Figure 3.34. One of its machine language instructions is an ANDM instruction that reads the contents of a memory location (specified by direct addressing), bitwise ANDs this data with the contents of a specified CPU register, then stores the result back in that same register. List and describe the sequence of steps that would have to be carried out under the supervision of the processor's control unit in order to implement this instruction.
18. What are the two principal design approaches for the control unit of a CPU? Describe each of them and discuss their advantages and disadvantages. If you were designing a family of high-performance digital signal processors, which approach would you use, and why?
19. In a machine with a microprogrammed control unit, why is it important to be able to do branching within the microcode?
20. Given the horizontal control word depicted in Figure 3.41 for our simple example machine, develop the microroutines required to fetch and execute the ANDM instruction using the steps you outlined in question 17.
21. Repeat question 20 using the vertical control word depicted in Figure 3.42.
22. Fill in the blanks below with the most appropriate term or concept discussed in this chapter:

    _____ The portion (bit field) of a machine language instruction that specifies the operation to be done by the CPU.

    _____ A type of instruction that modifies the machine's program counter (other than by simply incrementing it).

    _____ A way of specifying the location of an operand in memory by adding a constant embedded in the instruction to the contents of a "pointer" register inside the CPU.

    _____ These would be characteristic of a stack-based instruction set.

_____ This type of architecture typically has instructions that explicitly specify only one operand.

_____ A feature of some computer architectures in which operate instructions do not have memory operands; their operands are found in CPU registers.

_____ Machines belonging to this architectural class try to bridge the semantic gap by having machine language instructions that approximate the functionality of high-level language statements.

_____ This part of a CPU includes the registers that store operands as well as the circuitry that performs computations.

_____ This type of addition circuit develops all carries in logic directly from the inputs rather than waiting for them to propagate from less significant bit positions.

_____ A structure composed of multiple levels of carry save adders, which can be used to efficiently implement multiplication.

_____ This type of notation stores signed numbers as though they were unsigned; it is used to represent exponents in some floating-point formats.

_____ In IEEE 754 floating-point numbers, a normalized mantissa with the leading 1 omitted is called this.

_____ This is the result when the operation 1.0/0.0 is performed on a system with IEEE 754 floating-point arithmetic.

_____ This holds the currently executing machine language instruction so its bits can be decoded and interpreted by the control unit.

_____ A sequence of microinstructions that fetches or executes a machine language instruction, initiates exception processing, or carries out some other basic machine-level task.

_____ A technique used in microprogrammed control unit design in which mutually exclusive control signals are not encoded into bit fields, thus eliminating the need for decoding microinstructions.

_____ This keeps track of the location of the next microword to be retrieved from microcode storage.

## chapter four

# Enhancing CPU performance

In the previous chapter, we discussed the basics of instruction set architecture and datapath and control unit design. By now, the reader should have a good understanding of the essentials of central processing unit (CPU) architecture and implementation; you may even feel (the author dares to hope) that, given enough time, you could generate the logical design for a complete, usable CPU. Once upon a time, that would have been a good enough product to sell. Now, however, to succeed in the marketplace, a processor (and the system containing it) must not only work, but must perform extremely well on the application of interest. This chapter is devoted to exploring implementation techniques that manufacturers have adopted to achieve the goal of making their CPUs process information as rapidly as possible. The most ubiquitous of these techniques is known as *pipelining*. As we shall see, virtually all high-performance computers utilize some form of pipelining.

Implementation technologies for computer hardware change rapidly over time. Transistors continue to become smaller and faster; circuits that were considered blindingly fast 4 or 5 years ago are hopelessly slow today and will be completely obsolete that far, or less, in the future. Over the history of computing devices, technology has always improved and continues to improve, although at some point we must approach the ultimate physical limits. Where will more speed come from when we have switching elements close to the size of individual atoms with signal propagation over such tiny distances still limited by the velocity of light? Perhaps the unique approach of quantum computing (to be discussed in Section 7.4) will allow us to do much more than is currently thought possible within the limits of physics. Only time will tell where new innovations in technology will take us as designers and users of computer systems.

What is always true is that at any given time, be it 1955 or 1984 or 2016, we (computer manufacturers, specifically) can only make a given piece of hardware so fast using available technology. What if that is not fast enough to suit our purposes? Then we must augment technology with clever design. If a given hardware component is not fast enough to do the work we need to do in the time we have, we build faster hardware if we can; if we cannot, we build more hardware and divide the problem. This approach is known as *parallelism* or *concurrency*. We can achieve concurrency within a single processor core and/or by using multiple processors in a system (either in the form of a multicore chip, multiple CPU chips in a system, or both). The former approach—concurrency within one core—is the subject of the rest of this chapter; the latter is addressed in Section 6.1 when we discuss multiprocessor systems.

## 4.1   Pipelining

The original digital computers and their successors for a number of years were all serial (sequential) processors, not only in their architecture, but also in their implementation. Not only did they *appear* to execute only one instruction at a time as the von Neumann model suggests; they actually *did* execute only one instruction at a time. Each machine instruction was processed completely before the next one was started. This sequential execution property was the underlying assumption for our previous treatment of both hardwired and microprogrammed control unit design. This approach allowed the control unit to be kept relatively simple, as it only had to generate the control signals for one instruction at a time.

The sequential execution paradigm that forms the basis of the von Neumann machine cycle is very simple and very effective, but it has one obvious flaw: It does not make very efficient use of the hardware. Executing a single machine instruction requires several steps: fetch the instruction from memory, decode it, retrieve its operands, perform the operation it specifies, and store its result. (Slightly different breakdowns of the machine cycle are possible, but this will suffice for discussion.) If the machine processes one instruction at a time, what is the arithmetic/ logic unit (ALU) doing while the instruction (or an operand) is being fetched? Probably nothing. What work are the instruction decoder, memory address register (MAR), memory data register (MDR), and various other parts of the CPU accomplishing while the ALU is busy performing a computation? Probably none.

For the most efficient use of all the system components in which we have invested design time, chip area, electrical power, and other valuable resources, we would ideally like to keep all of these components as busy as possible as much of the time as possible. A component that is unused part of the time is not giving us our money's worth; designers should search for a way to make more use of it. Conversely, a component that is overused (needed more often than it is available) creates a *structural hazard*; it will often have other components waiting on it and will thus become a bottleneck, slowing down the entire system. Designers may need to replicate such a component to improve throughput. The art of designing a modern processor involves balancing the workload on all the parts of the CPU such that they are kept busy doing useful work as much of the time as possible without any of them clogging up the works and making the other parts wait. As the reader might expect, this balancing act is not a trivial exercise. Pipelining, which we are about to investigate, is an essential technique for helping bring about this needed balance.

*Pipelining*, in its most basic form, means breaking up a task into smaller subtasks and overlapping the performance of those subtasks for different instances of the task. (The same concept, when applied to the manufacture of automobiles or other objects, is called an assembly line.) To use terms more specifically related to computing, pipelining means dividing a computational operation into steps and overlapping those steps over successive computations. This approach, although much more common in today's computers than it was 30 or 40 years ago, is hardly new. The first use of pipelining in computers dates back to the IBM Stretch and Univac LARC machines of the late 1950s. Pipelining, as we shall see, improves the performance of a processor in much the same way that low-order interleaving improves the performance of main memory, while being subject to many of the same considerations and limitations.

To understand how pipelining works, consider a task that can be broken down into three parts performed sequentially. Let us refer to these parts as step 1, step 2, and step 3 (see Figure 4.1). The time taken to perform step 1 is represented as $t_1$, and $t_2$ and $t_3$ represent the times required to perform steps 2 and 3. Because the three steps (subtasks) are performed sequentially, the total time to perform the task is given by

$$t_{TASK} = t_1 + t_2 + t_3$$

Without pipelining, the time to perform three iterations of this task is $3 \times t_{TASK}$, and the time to perform 50 iterations is $50 \times t_{TASK}$. The time required is directly proportional to the number of iterations to be performed; there is no advantage to be gained by repeated performance of the task.

Now suppose that we separate the hardware that performs steps 1, 2, and 3 in such a way that it is possible for them to work independently of each other. (We shall shortly see how this can be accomplished in computer hardware.) Figure 4.2 illustrates this concept. We begin the first iteration of the task by providing its inputs to the hardware that performs step 1 (call this stage 1). After $t_1$ seconds, step 1 (for the first iteration of

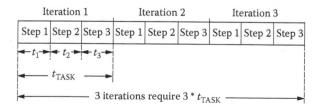

*Figure 4.1* Subdividing a task into sequential subtasks.

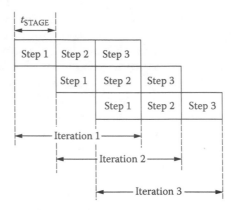

*Figure 4.2* Basic pipelining concept.

the task) is done, and the results are passed along to stage 2 for the performance of step 2. Meanwhile, we provide the second set of inputs to stage 1 and begin the second iteration of the task before the first iteration is finished. When stage 2 is finished processing the first iteration and stage 1 is finished processing the second iteration, the outputs from stage 2 are passed to stage 3, and the outputs from stage 1 are passed to stage 2. Stage 1 is then provided a new, third set of inputs—again, before the first and second iterations are finished. At this point, the pipeline is full; all three stages are busy working on something. When all are done this time, iteration 1 is complete, iteration 2 moves to stage 3, iteration 3 moves to stage 2, and the fourth iteration is initiated in stage 1. This process can continue as long as we have more iterations of the task to perform. All stages will remain busy until the last iteration leaves stage 1 and eventually "drains" from the pipeline (completes through all the remaining stages).

Obviously, this approach works best if the three stage times $t_1$, $t_2$, and $t_3$ are all equal to $t_{TASK}/3$. Let us refer to this as the stage time, $t_{STAGE}$. If $t_1$, $t_2$, and $t_3$ are not all equal, then $t_{STAGE}$ must be set equal to the greatest of the three values; in other words, we cannot advance the results of one stage to the next until the *slowest* stage completes its work. If we try to make $t_{STAGE}$ smaller, at least one of the stages will be given new inputs before it finishes processing its current inputs, and the process will break down (generate incorrect results).

What are the performance implications of this pipelined (or assembly line) approach? Pipelining does nothing to enhance the performance of a *single* computation. A single iteration of the task still requires at least $t_{TASK}$ seconds. In fact, it may take somewhat longer if the stage times are mismatched (we know that $3 \times t_{STAGE} \geq t_1 + t_2 + t_3$). The performance advantage occurs only if we perform two or more successive iterations of the task. Although the first iteration takes just as long as ever (or perhaps slightly longer), the second and all subsequent iterations are completed in $t_{STAGE}$ rather than the $3 \times t_{STAGE}$ taken by the first iteration. Two iterations can be completed in $4 \times t_{STAGE}$, three iterations in $5 \times t_{STAGE}$, four in $6 \times t_{STAGE}$, and so on. In general, we can define the time taken to perform $n$ iterations of the task using this three-stage pipeline as

$$t_{TOTAL} = [3 \times t_{STAGE}] + [(n - 1) \times t_{STAGE}] = [(n + 2) \times t_{STAGE}]$$

If $t_{STAGE}$ is equal (or even reasonably close) to $t_{TASK}/3$, then a substantial speedup is possible compared with the nonpipelined case; larger values of $n$ lead to a greater advantage for the pipelined implementation. Let us suppose for the sake of simplicity that there is no hardware overhead (we will address this topic later) and that $t_{STAGE}$ equals 1 ns and $t_{TASK}$ equals 3 ns. Five iterations of the task would take $(5 \times 3) = 15$ ns without pipelining, but only $(7 \times 1) = 7$ ns using the pipelined approach. The speed ratio in this case is 2.143 to 1 in favor of pipelining. If we consider 10 consecutive iterations, the total times required are 30 ns (nonpipelined) versus 12 ns (pipelined) with pipelining yielding a speedup factor of 2.5. For 50 iterations, the numbers are 150 ns and 52 ns, respectively, for a speedup of 2.885. In the limit, as $n$ grows very large, the speedup factor of the pipelined implementation versus the nonpipelined implementation approaches 3, which is—not coincidentally—the number of stages in the pipe.

Most generally, for a pipeline of $s$ stages processing $n$ iterations of a task, the time taken to complete all the iterations may be expressed as

$$t_{TOTAL} = [s \times t_{STAGE}] + [(n - 1) \times t_{STAGE}] = [(s + n - 1) \times t_{STAGE}]$$

The $[s \times t_{STAGE}]$ term represents the *flow-through time*, which is the time for the first result to be completed; the $[(n - 1) \times t_{STAGE}]$ term is the time required for the remaining results to emerge from the pipe. The time taken for the same number of iterations without pipelining is $n \times t_{TASK}$. In the ideal case of a perfectly balanced pipeline (in which all stages take the same time), $t_{TASK} = s \times t_{STAGE}$, so the total time for the nonpipelined implementation would be $n \times s \times t_{STAGE}$. The best case speedup obtainable by using an $s$-stage pipeline would thus be $(n \times s)/(n + s - 1)$, which, as $n$ becomes large, approaches $s$ as a limit.

From this analysis, it would appear that the more stages into which we subdivide a task, the better. This does not turn out to be the case for several reasons. First, it is generally only possible to break down a given task so far. (In other words, there is only so much *granularity* inherent in the task.) When each stage of the pipeline represents only a single level of logic gates, how can one further subdivide operations? The amount of logic required to perform a given task thus places a fundamental limitation on the *depth* of (number of stages in) a pipelined implementation of that task.

Another limiting factor is the number of consecutive, uninterrupted iterations of the task that are likely to occur. For example, it makes little sense to build a 10-stage multiplication pipeline if the number of multiplications to be done in sequence rarely exceeds 10. One would spend most of the time filling and draining the pipeline—in other words, with less than the total number of stages doing useful work. Pipelines only improve performance significantly if they can be kept full for a reasonable length of time. Mathematically speaking, achieving a speedup factor approaching $s$ (the number of pipeline stages) depends on $n$ (the number of consecutive iterations being processed) being large, and "large" is defined relative to $s$. "Deep" pipelines, which implement a *fine-grained* decomposition of a task, only perform well on long, uninterrupted sequences of task iterations.

Yet another factor that limits the speedup that can be achieved by subdividing a task into smaller subtasks is the reality of hardware implementation. Constructing a pipeline with actual hardware requires the use of a *pipeline register* (a parallel-in, parallel-out storage register composed of a set of flip-flops or latches) to separate the combinational logic used in each stage from that of the following stage (see Figure 4.3 for an illustration). The pipeline registers effectively isolate the outputs of one stage

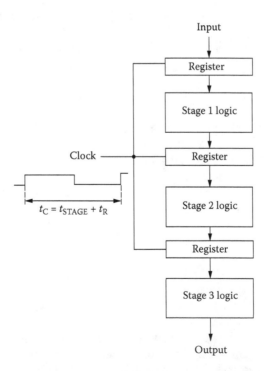

*Figure 4.3* Pipeline construction showing pipeline registers.

from the inputs of the next, advancing them only when a clock pulse is received; this prevents one stage from interfering with the operation of those preceding and following it. The same clock signal is connected to all the pipeline registers so that the outputs of each stage are transferred to the inputs of the next simultaneously.

Each pipeline register has a cost of implementation: It consumes a certain amount of power, takes up a certain amount of chip area, etc. Also, the pipeline registers have finite propagation delays that add to the propagation delay of each stage; this reduces the performance of the pipeline somewhat compared to the ideal theoretical case. The clock cycle time $t_C$ of the pipeline can be no smaller than $t_{STAGE}$ (the longest of the individual stage logic delays) plus $t_R$ (the delay of a pipeline register). The register delays represent an extra cost or "overhead factor" that takes away somewhat from the advantage of a pipelined implementation. If $t_R$ is small compared to $t_{STAGE}$, as is usually the case, the pipeline will perform close to theoretical limits, and certainly much better than a single-stage (purely combinational) implementation of the same task. If we try to divide the task into very small steps, however, $t_{STAGE}$ may become comparable to or even smaller than $t_R$, significantly reducing the advantage of pipelining. We cannot make $t_C$ smaller than $t_R$ no matter how finely we subdivide the task. (Put another way, the maximum clock frequency of the pipeline is limited to $1/t_R$ even if the stages do no work at all.) Thus, there is a point of diminishing returns beyond which it makes little sense to deepen the pipeline. The best design would probably be one with a number of stages that maximizes the ratio of performance to cost, where cost may be measured not just in dollars but in chip area, transistor count, wire length, power dissipation, and/or other factors.

CPU pipelines generally fall into one of two categories: *arithmetic pipelines* or *instruction unit pipelines*. Arithmetic pipelines are generally found in vector supercomputers, in which the same numerical (usually floating-point) computation(s) must be performed on many values in succession. Instruction unit pipelines, which are used to execute a variety of scalar instructions at high speed, are found in practically all modern general-purpose microprocessors. We will examine the characteristics of both types of pipelines in the following pages.

## 4.2   Arithmetic pipelines

High-performance computers intended for scientific and engineering applications generally place a premium on high-speed arithmetic computations above all else. In Chapter 3, we considered the design of circuitry to perform the basic arithmetic operations on integer and floating-point numbers. The circuits we discussed can be readily used to perform scalar arithmetic operations—that is, isolated operations on individual variables. Further optimization becomes possible (and highly desirable) when we perform computations on vectors. Vector manipulations generally involve performing the same operation on each of a long list of elements; thus, they lend themselves to pipelined implementation of those operations.

Recall the circuit of Figure 3.23 that used a tree of carry save adders to multiply two 4-bit integers. This circuit was constructed from a set of AND gates used to generate the partial products plus two sets of carry save adders and a final carry lookahead adder used to combine the sum and carry bits to obtain the product. As shown in Figure 3.23, this is a "flow-through" or purely combinational logic circuit. The two numbers to be multiplied are input, and after sufficient time elapses to satisfy the total propagation delay of the components (specifically, the longest path from any input to an output that depends on its value), the product is available. Once the product is saved or used in another computation, new inputs can be provided to the circuit and another multiplication can be done.

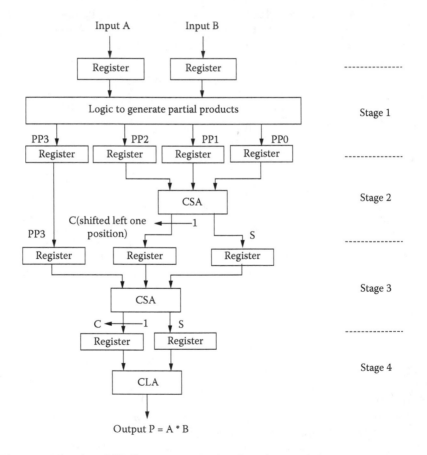

**Figure 4.4** Pipelined Wallace tree multiplier for 4-bit numbers.

Now suppose we have an application with a requirement that a large number of 4-bit multiplications must be done consecutively. We could use the circuit of Figure 3.23 to perform each multiplication sequentially, but performing $n$ multiplications would take $n$ times the propagation delay of the entire circuit. To speed things up, we could add pipeline registers and convert the circuit to the four-stage pipelined implementation seen in Figure 4.4. At successive time steps, the first pair of numbers would go through partial product generation, carry save adder stage 1, carry save adder stage 2, and finally the carry lookahead adder stage. Once the pipeline is full, a new result is produced every cycle, and the time for that cycle is much shorter than the time to do a complete multiplication. In the limit, many consecutive multiplications can be done with a speedup approaching a factor of 4 (the number of stages in the pipeline).

For another example, consider the floating-point adder/subtractor depicted in Figure 3.33. This circuit consists of circuitry to compare exponents, align mantissas, perform the addition or subtraction, and then renormalize the results. By inserting pipeline registers, we could subdivide the function of this circuit into three stages as shown in Figure 4.5. Once again, this could lead to a speedup of nearly 3 (the number of stages)

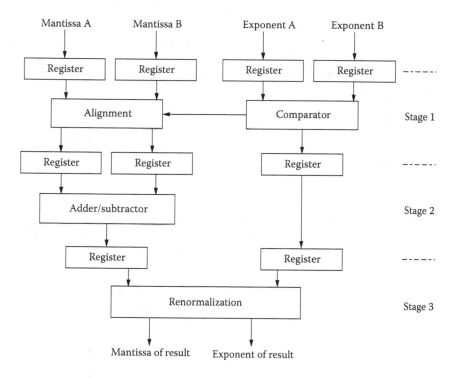

*Figure 4.5* Pipelined floating-point adder/subtractor.

in the best case, which is equal delay for all stages and a long sequence of floating-point additions and subtractions to be done.

We could go into much more detail on arithmetic pipelines, but the vector supercomputers that make such extensive use of them are a dying breed. Although in June 1993 approximately 65% of the world's top 500 supercomputers used a vector architecture, by November 2007 (early in the sixth generation) that figure had dropped to less than 1% and has now reached zero. The last vector machine remaining on the Top 500 list was Japan's Earth Simulator 2, an NEC SX-9/E system that debuted at number 22 in June 2009 but dropped to 217th place by November 2012, 472nd by November 2013, and disappeared from the top 500 in June 2014. So we now turn our attention to a different application of pipelining that is much more widely used in modern computing systems.

## 4.3   Instruction unit pipelines

Arithmetic pipelines like those described in the previous section are very useful in applications in which the same arithmetic operation (or some small set of such operations) must be carried out repeatedly. The most common example of this is in scientific and engineering computations carried out on vector supercomputers. Although this is (or was) an important segment of the overall computer market, it has always been a relatively small one. Arithmetic pipelines have limited usefulness in the general-purpose machines that are much more widely used in business and other applications. Does this mean only machines intended for scientific number-crunching use pipelining? Not at all; rather, almost all general-purpose machines being built today make use of one or more *instruction unit pipelines*.

Instruction unit pipelines are pipelines that are used to execute a machine's scalar instruction set (the instructions that operate on only one or two operands to produce a single result, which make up most if not all of the instruction set of a typical general-purpose machine). As we discussed in Section 3.3, the execution of each machine instruction can be broken into several steps. Often, particularly in a machine that has simple instructions, most of these steps are the same for many (possibly most or all) of the instructions. For example, all of the computational instructions may be done with the same sequence of steps with the only difference being the operation requested of the ALU once the operands are present. With some effort, the data transfer and control transfer instructions may also be implemented with the same, or a very similar, sequence of steps. If most or all of the machine's instructions can be accomplished with a similar number and type of steps, it is relatively easy to pipeline the execution of those steps and thus improve the machine's performance considerably over a completely sequential design.

## 4.3.1   Basics of an instruction pipeline

The basic concept flows from the principles that were introduced at the beginning of Section 4.1. Pipelining as we defined it means breaking up a computational task into smaller subtasks and overlapping the performance of those subtasks for different instances of the task. In this case, the basic task is the execution of a generic, scalar instruction, and the subtasks correspond to subdivisions of the von Neumann execution cycle, which the machine must perform in order to execute the instruction. The von Neumann cycle may be broken into more or fewer steps depending on the designer's preference and the amount of logic required for each step. (Remember that for the best pipelined performance, the overall logic delay should be divided as equally as possible among the stages.) As an example, let us consider a simple instruction unit pipeline with four stages as follows:

F: Fetch instruction from memory
D: Decode instruction and obtain operand(s)
E: Execute required operation on operand(s)
W: Write result of operation to destination

A sequential (purely von Neumann) processor would perform these four steps one at a time for a given instruction, then go back and perform them one at a time for the next instruction, and so on. If each step required one clock cycle, then each machine instruction would require four clock cycles. Two instructions would require eight cycles, three would require 12 cycles, etc. This approach is effective but slow, and does not make very efficient use of the hardware.

In an instruction-pipelined processor, we break down the required hardware into smaller pieces, separated by pipeline registers as described previously. Our example divides the hardware into four stages as shown in Figure 4.6. By splitting the hardware into four stages, we can overlap the execution of up to four instructions simultaneously. As Table 4.1 shows, we begin by fetching instruction $I_1$ during the first time step $t_0$. During the next time step $(t_1)$, $I_1$ moves on to the decoding/operand fetch stage while, simultaneously, instruction $I_2$ is being fetched by the first stage. Then, during step $t_2$, $I_1$ is in the execute (E) stage while $I_2$ is in the D stage and $I_3$ is being fetched by the F stage. During step $t_3$, $I_1$ moves to the final stage (W) while $I_2$ is in E, $I_3$ is in D, and $I_4$ is in F. This process continues indefinitely for subsequent instructions.

As in the nonpipelined case, it still takes four clock cycles to execute the first instruction $I_1$. However, because of the overlap of operations in the pipeline, $I_2$ is completed one clock cycle later; $I_3$ is completed one cycle after that, and likewise for subsequent instructions. Once the pipeline is

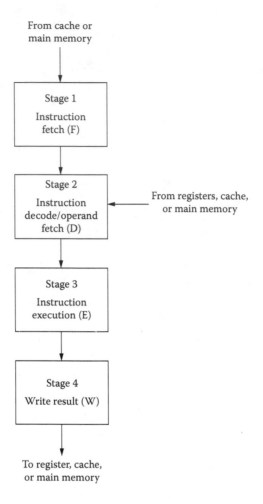

*Figure 4.6* Stages of a typical instruction pipeline.

*Table 4.1* Execution of instructions in a four-stage pipeline

|              | $t_0$ | $t_1$ | $t_2$ | $t_3$ | $t_4$ | $t_5$ | $t_6$ |
|--------------|-------|-------|-------|-------|-------|-------|-------|
| Stage 1 (F)  | $I_1$ | $I_2$ | $I_3$ | $I_4$ | $I_5$ | $I_6$ | $I_7$ |
| Stage 2 (D)  |       | $I_1$ | $I_2$ | $I_3$ | $I_4$ | $I_5$ | $I_6$ |
| Stage 3 (E)  |       |       | $I_1$ | $I_2$ | $I_3$ | $I_4$ | $I_5$ |
| Stage 4 (W)  |       |       |       | $I_1$ | $I_2$ | $I_3$ | $I_4$ |

full, we achieve a steady-state throughput of one instruction per clock cycle rather than one instruction per four cycles as in the nonpipelined case. Thus, in the ideal case, the machine's performance may increase by nearly a factor of four.

Of course, the ideal case is not always the way things work out in the real world. It appears from an initial examination that we can easily achieve a pipeline throughput of one instruction per cycle. (This was the goal of the original reduced instruction set computer [RISC] designers and is the goal of any pipelined processor design.) However, a sustained throughput of 1.0 instructions per cycle is never attainable in practice using a single instruction pipeline (although in some cases, we can come fairly close). Why not? For the same reason we can never achieve maximum throughput in an arithmetic or any other type of pipeline: Something happens to "break the chain" or temporarily interrupt the operation of the pipeline. For a vector arithmetic pipeline, this might include having to change a dynamic pipeline to another configuration, or simply reaching the end of the current vector computation (i.e., running out of operands). Anytime we miss initiating a new operation into the pipe for even a single cycle, we will correspondingly miss completing one operation per cycle at some later time and thus not average one operation per cycle.

Neither of the above scenarios applies directly to an instruction unit pipeline. Although different instructions are executed, they all use the same sequence of steps; thus, the pipeline structure is never reconfigured. And, of course, the CPU is always executing *some* instructions. They may be part of a user program or the operating system; they may be computational instructions, data transfers, or even no-operation instructions (NOPs), but the processor never stops executing instructions as long as it is "up." (We are ignoring the special case of some embedded processors that have sleep or standby modes in which they halt execution to save power.) If these types of situations are not the problem, what can and does happen to hold up the operation of a pipelined instruction unit and keep it from achieving its theoretical throughput of one instruction per cycle? We shall explore several possibilities, but the most fundamental difficulty is that execution of instructions is not always sequential.

## 4.3.2   Control transfers and the branch penalty

It is control transfer instructions, which are very important to all practical programs, that cause the most obvious problem with respect to pipelined instruction execution. Pipelining instructions that are sequentially stored and sequentially executed is relatively easy, but sooner or later in any useful program, we must make a decision (based on data input or retrieved from memory or on the results of computations performed on data) as to what to do next. This decision-making process is typically done using a comparison and conditional branching technique. The program performs a comparison (or some other arithmetic or logic operation), and then a conditional branch instruction is executed. This branch tests some condition related to the operation just performed and either *succeeds* or *fails* based on whether or not the specified condition is true. A conditional branch that succeeds means the next instruction executed is the one at the specified target address, while one that fails means the next instruction executed is the next one in program sequence.

If it were possible to know in advance which of these events would occur, handling the situation would pose few problems for the pipelined instruction execution unit; but if it were possible to always know ahead of time that a given branch would be taken or not, it would not have to be encoded as a branch. In fact, the CPU often (particularly in the case of program loops) encounters a particular branch repeatedly over time, and depending on the data being processed, it may succeed on some occasions and fail on others. Conditional transfers of control are an unavoidable fact of life in the logic of useful programs built on the von Neumann execution model, and as we shall see, these control transfers do pose problems, known as *control hazards*, for pipelined execution of instructions.

The problems we encounter with branching occur because a single pipeline can only process one sequence of instructions. There may indeed be only one sequence of instructions leading up to a given conditional branch, but there are always two possible sequences of instructions following it. There is the sequence of instructions following the branch instruction in memory, to be executed if it fails; there is also the sequence of instructions beginning at the branch target location in memory, to be executed if it succeeds. The pipeline can only process one of these sequences of instructions at a time. What if it is the wrong one?

If the pipeline control logic assumes that a given conditional branch will fail and begins working on the sequential instructions, and it turns out that the branch actually fails, then there is no problem, and the pipeline can continue completing instructions at the rate of one per clock cycle. Likewise, if the pipeline logic anticipates that the branch will succeed and begins processing at the target location, the pipeline can be kept full or nearly so. (Depending on the pipeline structure and the memory interface, there may be a slight delay as instructions must be obtained from a different part of memory.) In either of the other two cases, in which a branch succeeds while sequential instructions have already started down the pipe or in which a branch fails while the processor assumed it would be taken, there is a definite problem that will interrupt the completion of instructions—possibly for several clock cycles.

When any given conditional branch instruction first enters a pipeline, the hardware has no way of knowing whether it will succeed or fail. Indeed, because the first stage of the pipeline always involves fetching the instruction from memory, there is no way the control unit can even know yet that it is a conditional branch instruction. That information is not available until the instruction is decoded (in our example, this occurs in the second pipeline stage). Even once the instruction is identified as a conditional branch, it takes some time to check the appropriate condition and make the decision of whether or not the branch will succeed (and then to update the program counter with the target location if the branch does succeed). By this time, one, two, or more subsequent instructions, which may or may not be the correct ones, may have entered the pipe.

Table 4.2 illustrates a possible scenario in which a branch that was assumed to fail instead succeeds. Assume that instruction $I_4$ is a conditional branch that implements a small program loop; its target is $I_1$, the top of the loop, and the sequential instruction $I_5$ will be executed on completion of the loop. It is not possible to know ahead of time how many times the loop will iterate before being exited, so either $I_1$ or $I_5$ may follow $I_4$ at any time.

$I_4$ is fetched by the first pipeline stage during cycle $t_3$. During the following cycle, $t_4$, it is decoded while the pipeline is busy fetching the following instruction, $I_5$. Sometime before the end of cycle $t_4$, the control unit determines that $I_4$ is indeed a conditional branch, but it still has to test the branch condition. Let us assume that this does not happen until sometime during cycle $t_5$, when the fetch of instruction $I_6$ has begun. At this point, the control unit determines that the branch condition is true and loads the address of $I_1$ into the program counter to cause the branch to take place. $I_1$ will thus be fetched during the next clock cycle, $t_6$, but by this time, two instructions ($I_5$ and $I_6$) are in the pipeline where they should not be. Allowing them to continue to completion would cause the program to generate incorrect results, so they are aborted or *nullified* by the control logic (meaning the results of these two instructions are not written to their destinations). This cancellation of the incorrectly fetched instructions is known as "flushing" the pipeline. Although the correctness of program operation can be retained by nullifying the effects of $I_5$ and $I_6$, we can never recover the two clock cycles that were wasted in mistakenly attempting to process them. Thus, this branch has prevented the pipeline from achieving its maximum possible throughput of one instruction per clock cycle.

*Table 4.2* Pipelined instruction execution with conditional branching

|              | $t_0$ | $t_1$ | $t_2$ | $t_3$ | $t_4$ | $t_5$ | $t_6$ | $t_7$ | $t_8$ |
|--------------|-------|-------|-------|-------|-------|-------|-------|-------|-------|
| Stage 1 (F)  | $I_1$ | $I_2$ | $I_3$ | $I_4$ | $I_5$ | $I_6$ | $I_1$ | $I_2$ | $I_3$ |
| Stage 2 (D)  |       | $I_1$ | $I_2$ | $I_3$ | $I_4$ | $I_5$ | –     | $I_1$ | $I_2$ |
| Stage 3 (E)  |       |       | $I_1$ | $I_2$ | $I_3$ | $I_4$ | –     | –     | $I_1$ |
| Stage 4 (W)  |       |       |       | $I_1$ | $I_2$ | $I_3$ | $I_4$ | –     | –     |

In this example, a successful conditional branch caused a delay of two clock cycles in processing instructions. This delay is known as the *branch penalty*. Depending on the details of the instruction set architecture and the way it is implemented by the pipeline, the branch penalty for a given design might be greater or less than two clock cycles. In the best case, if the branch condition could be tested and the program counter modified in stage 2 (before the end of cycle $t_4$ in our example) the branch penalty could be as small as one clock cycle. If determining the success of the branch, modifying the program counter (PC), and obtaining the first instruction from the target location took longer, the branch penalty might be three or even four cycles (the entire depth of the pipeline, meaning its entire contents, would have to be flushed). The number of lost cycles may vary from implementation to implementation, but branches in a program are never good for a pipelined processor.

It is worth noting that branching in programs can be a major factor limiting the useful depth of an instruction pipeline. In addition to the other reasons mentioned earlier to explain the diminishing returns we can achieve from pipelines with many stages, one can readily see that the deeper an instruction-unit pipeline, the greater the penalty that may be imposed by branching. Rather than one or two instructions, a fine-grained instruction pipe may have to flush several instructions on each successful branch. If branching instructions appear frequently in programs, a pipeline with many stages may perform no better, or even worse, than one with a few stages.

It is also worth mentioning that conditional branch instructions are not the only reason the CPU cannot always initiate a new instruction into the pipeline every cycle. Any type of control transfer instruction, including unconditional jumps, subprogram calls and returns, etc., may cause a delay in processing. Although there is no branch condition to check, these other instructions must still proceed a certain distance into the pipeline before being decoded and recognized as control transfers, during which time one or more subsequent instructions may have entered the pipe. Exception processing (including internally generated traps and external events such as interrupts) also requires the CPU to suspend execution of the·sequential instructions in the currently running program, transfer control to a handler located somewhere else in memory, then return and resume processing the original program. The pipeline must be drained and refilled upon leaving and returning, once again incurring a penalty of one or more clock cycles in addition to other overhead, such as saving and restoring register contents.

### 4.3.3   Branch prediction

*Branch prediction* is one approach that can be used to minimize the performance penalty associated with conditional branching in pipelined processors. Consider the example presented in Table 4.2. If the control unit could somehow be made aware that instructions $I_1$ through $I_4$ make up a program loop, it might choose to assume that the branch would succeed and fetch $I_1$ (instead of $I_5$) after $I_4$ each time. With this approach, the full branch penalty would be incurred only once (upon exiting the loop) rather than each time through the loop. One could equally well envision a scenario in which assuming that the branch would fail would be the better course of action; but how, other than by random guessing (which, of course, is as likely to be wrong as it is to be right) can the control unit predict whether or not a branch will be taken?

Branch prediction, which dates back to the IBM Stretch machine of the late 1950s, can be done either *statically* (before the program is run) or *dynamically* (by the control unit at run time) or as a combination of both techniques. The simplest forms of static prediction either assume all branches succeed (assuming they all fail would be equivalent to no prediction at all) or assume that certain types of branches always succeed and others always fail. As the reader might imagine, these primitive schemes tend not to fare much better than random guessing.

A better way to do static prediction, if the architecture supports it, is to let the compiler do the work. Version 9 of Sun Microsystems' (now Oracle's) Scalable Processor ARChitecture (SPARC) provides a good example of this approach. Each SPARC V9 conditional branch instruction has two different op codes: one for "branch probably taken" and one for "branch probably not taken." The compiler analyzes the structure of the high-level code and chooses the version of the branch it expects will be correct most of the time; when the program runs, the processor uses the op code as a hint to help it choose which instructions to fetch into the pipeline. If the compiler is right most of the time, this technique will improve performance. However, even this more sophisticated form of static branch prediction has not proven to be especially effective in most applications when used alone (without run-time feedback).

Dynamic branch prediction relies on the control unit keeping track of the behavior of each branch encountered. This may be done by the very simple means of using a single bit to remember the behavior of the branch the last time it was executed. For more accurate prediction, two bits may be used to record the action of the branch for the last two consecutive times it was encountered. If it has been taken, or not taken, twice in succession, that is considered a strong indication of its likely behavior the next time, and instructions are fetched accordingly. If it was taken once and not taken once, the control unit decides randomly which way to predict the branch. Another even more sophisticated dynamic prediction technique associates a counter of two or more bits with each branch. The counter is initially set to a threshold value in the middle of its count range. Each time the branch succeeds, the counter is incremented; each time it fails, the counter is decremented. As long as the current count is greater than or equal to the threshold, the branch is predicted to succeed; otherwise, it is predicted to fail. The Intel Pentium processor used a two-bit, four-state counter like this to predict branches. Note that for this technique to work properly the counter must saturate or "stick" at its upper and lower limits rather than rolling over from the maximum count to zero or vice versa. Even more elaborate schemes (including hybrid, adaptive, and two-level mechanisms) are possible, but there is a point of diminishing returns beyond which the cost of additional hardware complexity is not justified by significantly better performance on typical compiled code.

Although dynamic branch prediction requires more hardware, in general it has been found to perform better than static prediction (which places greater demands upon the compiler). Because dynamic prediction is dependent upon the details of a particular implementation rather than the features of the instruction set architecture, it also allows compatibility with previous machines to be more easily maintained. Of course, the two approaches can be used together, with the compiler's prediction being used to initialize the state of the hardware predictor and/or as a tiebreaker when the branch is considered equally likely to go either way.

What are the performance effects of branch prediction? Performance is very sensitive to the success rate of the prediction scheme used. Highly successful branch prediction can significantly improve performance, and particularly poor prediction may do no better (or possibly even worse) than no prediction at all. Although a branch correctly predicted to fail may cost no cycles at all and one correctly predicted to succeed may incur only a minimal penalty (if any), a branch that is mispredicted either way will require the machine to recover to the correct execution path and thus may incur a branch penalty as severe as (or even more severe than) if the machine had no prediction scheme at all.

For comparison, let us first consider mathematically the throughput of a pipelined processor without branch prediction. Any instruction in a program is either a branching instruction or a nonbranching instruction. Let $p_b$ be the probability of any instruction being a branch; then $(1 - p_b)$ is the probability of it not being a branch. Let $C_B$ be the average number of cycles per branch instruction and $C_{NB}$ be the average number of cycles per nonbranching instruction. The average number of clock cycles per instruction is thus given by

$$C_{AVG} = p_b C_B + (1 - p_b)C_{NB}$$

In a pipelined processor, nonbranching instructions execute in one clock cycle, so $C_{NB} = 1$. $C_B$ depends on the fraction of branches taken (to the target) versus the fraction that is not taken and result in sequential execution. Let $p_t$ be the probability that a branch is taken and $(1 - p_t)$ be the probability it is not taken; let $b$ be the branch penalty in clock cycles. Because no prediction is employed, failed branches execute in 1 cycle, and successful branches require $(1 + b)$ cycles. The average number of cycles per branch instruction is thus given by

$$C_B = p_t(1 + b) + (1 - p_t)(1) = 1 + p_t b$$

We can substitute this expression for $C_B$ into the previous equation to determine the average number of cycles per instruction as follows:

$$C_{AVG} = p_b(1 + p_t b) + (1 - p_b)(1) = 1 + p_b p_t b$$

The throughput of the pipeline (the average number of instructions completed per clock cycle) is simply the reciprocal of the average number of cycles per instruction:

$$H = 1/C_{AVG} = 1/(1 + p_b p_t b)$$

The probabilities $p_b$ and $p_t$ will vary for different programs. Typical values for $p_b$, the fraction of branch instructions, have been found to be in the range of 0.1 to 0.3. The probability of a branch succeeding, $p_t$, may vary widely, but values in the 0.5 to 0.8 range are reasonable. As a numerical example, suppose for a given program that the branch penalty is three cycles, $p_b = 0.22$, and $p_t = 0.7$. The average number of cycles per instruction would be $1 + (0.22)(0.7)(3) = 1.462$, and the pipeline throughput would be approximately 0.684 instructions per cycle.

Another way to compute this result without memorizing the formula for $C_{AVG}$ is to construct a simple probability tree diagram as shown in Figure 4.7. To obtain $C_{AVG}$, it is simply necessary to multiply the number of cycles taken in each case times the product of the probabilities leading to

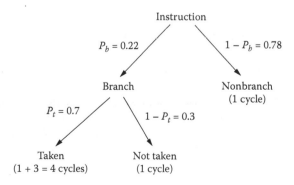

*Figure 4.7* Probability tree diagram for conditional branching example.

that case and then sum the results. Thus, we obtain $C_{AVG} = (0.22)(0.7)(4) + (0.22)(0.3)(1) + (0.78)(1) = 1.462$ as before, once again giving $H \approx 0.684$.

Now suppose the pipelined processor employs a branch prediction scheme to try to improve performance. Let $p_c$ be the probability of a correct prediction, and let $c$ be the reduced penalty associated with a correctly predicted branch. (If correctly predicted branches can execute as quickly as sequential code, $c$ will be equal to zero.) Branches that are incorrectly predicted (either way) incur the full branch penalty of $b$ cycles. In this scenario, the average number of cycles per branch instruction can be shown to be

$$C_B = 1 + b - p_c b + p_t p_c c$$

Substituting this into our original equation, $C_{AVG} = p_b C_B + (1 - p_b)(1)$, we find that with branch prediction, the average number of cycles per instruction is given by

$$C_{AVG} = 1 + p_b b - p_b p_c b + p_b p_t p_c c$$

Returning to our numerical example, let us assume that $b$ is still three cycles and $p_b$ and $p_t$ are still 0.22 and 0.7, respectively. Let us further assume that c = 1 cycle and the probability of a correct branch prediction, $p_c$, is 0.75. Substituting these values into the first equation, we find that the average number of cycles per branch instruction is $1 + 3 - (0.75)(3) + (0.7)(0.75)(1) = 2.275$. The second equation gives the overall average number of cycles per instruction as $1 + (0.22)(3) - (0.22)(0.75)(3) + (0.22)(0.7)(0.75)(1) = 1.2805$. The pipeline throughput $H$ with branch prediction is 1/1.2805, or approximately 0.781, a significant improvement over the example without branch prediction.

Again, if one does not wish to memorize formulas, the same result could be obtained using the probability tree diagram shown in Figure 4.8. Multiplying the number of cycles required in each case times the probability of occurrence and then summing the results, we get $C_{AVG} = (0.22)(0.7)(0.75)(2) + (0.22)(0.7)(0.25)(4) + (0.22)(0.3)(0.75)(1) + (0.22)(0.3)(0.25)(4) + (0.78)(1) = 1.2805$, the same result obtained from our equation.

The performance benefits derived from branch prediction depend heavily on the success rate of the prediction scheme. To illustrate this, suppose that instead of $p_c = 0.75$, our branch prediction scheme achieved only 50% correct predictions ($p_c = 0.5$). In that case, the average number of cycles per instruction would be $1 + (0.22)(3) - (0.22)(0.5)(3) + (0.22)(0.7)(0.5)(1) = 1.407$ for a throughput of approximately 0.711—not much better than the results with no prediction at all. If the prediction scheme performed very poorly, for example, with $p_c = 0.3$, the pipeline throughput could be even worse than with no prediction—in this case, $1/1.5082 \approx 0.663$ instructions per cycle.

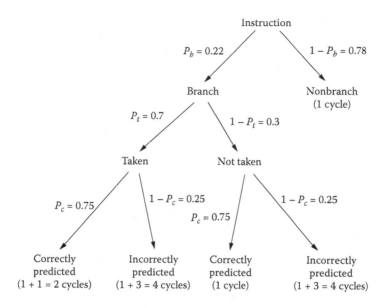

**Figure 4.8** Probability tree diagram for example with branch prediction.

To facilitate branch prediction, it is common for modern processors to make use of a *branch target buffer* (also known as a *branch target cache* or *target instruction cache*) to hold the addresses of branch instructions, the corresponding target addresses, and the information about the past behavior of the branch. Any time a branch instruction is encountered, this buffer is checked, and if the branch in question is found, the relevant history is obtained and used to make the prediction. A *prefetch queue* may be provided to funnel instructions into the pipeline. To further improve performance, some processors make use of dual-instruction prefetch queues to optimize branches. Using this *multiple prefetch* approach, the processor fetches instructions from both possible paths (sequential and target). By replicating the prefetch queue (and possibly even the first few stages of the pipeline itself), the processor can keep both possible execution paths "alive" until the branch decision is made. Instructions from the correct path continue on to execution, and those from the incorrect path are discarded. For this approach to work, there must be sufficient space on the chip for the prefetch queues and sufficient memory (particularly instruction cache) bandwidth to allow for the simultaneous prefetching of both paths.

## 4.3.4 Delayed control transfers

*Delayed control transfers* are another approach some designers have adopted to eliminate, or at least minimize, the penalty associated with control transfers (both conditional and unconditional) in pipelined processors. A *delayed branch* instruction is unlike any control transfer instruction the reader has likely encountered before. The branch instructions in most computer architectures take effect immediately. That is, if the branch condition is true (or if the instruction is an unconditional branch), the next instruction executed after the branch itself is the one at the target location. A delayed branch does not take effect immediately. The instruction sequentially following the branch is executed whether or not it succeeds. Only after the instruction following the control transfer instruction, which is said to be in the *delay slot*, is executed, will the target instruction be executed with execution continuing sequentially from that point.

An example of delayed branching is shown in Figure 4.9. Let us assume instructions $I_1$ through $I_4$ form a program loop with $I_4$ being a conditional branch back to $I_1$. In a "normal" instruction set architecture, $I_5$ would never be executed until the loop had finished iterating. If this instruction set were implemented with a pipelined instruction unit, $I_5$ would have to be flushed from the pipeline every time the branch $I_4$ succeeded. If the number of loop iterations turned out to be large, many clock cycles would be wasted fetching and flushing $I_5$.

The idea of delayed branching, which only makes sense in an architecture designed for pipelined implementation, comes from the desire not to waste the time required to fetch $I_5$ each time the branch ($I_4$) is taken. "Because $I_5$ is already in the pipeline," the argument goes, "why not go ahead and execute it instead of flushing it?" To programmers used to working in high-level languages or most assembly languages, this approach invariably seems very strange, but it makes sense in terms of efficient hardware implementation. (The delayed branch scheme appears to an uninitiated assembly programmer as a bug, but as all computing professionals should know, any bug that is documented becomes a feature.) $I_5$ appears to be—and is physically located in memory—outside the loop; but logically, in terms of program flow, it is part of the loop. The instructions are stored in the sequence $I_1$, $I_2$, $I_3$, $I_4$, $I_5$, but they are logically executed in the sequence $I_1$, $I_2$, $I_3$, $I_5$, $I_4$.

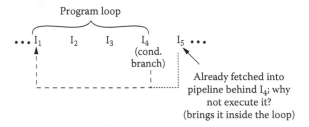

*Figure 4.9* Delayed branching example.

The trick to using this feature lies in finding an instruction that logically belongs before the branch but that is independent of the branch decision—neither affecting the branch condition nor being affected by it. (If the control transfer instruction is unconditional, any instruction that logically goes before it could be placed in its delay slot.) If such an instruction can be identified, it may be placed in the delay slot to make productive use of a clock cycle that would otherwise be wasted. If no such instruction can be found, the delay slot may simply be filled with a time-wasting instruction such as a NOP. This is the software equivalent of flushing the delay slot instruction from the pipe.

Delayed control transfer instructions are not found in complex instruction set computer (CISC) architectures, which, by and large, trace their lineage to a time before pipelining was widely used. However, they are often featured in RISC architectures, which were designed from the ground up for a pipelined implementation. CISC architectures were intended to make assembly language programming easier by making the assembly language look more like a high-level language. RISC architectures, however, were not designed to support assembly language programming at all, but rather to support efficient code generation by an optimizing compiler. (We discuss the RISC philosophy in more detail in Section 4.4.) The fact that delayed branches make assembly programming awkward is not really significant as long as a compiler can be designed to take advantage of the delay slots when generating code.

There is no reason that an architecture designed for pipelined implementation must have only one delay slot after each control transfer instruction. In fact, the example presented earlier in Table 4.2 is of a machine that could have two delay slots. (Machines with very deep pipelines might have even more.) Because instructions $I_5$ and $I_6$ will both have entered the pipeline before the branch at $I_4$ can be taken, they could both be executed before the target instruction if the architecture so specifies (see Figure 4.10). The more delay slots that exist after a conditional branch instruction, the more difficult it will be for the compiler (or masochistic assembly

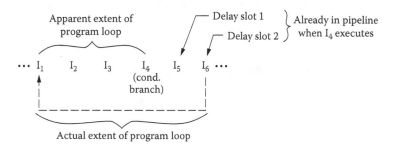

*Figure 4.10* Example of control transfer with two delay slots.

programmer) to find useful, independent instructions with which to fill them. In the worst case, they can still be filled with NOPs, which keeps the hardware simpler than it would be if it had to be able to recognize this situation and flush the instructions itself. This is one more example of the hardware–software trade-offs that are made all the time in computer systems design.

## 4.3.5   Memory accesses: delayed loads and stores

Control transfers are not the only occurrences that can interrupt or slow processing in a pipelined processor. Another potentially costly activity (in terms of performance) is accessing memory for operands. Register operands are generally accessible very quickly, such that they can be used for an arithmetic computation within the same clock cycle. Memory operands, even if they reside in cache, generally require at least one cycle to access before any use can be made of them. RISC architectures in particular attempt to minimize this problem by clearly subdividing their instruction sets into computational instructions (which operate only on register contents) and memory access (or load and store) instructions that do not perform computations on data. Even so, a problem may arise when a subsequent instruction tries to perform a computation on a data value being loaded by a previous instruction. For example, suppose the following two instructions appear consecutively in a program:

```
LOAD   VALUE, R5
ADD    R5, R4, R3
```

Given the slower speed of memory relative to most CPUs, the variable VALUE being loaded into register R5 might not be available by the time it is needed by the ADD instruction. One obvious solution to the problem would be to build in some sort of hardware interlock that would freeze or stall the pipeline until the data had been retrieved from memory and placed in R5. Then and only then would the ADD instruction be allowed to execute. The hardware is thus made responsible for the correct operation of the software—a typical, traditional computer design approach.

Another way of approaching this problem is simply to document that loads from memory always take at least one extra cycle to occur. The compiler or the assembly language programmer would be made aware that every load instruction has a "load delay slot." In other words, the instruction immediately following a load must not make use of the value being loaded. (In some architectures, the following instruction may also not be another load.) If the load's destination register is referenced by the following instruction, it is known that the value obtained will be the old value, not the new one being loaded from memory. This is another example of documenting what might appear to the casual user to be a bug and thus enshrining it as a feature of the instruction set. Instead of the hardware compensating for delays in loading data and ensuring the correct operation of the software, the software is simply informed of the details of the hardware implementation and forced to ensure correctness on its own. (Note that a hardware interlock will still be required to detect data cache misses and stall the pipeline when they occur, because main memory access may take several clock cycles rather than one.) In the example above, correctness of operation is ensured by simply inserting an unrelated but useful instruction (or a NOP if no such instruction can be found) between the LOAD and the ADD (or any other operational instruction) that uses the results of the LOAD.

Store operations (writes of data to memory) are less problematic than loads because the CPU is unlikely to need to retrieve the stored information from memory soon enough to pose a problem. (Because stores write data from a register to memory, presumably the CPU can use the copy of the data still in a register if it is needed.) However, back-to-back memory accesses may pose a problem for some machines because of the time required to complete an access. If two consecutive load or store operations are executed, it may be necessary to stall the pipeline. For that reason, stores in some architectures are also sometimes said to have a "delay slot" that should not contain another load or store instruction. If it does, the pipeline may have to be stalled for one cycle to allow the first store to complete before a second memory access is done.

## 4.3.6    Data dependencies and hazards

Dependency relations among computed results, giving rise to pipeline *data hazards*, may also hold up operations in a pipelined CPU and keep it from approaching its theoretical throughput. In a nonpipelined processor, each instruction completes before execution of the next instruction begins. Thus, values computed by a previous instruction are always available for use in subsequent instructions, and the program always obtains the results anticipated by the von Neumann sequential execution model. Because a pipelined instruction unit overlaps the execution of several instructions, however, it becomes possible for results to become sensitive to the timing of instructions rather than just the order in which they appear in the program. This is obviously not a desirable scenario and must be corrected (or at least accounted for) if we want our pipelined machine to compute the same results as a purely sequential machine.

For an example of a common data dependency problem in a pipelined machine, consider the following situation (illustrated in Figure 4.11).

| Cycle | Stage 1 | Stage 2 | Stage 3 | Stage 4 |
|-------|---------|---------|---------|---------|
| 1 | $I_1$ | | | |
| 2 | $I_2$ | $I_1$ | | |
| 3 | $I_3$ | $I_2$ | $I_1$ | |
| 4 | $I_4$ | $I_3$ | $I_2$ | $I_1$ |
| 5 | $I_5$ | $I_4$ | $I_3$ | $I_2$ |

(a)

(Stall)

| Cycle | Stage 1 | Stage 2 | Stage 3 | Stage 4 |
|-------|---------|---------|---------|---------|
| 1 | $I_1$ | | | |
| 2 | $I_2$ | $I_1$ | | |
| 3 | $I_3$ | $I_2$ | $I_1$ | |
| 4 | $I_3$ | $I_2$ | — | $I_1$ |
| 5 | $I_4$ | $I_3$ | $I_2$ | — |

(b)

*Figure 4.11* Data dependency problem in pipelined CPU. (a) Incorrect operation stage 3 needs result of $I_1$ at beginning of cycle 4; not available until end of cycle 4. (b) Correct operation due to stalling pipeline for 1 cycle after cycle 3; stage 3 will now have updated value for use in executing $I_2$.

Suppose a CPU with a four-stage pipeline such as the one in our previous example executed the following sequence of instructions:

$I_1$: ADD        R1, R2, R3
$I_2$: SUB        R3, R4, R6
$I_3$: XOR        R1, R5, R3

The last operand listed for each instruction is the destination. Thus, it can be seen that instruction $I_2$ uses the result computed by $I_1$, but will that result be available in time for $I_2$ to use it?

In Figure 4.11a, we can see how instruction $I_1$ proceeds through the pipeline with $I_2$ and $I_3$ following it stage by stage. The result from $I_1$ is not computed by stage 3 until the end of the third clock cycle and is not stored back into the destination register (R3) by stage 4 until the end of the fourth clock cycle. However, stage 3 needs to read R3 and obtain its new contents at the beginning of the fourth clock cycle so that they can be used to execute the subtraction operation for $I_2$. If execution proceeds unimpeded as shown, the previous contents of R3 will be used by $I_2$ instead of the new contents, and the program will operate incorrectly. This situation, in which there is a danger of incorrect operation because the behavior of one instruction in the pipeline depends on that of another, is known as a *data hazard*. This particular hazard is known as a *true data dependence* or (more commonly) a *read after write* (RAW) *hazard* because $I_2$, which reads the value in R3, comes after $I_1$, which writes it.

To avoid the hazard and ensure correct operation, the control unit must make sure that $I_2$ actually reads the data after $I_1$ writes it. The obvious solution is to stall $I_2$ (and any following instructions) for one clock cycle in order to give $I_1$ time to complete. Figure 4.11b shows how this corrects the problem.

$I_2$ now does not reach stage 3 of the pipeline until the beginning of the fifth clock cycle, so it is executed with the correct operand value from R3.

RAW hazards are the most common data hazards and, in fact, the only possible type in a machine such as we have been considering so far, with a single pipelined instruction unit in which instructions are always begun and completed in the same order and only one stage is capable of writing a result. Other types of data hazards, known as *write after read* (WAR) and *write after write* (WAW) hazards, are only of concern in machines with multiple pipelines (or at least multiple execution units, which might be "fed" by a common pipeline) or in situations in which writes can be done by more than one stage of the pipe. In such a machine, instructions may be completed in a different order than they were fetched (*out-of-order execution* or OOE). OOE introduces new complications to the process of ensuring that the machine obtains the same results as a nonpipelined processor.

Consider again the sequence of three instructions introduced above. Notice that $I_1$ and $I_3$ both write their results to register R3. In a machine with a single pipeline feeding a single ALU, we know that if $I_1$ enters the pipeline first (as it must), it will also be completed first. Later, its result will be overwritten by the result calculated by $I_3$, which of course is the behavior expected under the von Neumann sequential execution model. Now suppose the machine has more than one ALU that can execute the required operations. The ADD instruction ($I_1$) might be executed by one unit while the XOR ($I_3$) is being executed by another. There is no guarantee that $I_1$ will be completed first or have its result sent to R3 first. Thus, it is possible that due to out-of-order execution R3 could end up containing the wrong value. This situation, in which two instructions both write to the same location and the control unit must make sure that the second instruction's write occurs after the first write, is known as an *output dependence* or WAW hazard.

Now consider the relationship between instructions $I_2$ and $I_3$ in the example code. Notice that $I_3$ writes its result to R3 after the previous value in R3 has been read for use as one of the operands in the subtraction performed by $I_2$. At least that is the way things are supposed to happen under the sequential execution model. Once again, however, if multiple execution units are employed, it is possible that $I_2$ and $I_3$ may execute out of their programmed order. If this were to happen, $I_2$ could mistakenly use the new value in R3 that had been updated by $I_3$ rather than the old value computed by $I_1$. This is one more situation, known as an *antidependence* or WAR hazard, that must be guarded against in a machine in which out-of-order execution is allowed.

There is one other possible relationship between instructions that reference the same location for data and might have their execution overlapped in a pipeline(s). This "read after read" situation is the only one that never creates a hazard. In our example, both $I_1$ and $I_3$ read R1 for use as a source operand. Because R1 is never modified (written), both instructions are guaranteed to read the correct value regardless of their order of execution. Thus, a simple rule of thumb is that for a data hazard to exist between instructions, at least one of them must modify a commonly used value.

## 4.3.7   Controlling instruction pipelines

Controlling the operation of pipelined processors in order to detect and correct for data hazards is a very important but very complex task, especially in machines with multiple execution units. In a machine with only one pipelined instruction execution unit, RAW hazards are generally the only ones a designer must worry about. (The exception would be in the very unusual case in which more than one stage of the pipeline is capable of writing a result.) Control logic must keep track of (or "reserve") the destination register or memory location for each instruction in progress and check it against the source operands of subsequent instructions as they enter the pipe. If a RAW hazard is detected, one approach is to simply stall the instruction that uses the operand being modified (as shown in Figure 4.11b) while allowing the instruction that modifies it to continue. When the location in question has been modified, the reservation placed on it is released and the stalled instruction is allowed to proceed. This approach is straightforward, but forced stalls impair pipeline throughput.

Another approach that can minimize or, in some cases, eliminate the need to stall the pipeline is known as *data forwarding*. By building in additional connectivity within the processor, the result just computed by an instruction in the pipeline can be forwarded to the ALU for use by the subsequent instruction at the same time it is being sent to the reserved destination register. This approach generally saves at least one clock cycle compared to the alternative of writing the data into the first instruction's destination register and then immediately reading it back out for use by the following, dependent instruction.

Finally, designers can choose not to build in control logic to detect and interlocks or forwarding to avoid these RAW hazards caused by pipelining. Rather, they can document pipeline behavior as an architectural feature and leave it up to the compiler to reorder instructions and insert NOPs to artificially stall subsequent instructions, allowing a sufficient number of cycles to elapse so that the value in question will definitely be written before it is read. This solution simplifies the hardware design but is not ideal for designing a family of computers because it ties the instruction set architecture closely to the details of a particular implementation, which may later be superseded. In order to maintain compatibility, more advanced future implementations may have to emulate the behavior of the relatively primitive earlier machines in the family. Still, the approach of handling data dependencies in software is a viable approach for designers who want to keep the hardware as simple as possible.

Machines with multiple execution units encounter a host of difficulties not faced by simpler implementations. Adding WAW and WAR hazards to the RAW hazards inherent to any pipelined machine makes the design of control logic much more difficult. Two important control strategies were devised in the 1960s for high-performance, internally parallel machines of that era. Variations of these methods are still used in the microprocessors of today. These design approaches are known as the *scoreboard method* and *Tomasulo's method*.

The scoreboard method for resource scheduling dates back to the CDC 6600 supercomputer, which was introduced in 1964. James Thornton and Seymour Cray were the lead engineers for the 6600 and contributed substantially to this method, which was used later in the Motorola 88000 and Intel *i*860 microprocessors. The machine had 10 functional units that were not pipelined but did operate concurrently (leading to the possibility of out-of-order execution). The CDC 6600 designers came up with the idea of a central clearinghouse or *scoreboard* to schedule the use of functional units by instructions. As part of this process, the scoreboard had to detect and control interinstruction dependencies to ensure correct operation in an out-of-order execution environment.

The scoreboard is a hardware mechanism—a collection of registers and control logic—that monitors the status of all data registers and functional units in the machine. Every instruction is passed through the scoreboard as soon as it is fetched and decoded in order to check for data dependencies and resource conflicts before the instruction is issued to a functional unit for execution. The scoreboard checks to make sure the instruction's operands are available and that the appropriate functional unit is also available; it also resolves any write conflicts so that correct results are obtained.

There are three main parts, consisting of *tables* or sets of registers with associated logic, to the scoreboard: *functional unit status, instruction status,* and *destination register status*. The functional unit status table contains several pieces of information about the status of each functional unit. This information includes a *busy* flag that indicates whether or not the unit is currently in use, two fields indicating the numbers of its *source registers,* one field indicating its *destination register* number, and *ready* flags for each of the source registers to indicate whether they are ready to be read. (A source register is deemed ready if it is not waiting to receive the results of a previous operation.) If a given functional unit can perform more than one operation, there might also be a bit(s) indicating which operation it is performing.

The instruction status table contains an entry for each instruction from the time it is first decoded until it completes (until its result is written to the destination). This table indicates the current status of the instruction with respect to four steps: (a) whether or not it has been issued, (b) whether or not its operands have been read, (c) whether or not its execution is complete, and (d) whether or not the result has been written to the destination.

The destination register status table is the key to detecting data hazards between instructions. It contains one entry for each CPU register. This entry is set to the number of the functional unit that will produce the result to be written into that register. If the register is not the destination of a currently executing instruction, its entry is set to a null value to indicate that it is not involved in any write dependencies. Tables 4.3 through 4.5 show examples of possible contents of the functional unit status, instruction status, and destination register status tables for a machine with four functional units and eight registers. The actual CDC 6600 was more complex, but this example will serve for illustrative purposes.

The tables show a typical situation with several instructions in some stage of completion. Some functional units are busy, and some are idle. Unit 0, the first adder/subtractor, has just completed the first ADD operation and sent its result to R6, so it is momentarily idle. Unit 1, the second adder/subtractor, has just completed the second ADD operation. However, the result has yet to be written into R5 (in this case, simply because not enough time has elapsed for the write operation to complete; in some situations, the delay could be a deliberate stall caused by a WAR hazard between this and a previous instruction). Therefore, this instruction is not complete, and the reservations on the functional unit and destination register have not been released. Unit 2, the multiplier, has the first MUL operation in progress. Notice that R0 is reserved in the destination register status table as the result register for unit 2. As long as this is true, the control logic will not issue any subsequent instruction that uses R0.

As new instructions are fetched and decoded, the scoreboard checks each for dependencies and hardware availability before issuing it to a functional unit. The second MUL instruction has not yet been issued because the only multiplier, functional unit 2, is busy. Hardware (unit 0) is available such that the third ADD instruction could be issued, but it is stalled due to data dependencies: Neither of its operands (in R0 and R5) are yet available. Only when both of these values are available can this instruction be issued to one of the adder/subtractor units. The new R5 value will be available soon, but until the first multiply completes (freeing unit 2 and updating R0) neither of the two instructions following the first MUL can be issued.

As long as scoreboard entries remain, subsequent instructions may be fetched and checked to see if they are ready for issue. (In Table 4.4, the last scoreboard entry is still available, and the first one is no longer needed now that the instruction has completed, so two additional instructions

*Table 4.3* Scoreboard example: functional unit status table

| Unit name | Unit number | Busy? | Destination register | Source register 1 | Ready? | Source register 2 | Ready? |
|---|---|---|---|---|---|---|---|
| Adder/subtractor 1 | 0 | No | – | – | – | – | – |
| Adder/subtractor 2 | 1 | Yes | 5 | 2 | Yes | 7 | Yes |
| Multiplier | 2 | Yes | 0 | 1 | Yes | 3 | Yes |
| Divider | 3 | No | – | – | – | – | – |

*Table 4.4* Scoreboard example: instruction status table

| Instruction | Instruction address | Instruction issued? | Operands read? | Execution complete? | Result written? |
|---|---|---|---|---|---|
| ADD R1,R4,R6 | 1000 | Yes | Yes | Yes | Yes |
| ADD R2,R7,R5 | 1001 | Yes | Yes | Yes | No |
| MUL R1,R3,R0 | 1002 | Yes | Yes | No | No |
| MUL R2,R6,R4 | 1003 | No | No | No | No |
| ADD R0,R5,R7 | 1004 | No | No | No | No |
| – | – | – | – | – | – |

*Table 4.5* Scoreboard example: destination register status table

| | R7 | R6 | R5 | R4 | R3 | R2 | R1 | R0 |
|---|---|---|---|---|---|---|---|---|
| Functional unit number | – | – | 1 | – | – | – | – | 2 |

could be fetched.) If all scoreboard entries are in use, no more instructions can be fetched until some instruction currently tracked by the scoreboard is completed. The limited number of scoreboard entries, along with the frequent stalls caused by all three types of data hazards and the fact that all results must be written to the register file before use, are the major limitations of the scoreboard approach.

Another important control strategy was first used in the IBM 360/91, a high-performance scientific computer introduced in 1967. Its multiple pipelined execution units were capable of simultaneously processing up to three floating-point additions or subtractions and two floating-point multiplications or divisions in addition to six loads and three stores. The operation of the floating-point registers and execution units (the heart of the machine's processing capability) was controlled by a hardware scheduling mechanism designed by Robert Tomasulo.

Tomasulo's method is essentially a refinement of the scoreboard method with some additional features and capabilities designed to enhance concurrency of operations. One major difference is that in Tomasulo's method, detection of hazards and scheduling of functional units are distributed, not centralized in a single scoreboard. Each data register has a *busy* bit and a *tag* field associated with it. The busy bit is set when an instruction specifies that register as a destination and cleared when the result of that operation is written to the register. The tag field is used to identify which unit will compute the result for that register. This information is analogous to that kept in the destination status register table of a scoreboard.

Another principal feature of Tomasulo's method that differs from the scoreboard method is the use of *reservation stations* to hold operands (and their tags and busy bits, plus an operation code) for the functional units. Each reservation station is essentially a set of input registers that are used to buffer operations and operands for a functional unit. The details of functional unit construction are not important to the control strategy. For example, if the machine has three reservation stations for addition/subtraction (as the 360/91 did), it does not matter whether it has three non-pipelined adders or one three-stage pipelined adder (as was actually the case). Each reservation station has its own unit number and appears to the rest of the machine as a distinct "virtual" adder. Likewise, if there are two reservation stations for multiplication, the machine appears to have two virtual multipliers regardless of the actual hardware used.

The tags associated with each operand in Tomasulo's method are very important because they specify the origin of each operand independently of the working register set. Although program instructions are written with physical register numbers, by the time an instruction is dispatched to a reservation station, its operands are no longer identified by their original register designations. Instead, they are identified by their tags, which indicate the number of the functional unit (actually the number of a virtual functional unit—one of the reservation stations) that will produce that operand. Operands being loaded from memory are tagged with the number of their *load buffer* (in a modern machine they might be tagged by cache location, but the 360/91 had no cache). Once an operand has been produced (or loaded from memory) and is available in the reservation station, its tag and busy bit are changed to 0 to indicate that the data value is present and ready for use. Any time a functional unit is ready to accept operands, it checks its reservation stations to see if any of them have all operands present; if so, it initiates the requested operation. Thus, although programs for the machine are written sequentially using the von Neumann model, the functional units effectively operate as *dataflow* machines (see Section 7.1) in which execution is driven by the availability of operands.

This use of tags generated on the fly rather than the original register numbers generated by the programmer or compiler to identify operands is known as *register renaming*. The register renaming scheme significantly reduces the number of accesses to data registers; not only do they not have to be read for operands if data are coming directly from a functional unit, but only the last of a series of writes to the same register actually needs to be committed to it. The intermediate values are just sent directly to the reservation stations as necessary. Tomasulo's register renaming scheme is instrumental in avoiding stalls caused by WAR and WAW hazards and thus achieves a significant advantage over the scoreboard method.

Another important feature of the 360/91 that helped to reduce or eliminate stalls caused by RAW hazards was the use of a *common data bus* (CDB) to forward data to reservation stations that need a just-calculated

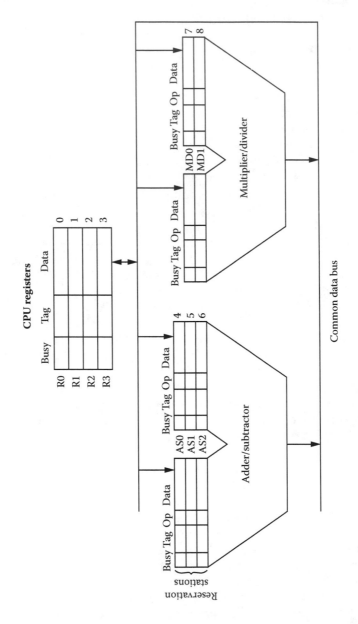

*Figure 4.12* Example architecture with a CDB and reservation stations.

result. Because of this data-forwarding mechanism (which goes hand in hand with the register renaming scheme), the reservation stations do not have to wait for data to be written to the architectural register file. If the tag of a value on the CDB matches the tag of an operand needed by any reservation station(s), the operand is captured from the bus and can be used immediately. Meanwhile, the register file also monitors the CDB, loading a new value into any register with a busy bit set whose tag matches the one on the bus. Figure 4.12 is a simplified view of a machine with Tomasulo scheduling and a CDB architecture like the IBM 360/91.

Tomasulo's method has some distinct disadvantages. Not only does it require complex hardware for control, but its reliance on a single, shared bus makes it hard to scale up for a machine with many registers and functional units (typical of modern CPUs). Given sufficient real estate on the chip, additional shared internal buses can be constructed to remove the CDB bottleneck. Likewise, the larger tags (and greater number of tag comparisons) needed in a more complex machine can be accommodated if sufficient space is available. However, the great selling point for Tomasulo's method is that it helps encourage concurrency of execution (to a greater degree than the scoreboard method) while preserving the dependency relationships inherent to programs written under the von Neumann sequential execution model. Tomasulo's method in particular helps get better performance out of architectures with multiple pipelines. Because of this, approaches based on or very similar to this method are used in many modern, *superscalar* (see Section 4.5.2) microprocessors including the more advanced members of the Alpha, MIPS, PA/RISC, Pentium, and PowerPC families.

## 4.4   Characteristics of RISC machines

In Section 3.1.6, we discussed some of the architectural features common to machines built around the RISC philosophy. These included an instruction set with a limited number of fixed-length instructions using as few instruction formats as possible, simple functionality of each instruction with optimization performed by the compiler, support for only a few simple addressing modes, and a load–store approach in which computational instructions operate only on registers or constants. Later, when discussing control unit design, we mentioned another characteristic of RISC machines that distinguishes them from CISCs: RISCs invariably use hardwired control.

Many of the other characteristics common to RISC architectures stem from one other guiding principle, which we alluded to earlier and the reader can now fully appreciate: RISCs are architecturally designed to accommodate a pipelined implementation. It is true that modern implementations of the remaining CISC architectures are almost always pipelined to some extent, but this pipelining is done in spite of their architectural complexity, not because of it. RISCs are designed with a simple pipeline implementation in mind, one that allows the hardware to be made as simple and fast as possible while consigning complex operations (or anything that cannot be done quickly in hardware) to software. One humorous definition of the RISC acronym, "Relegate Impossible Stuff to Compilers," expresses this philosophy quite well.

Other characteristics common to RISC architectures are essentially secondary traits that are implied by one or more of the attributes already mentioned. The three-operand instructions common to RISC architectures are designed to make things easier for compilers. Also, because of their load–store instruction set architecture, RISCs need a large register set to achieve good performance. This large register set (as well as other enhancements such as on-chip cache memory and floating-point hardware) is made possible by the use of a hardwired control unit, which takes up much less space on the chip than would a microprogrammed control unit. (A typical hardwired control unit may consume only about 10% or less of the available chip area, while the microprogrammed control units typical of fourth-generation CISC microprocessors often occupied 50% or more of the silicon.) Single-cycle instruction execution and the existence of delay slots behind control transfer and memory access instructions are direct consequences of a design that supports pipelining. Likewise, a pipelined machine benefits greatly from a Harvard architecture that keeps memory accesses for data from interfering with instruction fetching. RISC machines almost always have a modified Harvard architecture, which manifests itself in the form of separate, on-chip data and instruction caches.

The main distinguishing characteristics of a typical RISC architecture and its implementation can be summarized briefly as follows:

- *Fixed-length instructions* are used to simplify instruction fetching.
- The machine has only a *few instruction formats* in order to simplify instruction decoding.
- A *load–store instruction set architecture* is used to decouple memory accesses from computations so that each can be optimized independently.
- *Instructions have simple functionality*, which helps keep the control unit design simple.
- A *hardwired control unit* optimizes the machine for speed.
- The architecture is *designed for pipelined implementation*, again to optimize for speed of execution.
- Only a *few, simple addressing modes* are provided because complex ones may slow down the machine and are rarely used by compilers.

- *Optimization of functions by the compiler* is emphasized because the architecture is designed to support high-level languages rather than assembly programming.
- *Complexity is in the compiler* (where it only affects the performance of the compiler), not in the hardware (where it would affect the performance of every program that runs on the machine).

Additional, secondary characteristics prevalent in RISC machines include the following:

- *Three-operand instructions* make it easier for the compiler to optimize code.
- A *large register set* (typically 32 or more registers) is possible because the machine has a small, hardwired control unit and desirable because of the need for the compiler to optimize code for the load–store architecture.
- *Instructions execute in a single clock cycle* (or at least most of them appear to, due to pipelined implementation).
- *Delayed control transfer instructions* are used to minimize disruption to the pipeline.
- *Delay slots behind loads and stores* help cover up the latency of memory accesses.
- A modified *Harvard architecture* is used to keep memory accesses for data from interfering with instruction fetching and thus keep the pipeline(s) full.
- *On-chip cache* is possible due to the small, hardwired control unit and necessary to speed instruction fetching and keep the latency of loads and stores to a minimum.

Not every RISC architecture exhibits every one of the above features, nor does every CISC architecture avoid all of them. They do, however, serve as reference points that can help us understand and classify a new or unfamiliar architecture. What distinguishes a RISC architecture from a CISC is not so much a checklist of features but the assumptions underlying its design. RISC is not a specific machine, but a philosophy of machine design that stresses keeping the hardware as simple and fast as reasonably possible while moving complexity to the software. (Note that "reduced instruction set" does not mean reduced to the absolute, bare minimum required for computation. It means reduced to only those features that contribute to increased performance. Any features that slow the machine in some respect without more than making up for it by increasing performance in some other way are omitted.) Given a suitable, limited set of machine operations, the RISC philosophy effectively posits that he who can execute the most operations in the shortest time wins. It is up to the compiler to find the best encoding of a high-level algorithm into machine language instructions executed by this fast-pipelined processor and thus maximize system performance.

It is interesting to note that although the acronym RISC was coined circa 1980 by David Patterson of the University of California at Berkeley (and took several years after that to make its way into common parlance), many of the concepts embodied in the RISC philosophy predated Patterson's work by more than 15 years. In hindsight, the CDC 6600 has been recognized by many computer scientists as the first RISC architecture. Although it was not pipelined, the 6600 did have multiple execution units and thus was sometimes able to approach a throughput of approximately one instruction per clock cycle (an unheard-of achievement in 1964). It displayed many of the other RISC characteristics described above, including a small instruction set with limited functionality, a load–store architecture, and the use of three operands per instruction. It did, however, violate our criterion of having set length instructions; some were 32 bits long, and others took up 64 bits. Although no one gave their architecture a catchy nickname, the 6600's designers Thornton and Cray were clearly well ahead of their time.

Another RISC architecture that existed before the name itself came about was the IBM 801. This project, which was code-named America at first but would eventually be given the number of the building that housed it, was started in 1975, several years before the work of Patterson at Berkeley and John Hennessy at Stanford. The 801 was originally designed as a minicomputer to be used as the control processor for a telephone exchange system—a project that was eventually scrapped. The 801 never made it to the commercial market as a standalone system, but its architecture was the basis for IBM's ROMP microprocessor, which was developed in 1981. The ROMP was used inside other IBM hardware (including the commercially unsuccessful PC/RT, which was the first RISC-based PC). The 801's simple architecture was based on studies performed by IBM engineer John Cocke, who examined code compiled for the IBM System/360 architecture and saw that most of the machine instructions were never used. Although it was not a financial success, the IBM 801 inspired a number of other projects, including the ones that led to IBM's development of the Performance Optimized With Enhanced RISC (POWER) architecture used in the RS/6000 systems, as well as the PowerPC microprocessor family.

It was only some time after the IBM 801 project was underway that Patterson began developing the RISC I microprocessor at Berkeley while, more or less simultaneously, Hennessy started the MIPS project at Stanford. (Both projects were funded by the Defense Advanced Research Projects Agency [DARPA].) These two architectures were refined over a period of several years in the early 1980s; derivatives of both were eventually commercialized. Patterson's second design, the RISC II, was adapted by Sun Microsystems (now Oracle) to create the SPARC architecture, and Hennessy created a company called MIPS Computer Systems (later MIPS Technologies, acquired in 2013 by Imagination Technologies) to market chips based on his architecture.

Patterson's design was particularly innovative in its use of a register remapping scheme known as *overlapping register windows* to reduce the performance cost associated with procedure calls. (Analysis done by the Berkeley research team indicated that calling and returning from procedures consumed much of a processor's time when running compiled code.) In this scheme, the CPU has many more hardware registers than the programmer can "see." Each time a procedure call occurs, the registers (except for a few that have global scope) are logically renumbered such that the called routine uses a different subset (or "window") of them. However, the remapping leaves a partial overlap of registers that can be used to pass arguments from the caller to the callee (see Figure 4.13). The previous mapping is restored when the called procedure is exited. By using registers rather than the stack for parameter passing, the CPU can avoid many memory accesses that would otherwise be necessary and thus achieve higher performance when executing a given program.

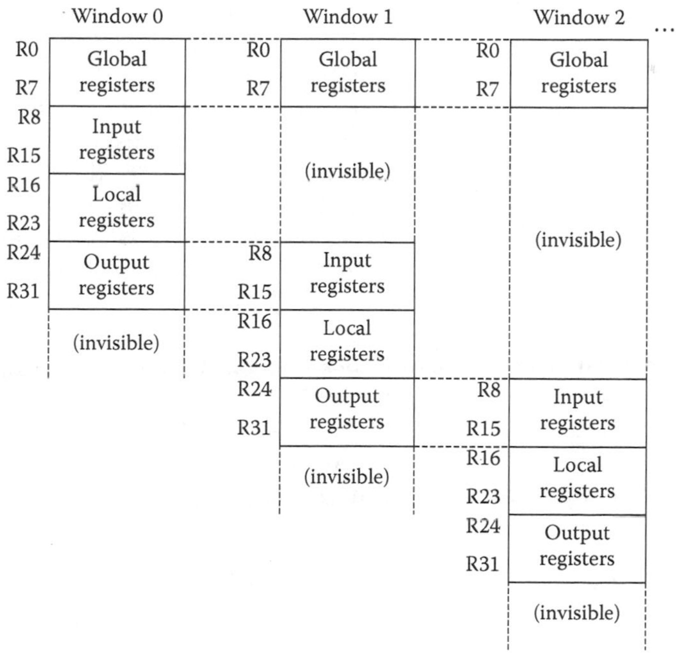

*Figure 4.13* Overlapping register window scheme.

Although these new chips took some time to gain widespread acceptance, they demonstrated the viability of the RISC approach. The commercial versions of Patterson's and Hennessy's pioneering architectures were used in high-performance engineering workstations successfully marketed by Sun Microsystems and Silicon Graphics; they eventually became two of the most successful RISC architectures to date and ultimately brought the RISC philosophy into the mainstream of computer architecture. By the 1990s, RISC had become the dominant philosophy of computer design.

What will be the design philosophy of the future? For quite some time the once-sharp distinctions between RISC and CISC machines have been blurred. The sixth generation (and the latter part of the fifth) may be considered the "post-RISC era." Many modern computer architectures borrow features from both RISC and CISC machines; their designers have found that by selectively adding back certain features that were eliminated from the original RISC architectures, performance on many applications (especially those that use graphics) can be improved. Meanwhile, computer engineers charged with the task of keeping legacy CISC architectures, such as the Intel x86 and its later enhancements, competitive have adopted many of the techniques pioneered by RISC designers. AMD's K5, K6, Athlon, Opteron and subsequent chips as well as Intel's Core processors emulate the x86 instruction set rather than executing it directly. These CPUs achieve high performance by breaking each CISC "macro" (x86 machine language) instruction down into simpler RISC-like instructions ($\mu$ops) that are then fed to multiple, highly pipelined execution units.

Considering that RISC and CISC are radically different philosophies of computer design, it makes sense to view the continued evolution of computer architecture in philosophical terms. The philosopher Georg Hegel believed that human thought progresses from an initial idea to its opposite and then to a new, higher concept that binds the two together and transcends them. These three steps are known as *thesis, antithesis,* and *synthesis*; through them, the truth develops and progress is achieved. Considering computer systems design in this light, CISC can be seen as a thesis put forward in the 1960s and 1970s with RISC, its antithesis, stepping forward to challenge it in the 1980s and 1990s. The most prominent computing architectures of the 21st century may thus be seen as the synthesis of CISC and RISC, with the demonstrated ability to go beyond the limitations of either approach to deliver new levels of computing performance.

## 4.5   *Enhancing the pipelined CPU*

We learned in the preceding sections that pipelining is an effective way to improve the processing performance of computer hardware. However, a simple pipeline can only improve performance by a factor approaching, but never reaching or exceeding, the small number of stages employed. In this section, we describe design approaches that have been adopted to achieve further performance improvements in pipelined CPUs.

## 4.5.1   Superpipelined architectures

The fundamental limitation on performance improvement using pipelining is the number of stages into which the task is subdivided. A three-stage pipeline can at best yield a speedup approaching a factor of three; a five-stage pipeline can only approach a 5:1 speed ratio. The simplest and most straightforward approach to achieving further performance improvement, then, is simply to divide the pipeline into more but smaller stages in order to clock it at a higher frequency. There is still only one pipeline, but by increasing the number of stages, we increase its *temporal parallelism*—it is working on more instructions at the same time. This use of a very deep, very high-speed pipeline for instruction processing is called *superpipelining*.

One of the first superpipelined processors was the MIPS R4000 (introduced in 1991), which had an eight-stage pipeline instead of the four- or five-stage design that was common in RISC architectures at the time. The eight stages were instruction fetch (first half), instruction fetch (second half), instruction decode and register fetch, execution, data cache access (first half), data cache access (second half), tag check, and write back. By examining this decomposition of the task, one can see that the speed of the cache had been a limiting factor in CPU performance. By splitting up memory access across two stages, the MIPS designers were able to better match that slower operation with the speed of internal CPU operations and thus balance the workload throughout the pipe.

The MIPS R4000 illustrated both the advantages and disadvantages of a superpipelined approach. On the plus side, it achieved a very high clock frequency for its time. The single pipelined integer functional unit was simple to control and took up little space on the chip, leaving room for more cache and other components, including a floating-point unit and a memory management unit. However, as with all deeply pipelined instruction execution units, branching presented a major problem. The branch penalty of the R4000 pipeline was "only" three cycles (one might reasonably expect an even greater penalty given an eight-stage implementation). However, the MIPS instruction set architecture had been designed for a shallower pipeline and thus only specified one delay slot following a branch. This meant there were always at least two stall cycles after a branch—three if the delay slot could not be used constructively. This negated much of the advantage gained from the chip's higher clock frequency. The R4000's demonstration of the increased branch delays inherent to a single superpipeline was so convincing that this approach has largely been abandoned. Most current processors that are superpipelined are also *superscalar* (see next section) to allow speculative execution down both paths from a branch.

### 4.5.2   *Superscalar architectures*

A *superscalar* machine is one that uses a standard, von Neumann–type instruction set but can issue (and thus often complete) multiple instructions per clock cycle. Obviously, this cannot be done with a single conventional pipeline, so an alternate definition of superscalar is a machine with a sequential programming model that uses multiple pipelines. Superscalar CPUs attempt to exploit whatever degree of *instruction level parallelism* (ILP) exists in sequential code by increasing *spatial parallelism* (building multiple execution units) rather than temporal parallelism (building a deeper pipeline). By building multiple pipelines, designers of superscalar machines can get beyond the theoretical limitation of one instruction per clock cycle inherent to any single-pipeline design and achieve higher performance without the problems of superpipelining.

Many microprocessors of the modern era have been implemented as superscalar designs. Superscalar CPUs are generally classified by the maximum number of instructions they can issue at the same time. (Of course, due to data dependencies, structural hazards, and branching, they do not actually issue the maximum number of instructions during every cycle.) The maximum number of instructions that can be simultaneously issued depends, as one might expect, on the number of pipelines built into the CPU. DEC's Alpha 21064 (a RISC processor introduced in 1992) was a two-issue chip, and its successor the 21164 (1995) was a four-issue processor. Another way of saying the same thing is to call the 21064 a *two-way superscalar* CPU, and the 21164 is said to be *four-way superscalar*. The MIPS R10000, R12000, R14000, and R16000 are also four-way superscalar processors. CISC architectures can also be implemented in superscalar fashion: As far back as 1993, Intel's first Pentium chips were based on a two-way superscalar design. The Pentium had essentially the same instruction set architecture as the 486 processor but achieved higher performance by using two execution pipelines instead of one. As we previously noted, more recent x86 family chips from both Intel and AMD achieve high performance by decoding architectural instructions into simpler RISC μops, which are then executed on a multiway superscalar core.

In order to issue multiple instructions per cycle from a sequentially written program and still maintain correct execution, the processor must check for dependencies between instructions using an approach like those we discussed previously (the scoreboard method or Tomasulo's method) and issue only those instructions that do not conflict with each other. Because out-of-order execution is generally possible, all types of hazards can occur; register renaming is often used to help solve the problems. Precise handling of *exceptions* (discussed in Chapter 5) is more difficult in a superscalar environment, too. The logic required to detect and resolve all of these problems is complex to design and adds significantly to the amount of chip area required. The multiple pipelines also take up more room, making superscalar designs very space-sensitive and thus more amenable to implementation technologies with small feature (transistor) sizes. (Superpipelined designs, by contrast, are best implemented with technologies that have short propagation delays.) These many difficulties of building a superscalar CPU are offset by a significant advantage: With multiple pipelines doing the work, clock frequency is not as critical in superscalar machines as it is in superpipelined ones. Because generating and distributing a high-frequency clock signal across a microprocessor is far from a trivial exercise, this is a substantial advantage in favor of the superscalar approach.

Superscalar and superpipelined design are not mutually exclusive. Many CPUs have been implemented with multiple, deep pipelines, making them both superpipelined and superscalar. Sun's UltraSPARC processor, introduced in 1995, was an early example of this hybrid approach: It was both superpipelined (nine stages) and four-way superscalar. The AMD Athlon (first introduced in 1999), Intel Pentium 4 (2000), IBM PowerPC 970 (2003), and ARM Cortex A8 (2005), among others, followed suit by combining superscalar and superpipelined design in order to maximize performance. Given sufficient chip area, superscalar design is a useful enhancement that makes superpipelining much more practical. When a branch is encountered, a superscalar/superpipelined machine can use one (or more) of its deep pipelines to continue executing sequential code while another pipeline executes speculatively down the branch target path. Whichever way the branch decision goes, at least one of the pipelines will have correct results; any that took the wrong path can be flushed. Some work is wasted, but processing never comes to a complete halt.

### 4.5.3 Very long instruction word (VLIW) architectures

Superscalar CPU designs have many advantages. However, there remains the significant problem of scheduling which operations can be done concurrently and which ones have to wait and be done sequentially after others. Determining the precise amount of instruction level parallelism in a program is a complex exercise that is made even more difficult by having to do it "on the fly" within the limitations of hardware design and timing constraints. How can we get the benefits of multiple instruction issue found in superscalar CPUs without having to build in complex, space- and power-consuming circuitry to analyze multiple types of data dependencies and schedule the initiation of operations? A look back at our discussion of RISC principles provides one possible answer: Let the compiler do the work. This idea is the foundation of a class of architectures known by the acronym (coined by Josh Fisher) *VLIW*, which stands for very long instruction word.

The centerpiece of a VLIW architecture is exactly what the name implies: a machine language instruction format that is fixed in length (as in a RISC architecture) but much longer than the 32- to 64-bit formats common to most conventional CISC or RISC architectures (see Figure 4.14 for an example). Each very long instruction contains enough bits to specify not just one machine operation, but several to be performed simultaneously. Ideally, the VLIW format includes enough *slots* (groups of bit fields) to specify operands and operations for every functional unit present in the machine. If sufficient instruction level parallelism is present in the high-level algorithm, all of these fields can be filled with useful operations, and maximum use will be made of the CPU's internal parallelism. If not enough logically independent operations can be scheduled at one time, some of the fields will be filled with NOPs, and thus some of the hardware will be idle while other units are working.

The VLIW approach has essentially the same effect as a superscalar design, but with most or all of the scheduling work done statically (offline) by the compiler rather than dynamically (at run time) by the control unit. It is the compiler's job to analyze the program for data and resource dependencies and pack the slots of each very long instruction word with as many concurrently executable operations as possible. Because the compiler has more time and resources to perform this analysis than would the control unit and because it can see a bigger picture of the code at once, it may be able to find and exploit more instruction-level parallelism than would be possible in hardware. Thus, a VLIW architecture may be able to execute more (equivalent) instructions per cycle than a superscalar machine.

Finding and expressing the parallelism inherent in a high-level program is quite a job for the compiler; however, the process is essentially the same whether it is done in hardware or software. Doing scheduling in software allows the hardware to be kept simpler and ideally faster. One of the primary characteristics of the original, single-pipeline RISC

| Operation 1 | Operand 1A | Operand 1B | Operand 1C | Operation 2 | Operand 2A | Operand 2B | Operand 2C | Operation 3 | Operand 3A | Operand 3B | Operand 3C | Operation 4 | Operand 4A | Operand 4B | Operand 4C |
|---|---|---|---|---|---|---|---|---|---|---|---|---|---|---|---|

*Figure 4.14* Example of very long instruction word format.

architectures was reliance on simple, fast hardware with optimization to be done by the compiler. VLIW is nothing more or less than the application of that same principle to a system with multiple pipelined functional units. Thus, in its reliance on moving pipeline scheduling from hardware to software (as well as in its fixed-length instruction format), VLIW can be seen as the logical successor to RISC.

The VLIW concept goes back further than most computing professionals realize. The first VLIW system was the Yale ELI-512 computer (with its Bulldog compiler) built in the early 1980s, just about the time that RISC architectures were being born. Josh Fisher and some colleagues from Yale University started a company named Multiflow Computer, Inc., in 1984. Multiflow produced several TRACE systems, named for the trace scheduling algorithm used in the compiler. Another company, Cydrome Inc., which was founded about the same time by Bob Rau, produced a machine known as the Cydra-5. However, the VLIW concept was clearly ahead of its time. (Remember, RISC architectures, which were a much less radical departure from conventional design, had a hard time gaining acceptance at first, too.) Both companies failed to thrive and eventually went out of business.

For some time after this, VLIW architectures were only found in experimental research machines. IBM got wind of the idea and started its own VLIW project in 1986. Most of the research involved the (very complex) design of the compilers; IBM's hardware prototype system was not constructed until about 10 years after the project began. This system was based on what IBM calls the Dynamically Architected Instruction Set from Yorktown (DAISY) architecture. Its instructions are 759 bits long; they can simultaneously specify up to eight ALU operations, four loads or stores, and seven branch conditions. The prototype machine contained only 4 MB of data memory and 64K VLIWs of program memory, so it was obviously intended more as a proof of the VLIW concept (and of the compiler) than as a working system.

Only a few general-purpose processors based on the VLIW idea have been built, and those have met with somewhat limited success. In 2005, the Moscow Center of SPARC Technologies introduced its Elbrus 2000 microprocessor, which featured 512-bit-wide instructions and could execute up to 20 operations (or more in later versions) per clock cycle. This chip and its successors (up through the Elbrus-8S, introduced in 2014) have been used in a limited number of Russian-built computers but have not been mass-produced or sold internationally. Special-purpose processors based on VLIW architectures have been somewhat more successful. For example, the ST200 family of processor cores, based on the Lx architecture jointly developed by Hewlett-Packard (HP) and STMicroelectronics, and which trace their lineage to Fisher's work at Multiflow, have been widely used in embedded media processing applications. Texas Instruments' popular

TMS320C6000 family of digital signal processing chips is also based on a VLIW architecture that bundles eight 32-bit RISC-like operations into 256-bit *fetch packets* that are executed in parallel.

The (relatively) most popular desktop–server architecture to date that incorporates elements of VLIW is the Itanium (formerly known as IA-64) architecture, which was jointly developed by Intel and Hewlett-Packard and initially implemented in the Itanium (Merced) processor. Subsequent chips in this family include the Itanium 2 (2002), Itanium 2 9000 series (2006), 9100 series (2007), 9300 series (code name Tukwila, 2010), and 9500 series (code name Poulson), introduced in 2012. Intel does not use the acronym VLIW, preferring to refer to its slightly modified version of the concept as EPIC (which stands for Explicitly Parallel Instruction Computing). However, EPIC's lineage is clear: VLIW pioneers Fisher and Rau went to work with HP after their former companies folded and contributed to the development of the new architecture. The Itanium instruction set incorporates 128-bit bundles (see Figure 4.15) containing three 41-bit RISC-type instructions; each bundle provides explicit dependency information. This dependency information (determined by the compiler) is used by the hardware to schedule the execution of the bundled operations. The purpose of this arrangement (with some of the scheduling done in hardware in addition to that done by software) is to allow binary compatibility to be maintained between different generations of chips. Ordinarily, each generation of a given manufacturer's VLIW processors would be likely to have a different number and types of execution units, and thus a different instruction format. Because hardware is responsible for some of the scheduling and because the instruction format is not directly tied to a particular physical implementation, the Itanium processor family is architecturally somewhere between a pure VLIW and a superscalar design (although it is arguably closer to VLIW).

The VLIW approach has several disadvantages, which IBM, HP, and Intel hoped would be outweighed by its significant advantages. Early VLIW machines performed poorly on branch-intensive code; IBM attempted to address this problem with the tree structure of its system, and the Itanium architecture addresses it with a technique called *predication* (as opposed to

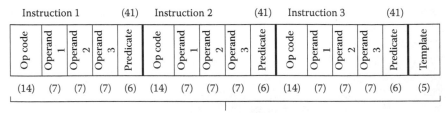

*Figure 4.15* Itanium EPIC instruction format.

prediction, which has been used in many RISC and superscalar processors). The predication technique uses a set of *predicate registers*, each of which can be used to hold a true or false condition. Where conditional branches would normally be used in a program to set up a structure such as if/then/ else, the operations in each possible sequence are instead *predicated* (made conditional) on the contents of a given predicate register. Operations from both possible paths (the "then" and "else" paths) then flow through the parallel, pipelined execution units, but only one set of operations (the ones with the predicate that evaluates to true) are allowed to write their results.

Another problem that negatively affected the commercial success of early VLIW architectures was the lack of compatibility with established architectures. Unless one had access to all the source code and could recompile it, existing software could not be run on the new architectures. IBM's DAISY addresses this problem by performing dynamic translation (essentially, run-time interpretation) of code for the popular PowerPC architecture. Early (pre-2006) EPIC processors also supported execution of the huge existing Intel IA-32 (x86) code base, as well as code compiled for the HP PA-RISC architecture, through hardware emulation at run time. However, performance proved to be poor compared to both native IA-64 code running on Itanium series chips as well as IA-32 code running on existing IA-32 processors. Thus, starting with the Itanium 2 9000 series (Montecito) processors, a software emulator known as the IA-32 Execution Layer was introduced to improve performance on legacy x86 code.

Yet another VLIW disadvantage is poor code density. Because not every VLIW instruction has useful operations in all its slots, some bit fields are inevitably wasted, making compiled programs take up more space in memory than they otherwise would. No real cure has been found for this problem, as the available hardware resources are constant and the ILP available in software varies. However, this issue has been considerably ameliorated by the fact that memory prices have trended downward over time.

Very long instruction word architectures have several advantages that would seem to foretell commercial success. The most obvious is the elimination of most or all scheduling logic from the hardware, freeing up more space for additional functional units, more on-chip cache, multiple cores, etc. Simplifying the control logic may also reduce overall delays and allow the system to operate at a higher clock frequency than a superscalar CPU built with the same technology. Also, as was previously mentioned, the compiler (relatively speaking) has "all day" to examine the entire program and can potentially uncover much more instruction-level parallelism in the programmed algorithm than could a real-time, hardware scheduler. This advantage, coupled with the presence of additional functional units, should allow VLIW machines to execute more operations per cycle than are possible with a superscalar approach. Because the previously steady increase in CPU clock speeds has flattened out in recent years, the ability of VLIW architectures to exploit ILP would seem to bode well for their success in the high-performance computing market.

However, in actual practice, Itanium family chips account for only a small fraction of Intel's overall sales. The architecture never caught on in the personal computer (desktop/laptop) market, and Itanium-based systems represent only a fraction of the overall server and high-performance computing segment (almost all of it in systems made by Hewlett-Packard). It appears that this trend is likely to continue; although VLIW-based chips will no doubt continue to be used in embedded systems and DSP applications, for the fore-seeable future it appears that the vast majority of high-end systems will be based on multicore and, in many cases, multithreaded (see the following section) implementations of more conventional architectures. In many cases, these will be supplemented by the increasingly powerful and popular graphics processing units (GPUs) that we will examine in Section 6.1.2.

### 4.5.4  *Multithreaded architectures*

Previously, in Section 4.3.7, we mentioned that Tomasulo's scheduling algorithm for machines using out-of-order execution was based on a dataflow approach. Although they outwardly perform as von Neumann machines and are programmed using the conventional, sequential programming paradigm, processors with Tomasulo schedulers operate internally as dataflow machines—a type of "non–von Neumann" architecture that we will examine in more detail in Section 7.1. Although pure dataflow machines have not become popular in their own right, they did inspire this and certain other innovations that have been applied to more traditional CPU designs.

Another way in which dataflow concepts have influenced conventional computer designs is in the area of multithreaded, superthreaded, and hyperthreaded processors. Consider the behavior of programs in conventional machines. We think of programs running on a traditional, von Neumann–style architecture as being comprised of one or more relatively coarse-grained processes. Each process is an instruction stream consisting of a number of sequentially programmed instructions. Multiple processes can run on a parallel system (of the kind we will discuss in Section 6.1) in truly concurrent fashion, while on a uniprocessor machine, they must run one after another in time-sliced fashion. In a true dataflow machine, each individual machine operation can be considered to be an extremely "lightweight" process of its own to be scheduled when it is ready to run (has all its operands) and when hardware (a processing element) is available to run it. This is as fine-grained a decomposition of processes as is possible, and lends itself well to a highly parallel machine implementation. However, for reasons to be discussed later, massively parallel dataflow machines are not always practical or efficient.

The same concept, to less of a degree (and much less finely grained) is used in multithreaded systems. Multithreaded architectures, at least in some aspects, evolved from the dataflow model of computation. The idea of dataflow, and multithreading as well is to avoid data and control hazards and thus keep multiple hardware functional units busy, particularly when latencies of memory access or communication are high and/or not a great deal of instruction-level parallelism is present within a process. Although each thread of execution is not nearly as small as an individual instruction (as it would be in a true dataflow machine), having more, "lighter weight" threads instead of larger, monolithic processes and executing these threads in truly concurrent fashion is in the spirit of a dataflow machine because it increases the ability to exploit replicated hardware.

Time-slice multithreading, or superthreading, is a dataflow-inspired implementation technique whereby a single, superscalar processor executes more than one process or thread at a time. Each clock cycle, the control unit's scheduling logic can issue multiple instructions belonging to a single thread, but on a subsequent cycle, it may issue instructions belonging to another thread. (Figure 4.16 illustrates this approach.) In

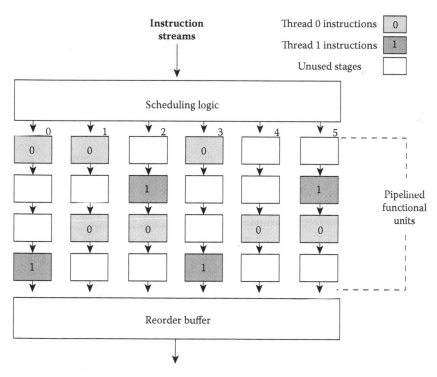

*Figure 4.16* Instruction execution in superthreaded processor.

effect, process/thread execution on a superthreaded machine is still time-sliced (as it is in traditional uniprocessor architectures), but the time slice can be as small as one clock cycle.

The superthreaded approach can lead to better use of hardware resources in a superscalar processor due to mitigation of dependency-based hazards, because instructions from one thread are unlikely to depend on results from instructions belonging to other threads. It can also reduce cycles wasted due to access latencies because instructions from other threads can still execute while a given thread is waiting on data from memory. However, utilization of hardware resources is still limited by the inherent instruction-level parallelism or lack thereof in the individual threads. (Studies have shown that, on average, a single thread only has sufficient ILP to issue about 2.5 instructions per clock cycle.) To address this limitation, simultaneous multithreading (SMT), also known as hyperthreading, goes a step beyond superthreading.

Compared to superthreading, hyperthreading (which is used in some of the latest generation CPUs from Intel, the IBM POWER5, POWER6, and POWER7 processors, and others) gives the scheduling logic more flexibility in deciding which instructions to issue at what time. In a hyper-threaded processor, it is not necessary that all the instructions issued during a given clock cycle belong to the same thread. For example, a six-issue superscalar machine could issue three instructions from thread 0 and three from thread 1 at the same time or four from thread 0 and two from thread 1, etc., as illustrated in Figure 4.17.

In effect, simultaneous multithreading internally splits a single, phys-ical superscalar processor into two or more logically separate CPUs, each executing its own instruction stream by sharing processor resources that might otherwise go unused if only a single thread were executing. (Of course, some CPU resources, including the program counter and certain other system registers, may have to be replicated to allow for hyperthread-ing.) Simultaneous multithreading implementation can significantly increase the efficiency of resource utilization and improve overall system performance as seen by the user when multiple tasks (or a single applica-tion coded with multiple threads) are running.

Of course, no approach to processor design comes without caveats. Some of the issues that may pose problems in multithreaded architec-tures include limitations on single-threaded performance when multiple threads are running and disadvantages associated with sharing of mem-ory resources, including cache and Translation Lookaside Buffers (TLBs). It is an observed fact that some applications perform noticeably worse on hyperthreaded processors due to contention with other threads for CPU resources, such as instruction decoding logic, execution units, branch predictors, etc. In each new design, manufacturers must struggle to bal-ance hardware resources with the anticipated needs of applications. As

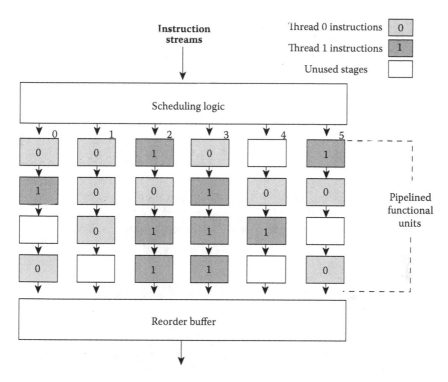

*Figure 4.17* Instruction execution in hyperthreaded processor.

an alternative, some multithreaded CPUs, for example, the UltraSPARC T4, provide a mechanism for multithreading to be effectively turned off, temporarily dedicating an entire core's hardware to running a single thread that had been suffering from a bottleneck, using the Solaris operating system's "Critical Thread Optimization" feature. More research and development are needed to provide convenient and efficient means for systems to detect when this is occurring and to use this information to switch back and forth between single-threaded and multithreaded modes for optimum overall performance.

Sharing of memory resources between threads running on a common core creates issues for both performance and security. Obviously, when the instruction and/or data caches belonging to a core are shared between two or more threads, the effective cache size available to each thread—and thus the hit ratio for each—is less than it would be if only one thread were running. Effective cache size may not be cut exactly in half if two running threads share some common code or data, but it is extremely unlikely that they would share everything; thus, some degradation of performance is almost inevitable. Important research questions here involve determining the optimum cache size, line size, and other design parameters for multiple threads to share a common cache with the least disruption. But perhaps an even more important topic is preventing threads from sharing the cache "too much"—in other words, spying on data belonging to other threads. As far back as 2005, researcher Colin Percival discovered that it was possible for a malicious thread running on a Pentium 4 CPU with hyperthreading to use cache access patterns to steal a cryptographic key from another thread running on the same core. Although operating systems can be configured to block this attack vector, manufacturers of multithreaded CPUs now know they must keep information security in mind when designing cache line replacement policies; this will surely be an ongoing concern as more processors of this type are designed. Despite these issues, however, because of its adoption by the largest manufacturer of microprocessors and other industry giants, SMT/hyperthreading could prove to be the most lasting legacy of the dataflow approach.

## 4.6   Chapter wrap-up

No part of a modern computer system is as complex as its CPU. (This explains why we have devoted two chapters, rather than just one, to CPU design topics.) In this chapter, we examined several advanced design concepts involving internal concurrency of operations within a single CPU core. (Multicore/multiprocessing systems are addressed in Section 6.1.) The concept of pipelining was the main focus of our attention because almost all commonly used methods for enhancing the performance of a CPU are based on pipelining and because, as a result, almost all modern microprocessors are internally pipelined to some extent. As the reader has seen, although the potential performance benefits of pipelining are obvious, "the devil is in the details," particularly when pipelines become very "deep" and/or when multiple pipelines are used in one processor core.

Of course, the examples provided above as illustrations of pipelined, superpipelined, superscalar, VLIW, and multithreaded architectures were kept relatively simple so as not to lose the reader in myriad implementation details—details that would have to be addressed in any practical, real-world design. Put another way, we have only scratched the surface of the intricacies of this subject. Computer engineers may spend their entire careers designing high-performance microprocessors and not learn everything there is to know about it, and even comprehensive knowledge of technical details can quickly become obsolete as technology changes. Certainly no single course or textbook can do full justice to this highly specialized discipline. Given that mastery of a vast body of ever-changing details is virtually impossible, however, it is all the more crucial for computing professionals to appreciate the history of CPU design and understand the techniques involved in extracting the best performance from an architecture.

It is likely that relatively few of the students using this book will subsequently be employed in the design of high-performance microprocessor chips. However, it is highly probable that at some point the reader will be called upon to provide system specifications or choose the best machine for a particular task. In order to be able to intelligently select a system for a desired application, it is important, among other things, to have a basic understanding of the performance-enhancing techniques used in microprocessor design. By studying and understanding the elements of processor architecture and implementation covered in this chapter and the one before it, the serious student will have helped himself or herself understand their performance implications and thus will be better equipped to make wise choices among the many competing computer systems in the constantly changing market of the present and the future. If, however, the reader's goal is to help design the next generation of high-performance CPUs, studying this material will have taken him or her the first few steps down that highly challenging career path.

## REVIEW QUESTIONS

1. Suppose you are designing a machine that will frequently have to perform 64 consecutive iterations of the same task (for example, a vector processor with 64-element vector registers). You want to implement a pipeline that will help speed up this task as much as is reasonably possible, but recognize that dividing a pipeline into more stages takes up more chip area and adds to the cost of implementation.
   a. Make the simplifying assumptions that the task can be subdivided as finely or coarsely as desired and that pipeline registers do not add a delay. Also assume that one complete iteration of the task takes 16 ns (thus, a nonpipelined implementation would take $64 \times 16 = 1024$ ns to complete 64 iterations). Consider possible pipelined implementations with 2, 4, 8, 16, 24, 32, and 48 stages. What is the total time required to complete 64 iterations in each case? What is the speedup (vs. a nonpipelined implementation) in each case? Considering cost as well as performance, what do you think is the best choice for the number of stages in the pipeline? Explain. (You may want to make graphs of speedup and/or total processing time vs. the number of stages to help you analyze the problem.)
   b. Now assume that a total of 32 levels of logic gates are required to perform the task, each with a propagation delay of 0.5 ns (thus, the total time to produce a single result is still 16 ns). Logic levels cannot be further subdivided. Also assume that each pipeline register has a propagation delay equal to that of two levels

of logic gates, or 1 ns. Reanalyze the problem; does your previous recommendation still hold? If not, how many stages would you recommend for the pipelined implementation under these conditions?

2. Considering the overall market for all types of computers, which of the following are more commonly found in today's machines: arithmetic pipelines (as discussed in Section 4.2) or instruction unit pipelines (Section 4.3)? Explain why this is so.

3. Why do control transfers, especially conditional control transfers, cause problems for an instruction-pipelined machine? Explain the nature of these problems and discuss some of the techniques that can be employed to cover up or minimize their effect.

4. A simple RISC CPU is implemented with a single scalar instruction-processing pipeline. Instructions are always executed sequentially except in the case of branch instructions. Given that $p_b$ is the probability of a given instruction being a branch, $p_t$ is the probability of a branch being taken, $p_c$ is the probability of a correct prediction, $b$ is the branch penalty in clock cycles, and $c$ is the penalty for a correctly predicted branch:

   a. Calculate the throughput for this instruction pipeline if no branch prediction is made given that $p_b = 0.16$, $p_t = 0.3$, and $b = 3$.

   b. Assume that we use a branch prediction technique to try to improve the pipeline's performance. What would be the throughput if $c = 1$, $p_c = 0.8$, and the other values are the same as above?

5. What are the similarities and differences between a delayed branch and a delayed load?

6. Given the following sequence of assembly language instructions for a CPU with multiple pipelines, indicate all data hazards that exist between instructions.

   $I_1$: Add R2, R4, R3  ;R2 = R4 + R3
   $I_2$: Add R1, R5, R1  ;R1 = R5 + R1
   $I_3$: Add R3, R1, R2  ;R3 = R1 + R2
   $I_4$: Add R2, R4, R1  ;R2 = R4 + R1

7. What are the purposes of the scoreboard method and Tomasulo's method of controlling multiple instruction execution units? How are they similar and how are they different?

8. List and explain nine common characteristics of RISC architectures. In each case, discuss how a typical CISC processor would (either completely or partially) not exhibit the given attribute.

9. How does the overlapping register windows technique, used in the Berkeley RISC and its commercial successor the Sun SPARC, simplify the process of calling and returning from subprograms?

10. You are on a team helping design the new Platinum V® processor for AmD$_e$l Corporation. Consider the following design issues:
    a. Your design team is considering a superscalar versus superpipeline approach to the design. What are the advantages and disadvantages of each option? What technological factors would tend to influence this choice one way or the other?
    b. Your design team has allocated the silicon area for most of the integrated circuit and has narrowed the design options to two choices: one with 32 registers and a 512-KB on-chip cache and one with 512 registers but only a 128-KB on-chip cache. What are the advantages and disadvantages of each option? What other factors might influence your choice?
11. How are VLIW architectures similar to superscalar architectures, and how are they different? What are the relative advantages and disadvantages of each approach? In what way can VLIW architectures be considered the logical successors to RISC architectures?
12. Is explicitly parallel instruction computing (EPIC) the same thing as a VLIW architecture? Explain why or why not.
13. Are superthreaded and hyperthreaded processors the same thing? If not, how do they differ?
14. Fill in the blanks below with the most appropriate term or concept discussed in this chapter:
    _____ The time required for the first result in a series of computations to emerge from a pipeline.
    _____ This is used to separate one stage of a pipeline from the next.
    _____ Over time, this tells the mean number of operations completed by a pipeline per clock cycle.
    _____ The clock cycles that are wasted by an instruction-pipelined processor due to executing a control transfer instruction.
    _____ A technique used in pipelined CPUs when the compiler supplies a hint as to whether or not a given conditional branch is likely to succeed.
    _____ The instruction(s) immediately following a conditional control transfer instruction in some pipelined processors, which are executed whether or not the control transfer occurs.
    _____ A technique used in pipelined CPUs when the instruction immediately following another instruction that reads a memory operand cannot use the updated value of the operand.
    _____ The most common data hazard in pipelined processors—also known as a true data dependence.

_____ Also known as an output dependence, this hazard can occur in a processor that utilizes out-of-order execution.

_____ A centralized resource scheduling mechanism for internally concurrent processors; it was first used in the CDC 6600 supercomputer.

_____ These are used by a Tomasulo scheduler to hold operands for functional units.

_____ A technique used in some RISC processors to speed up parameter passing for high-level language procedure calls.

_____ This type of processor architecture maximizes temporal parallelism by using a very deep pipeline with very fast stages.

_____ This approach to high-performance processing uses multiple pipelines with resolution of interinstruction data dependencies done by the control unit.

_____ The "architecture technology" used in Intel's IA-64 (Itanium) chips.

_____ The Itanium architecture uses this approach instead of branch prediction to minimize the disruption caused by conditional control transfers.

_____ A machine using this technique can issue instructions from more than one thread of execution during the same clock cycle.

# chapter five

# Exceptions, interrupts, and input/output systems

We have studied two of the three main components of a modern computer system: the memory and the central processing unit. The remaining major component, the *input/output* (I/O) system, is just as important as the first two. It is always necessary for a useful computer system to be able to receive information from the outside world, whether that information happens to be program instructions, interactive commands, data to be processed, or (usually) some combination of these things. Likewise, regardless of the memory capacity or computational capabilities of a system, it is essentially useless unless the results of those computations can be conveyed to the human user (or to another machine for further processing, display, storage, etc.). Anyone who has ever bought an expensive audio amplifier knows that the resulting sound is only as good as the signal source fed to the inputs and the speaker system that converts the amplified signal into sound waves for the listener to hear. Without a clean sound source and high-fidelity speakers, the amplifier is nothing more than an expensive boat anchor. Similarly, without an I/O system able to quickly and effectively interact with the user and other devices, a computer might just as well be thrown overboard with a stout chain attached.

In this chapter, we also study how computer systems use *exceptions* (including traps, hardware interrupts, etc.) to alert the processor to various conditions that may require its attention. The reader who has had a previous course in basic computer organization will no doubt recall that exceptions, both hardware and software related, are processed by the central processing unit (CPU). Thus, it may have seemed logical to cover that material in the previous chapters on CPU design. Because those chapters were already filled with so many other important concepts related to CPU architecture and implementation and, even more importantly, because most exceptions are related directly or indirectly to I/O operations, we have reserved the discussion of this important topic to its logical place in the context of the function of the I/O system.

## 5.1   Exceptions

An *exception*, in generic terms, can be any synchronous or asynchronous system event that occurs and requires the attention of the CPU for its resolution. The basic idea is that while the CPU is running one program (and it is always running some program unless it has a special low-power or sleep mode), some condition arises inside or outside the processor that requires its intervention. In order to attend to this condition, the CPU suspends the currently running process, then locates and runs another program (usually referred to as a *service routine* or *handler*) that addresses the situation. Whether it is a hardware device requesting service, an error condition that must be investigated and fixed if possible, or some other type of event, the handler is responsible for servicing the condition that caused the exception. Once the handler runs to completion, control is returned (if possible) to the previously running program, which continues from the point where it left off.

Exceptions can be divided into two general types: those caused by hardware devices (which may be on- or off-chip, but in either case are outside the CPU proper) and those that are caused by the software running on the CPU. There are many similarities between hardware- and software-related exceptions, plus a few key differences that we examine in this section. Because hardware-related exceptions, more commonly known as interrupts, are a bit more straightforward concept and are usually more directly related to I/O, we examine their characteristics first.

### 5.1.1   Hardware-related exceptions

Exceptions caused by hardware devices (often I/O devices) in a system are generally referred to as *interrupts*. The obvious origin of this term lies in the mechanism employed by external devices to get the attention of the CPU. Because the processor is busy executing instructions that likely have nothing to do with the device in question, the device interrupts the execution of the current task by sending a hardware signal (either a specified logic level 0 or 1, or the appropriate transition of a signal from 1 to 0 or from 0 to 1) over a physical connection to the CPU. In Figure 5.1, this physical connection is shown as an active low interrupt request (IRQ)

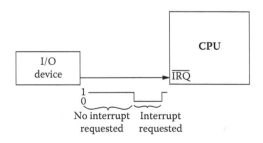

*Figure 5.1* Interrupt request from device to CPU.

input from the device to the CPU; when this input is logic 0, the device is requesting service. Assuming interrupts are *enabled* (meaning the CPU is listening to devices) and the request is of sufficient priority (we will discuss priority schemes below), the currently running program will be suspended and the appropriate interrupt handler will be executed to service the source of the interrupt.

It is important to recognize that because interrupts come from hardware devices outside the CPU, they are asynchronous events. Interrupts may occur at any time without regard to anything else the processor may be doing. Because they may fall anywhere within the instruction stream the CPU is executing, it is crucial that interrupts must be processed *transparently*. That is, all relevant CPU state information, including the contents of all registers and status and condition code flags, must be saved before interrupt processing begins and then restored before the interrupted program is resumed in order to guarantee correct operation.

To illustrate how critical it is for all relevant state information to be maintained, consider the situation illustrated in Figure 5.2. An interrupt is recognized after the CMP (compare) instruction. The CPU suspends execution at this point; after the interrupt handler completes, it will resume execution with the BEQ (branch if equal) instruction. This conditional branch instruction bases its branch decision on whether or not the previous comparison showed equality. (This is generally accomplished by setting a status flag, such as a zero or Z flag, to reflect the result of the CMP instruction.) If no interrupt had occurred, no other arithmetic or logical operation could possibly intervene between the CMP and the BEQ, so the branch would execute properly. However, if an interrupt occurs as indicated in Figure 5.2, it is possible that some arithmetic or logic instruction that is part of the interrupt handler could change the state of the Z flag and cause the branch to operate incorrectly when the original program is resumed. Clearly, this would not be a desirable situation. Thus, it is critical that the complete CPU state be maintained across any interrupt service routine.

Partly in order to reduce the amount of processor state information that must be saved (and also because it simplifies the design of the control unit), most CPUs only examine their interrupt request inputs at the conclusion of an instruction. Even if the processor notices that an interrupt is pending, it will generally complete the current instruction before

Interrupt occurs ⟶ CMP   R1, R2   ; compare two values and set status bits
                    BEQ   Target   ; conditional branch, succeeds if operands were equal

*Figure 5.2* Occurrence of an interrupt between related instructions.

acknowledging the interrupt request or doing anything to service it. Pipelined CPUs may even take the time to drain the pipeline by completing all instructions in progress before commencing interrupt processing. This takes a little extra time but greatly simplifies the process of restarting the original instruction stream.

The usual mechanism employed for saving the processor state is to push all the register contents and status flag/condition code values on a stack in memory. When interrupt processing is complete, a special instruction (usually with a mnemonic, such as RTI or IRET [return from interrupt] or RTE [return from exception]) pops all the status bits and register contents (including the program counter) off the stack in reverse order, thus restoring the processor to the same state it was in before the interrupt occurred. (Because interrupt handlers generally run with system or supervisor privileges, although the interrupted program may have been running at any privilege level, restoring the previous privilege level must be part of this process.) Some architectures specify that only absolutely essential information (such as the program counter and flag values) are automatically preserved on the stack when an interrupt is serviced. In this case, the handler is responsible for manually saving and restoring any other registers it uses. If the machine has many CPU registers and the interrupt service routine uses only a few of them, this method will result in reduced latency when responding to an interrupt (an important consideration in some applications). An even faster method, which has been used in a few designs in which sufficient chip space was available, is for the processor to contain a duplicate set of registers that can be used for interrupt processing. After doing its job, the service routine simply switches back to the original register set and returns control to the interrupted program.

### 5.1.1.1   Maskable interrupts

It is usually desirable for the operating system to have some way of enabling and disabling the processor from responding to interrupt requests. Sometimes the CPU may be engaged in activities that need to be able to proceed to completion without interruption or are of higher priority than some or all interrupt sources. If this is the case, there may be a need to temporarily *mask* or disable some or all interrupts. *Masking* an interrupt is the equivalent of putting a mask over a person's face; it prevents the interrupt request from being "seen" by the CPU. The device may maintain its active interrupt request until a later time when interrupts are re-enabled.

Masking is generally accomplished by manipulating the status of one or more bits in a special CPU configuration register (often known by a name such as the status register, processor state register, or something similar). There may be a single bit that can be set or cleared to disable or enable all interrupts, individual bits to mask and unmask specific interrupt types, or even a field of bits used together to specify a priority threshold below which interrupts will be ignored. In systems with separate supervisor and user modes (this includes most modern microprocessors), the instructions that affect the mask bits are usually designated as privileged or system instructions. This prevents user programs from manipulating the interrupt mechanism in ways that might compromise system integrity. The specifics of the interrupt masking scheme are limited only by the area available for interrupt logic, the intended applications for the system, and the creativity of the CPU designers.

There are often several devices that can cause interrupts in a system. Some of these devices may be more important or may have more stringent timing requirements than others. Thus, it may be important to assign a higher priority to certain interrupt requests so that they will be serviced first in case more than one device has a request pending. Most architectures support more than one—often several—priority levels for interrupts. To take two well-known examples, the Motorola 68000 supports seven interrupt priority levels, and Sun's SPARC architecture supports 15. The various levels are distinguished by a priority encoder, which can be an external chip or may be built into the CPU.

Consider the case of a CPU that has three interrupt request lines: IRQ2, IRQ1, and IRQ0 (see Figure 5.3). One very simple approach would be to treat these as three separate interrupt requests with different priorities (which would be established by priority encoding logic inside the CPU). Each interrupt request input could be driven by a separate device (thus three types of devices could be supported). If the system has more than three devices capable of generating interrupts, multiple devices will have to share an IRQ input, and additional logic will be required to distinguish among them.

Any time none of the three IRQ inputs are active, the CPU proceeds with normal processing. If one or more interrupt requests become active, logic inside the CPU compares the highest priority active input with the current priority level to see whether or not this request is masked. If it is unmasked, the CPU acknowledges the interrupt request and begins servicing it; if it is masked, the CPU ignores it and continues processing the current program.

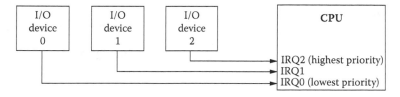

*Figure 5.3* CPU with simple interrupt priority scheme.

A slightly more complex but more flexible scheme could be devised using the same three interrupt request inputs: IRQ2, IRQ1, and IRQ0. Instead of considering them as three separate interrupt requests, the logic inside the CPU could be constructed to interpret them as a 3-bit field indicating the current interrupt request level. Level 0 (000 binary) could indicate that no device is currently requesting an interrupt, and levels 1 through 7 (001 through 111 binary) could be interpreted as distinct interrupt requests with different priorities (for example, 1 lowest through 7 highest). In such a scheme, the devices would not be connected directly to the IRQ inputs of the CPU. Instead, the device interrupt request lines would be connected to the inputs of a priority encoder with the outputs of the encoder connected to the interrupt request lines (see Figure 5.4).

This scheme allows for up to seven interrupting devices, each with its own request number and priority level. (Again, if more than seven devices were needed, additional hardware could be added to allow multiple devices to share a single interrupt request level.) This type of arrangement could also be realized completely with logic inside the CPU chip, but that would require the use of seven pins on the package (rather than just three) for interrupt requests.

In a system such as this one with several priority levels of interrupts, the masking scheme would probably be designed such that interrupts of a given priority or higher are enabled. (It would be simpler to mask and unmask all levels at once, but that would defeat most of the advantage of having several interrupt types.) All this requires is a bit field (in this case 3 bits), either in a register by itself or as some subset of a CPU control register, that can be interpreted as a threshold—a binary number specifying the lowest interrupt priority level currently enabled. (Alternatively, it could represent the highest priority level currently masked.) A simple magnitude comparator (see Figure 5.5) could then be used to compare the current interrupt request level with the priority threshold; if the current request is greater than or equal to (or in the second case, simply greater

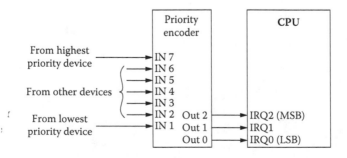

*Figure 5.4* System with seven interrupt priority levels.

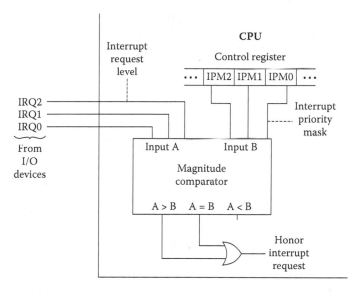

**Figure 5.5** Circuit for comparing interrupt request level and mask threshold level.

than) the threshold value, the interrupt request is honored. If not, it is ignored until the threshold is changed to a lower value. Many architectures specify an interrupt masking mechanism similar to this.

### 5.1.1.2 Nonmaskable interrupts

Nonmaskable interrupts are supported by almost all architectures. As the name implies, a *nonmaskable interrupt* is one that cannot be disabled. Whenever it is activated, the processor will recognize it and suspend the current task to service it. Thus, a nonmaskable interrupt is the highest priority hardware interrupt. Interrupt priority level 7 is the nonmaskable interrupt in a Motorola 68000 system, and level 15 is the nonmaskable interrupt in a SPARC-based system. In the example of Figure 5.3, the CPU designers could either designate one of the interrupt request inputs (for example, IRQ2) as a nonmaskable interrupt or provide a separate nonmaskable interrupt pin for that purpose. Similarly, in the example of Figure 5.4, one of the levels (for example, level 7 as in the 68000) could be set up as nonmaskable, or a separate input pin could be used for the nonmaskable interrupt (making eight types of interrupts in all).

We have mentioned that it may sometimes be necessary for some or all interrupts to be temporarily disabled; the reader might wonder what sort of scenario might require the use of a nonmaskable interrupt. One classic example is the use of a power fail sensor to monitor the system power source. If the sensor detects a voltage dropout, it uses an interrupt to warn the CPU that system power is about to be lost. (When line voltage is lost, power supply capacitors will typically maintain the output voltage for a few milliseconds, perhaps just long enough for the CPU to save crucial information and perform an orderly system shutdown.) Given that no system activity of any priority level will be possible without power, it makes sense to assign to this power fail sensor an interrupt request line that will never be ignored. One can also envision the use of an embedded microcontroller in a piece of medical equipment, such as a heart monitor. If the sensor that detects the patient's pulse loses its signal, that would probably be considered an extremely high-priority event requiring immediate attention regardless of what else might be going on.

### 5.1.1.3 Watchdog timers and reset

Watchdog timers and reset are special types of interrupts that can be used to reinitialize a system under certain conditions. All microprocessors have some sort of reset capability. This typically consists of a hardware connection that under normal conditions is always maintained at a certain logic level (for example, 1). If this pin is ever driven by external hardware to the opposite logic level, it signals the CPU to stop processing and reinitialize itself to a known state from which the operating system and any other necessary programs can be reloaded. This pin is often connected to a hardware circuit including a button (switch) that can be pressed by a user when the system crashes and all other attempts at recovery have failed. Some architectures also have RESET instructions that use internal hardware to pulse the same signal, resetting the processor and external devices to a known state. The reset mechanism is very much like a nonmaskable interrupt except for one thing: On a reset, no machine state information is saved because there is no intention of resuming the task that was running before the reset occurred.

Processors that are intended for use in *embedded* systems (inside some other piece of equipment) also generally have an alternate reset mechanism known as a *watchdog timer*. (Some manufacturers have proprietary names, such as computer operating properly [COP], for this type of mechanism.) A watchdog timer is a counter that runs continuously, usually driven by the system clock or some signal derived from it. Periodically, the software running on the CPU is responsible for reinitializing the watchdog timer to its original value. (The initial value chosen and the frequency of the count pulses determine the timeout period.) As long as the timer is reinitialized before going through its full count range, the system operates as normal. However, if the watchdog timer ever rolls over its maximum count (or reaches zero, depending on whether it is an up or down counter), it can be assumed that somehow the system software has locked up and must be restarted. A hardware circuit detects the timer rollover and generates a system reset. This type of monitoring mechanism is often necessary in embedded systems because it may be inconvenient or impossible for a user to manually reset the system when trouble occurs. Not all computers are desktop workstations with convenient reset buttons. The processor may be buried deep within an automobile transmission or, worse, an interstellar space probe.

### 5.1.1.4   Nonvectored, vectored, and autovectored interrupts

Now that we have discussed various types of hardware interrupts and the approaches that can be used to assign priorities to those that are maskable, we should consider the process required to identify the correct service routine to run for a given type of interrupt. The process of identifying the source of the interrupt and locating the service routine associated with it is called interrupt *vectoring*. When an interrupt is recognized, the CPU suspends the currently running process and executes a service routine or handler to address the condition causing the interrupt. How does the CPU go about finding the handler?

The simplest approach would be to simply assign a fixed address in memory to the hardware interrupt request line. (If there are multiple interrupt request lines, a different fixed address could be assigned to each.) Whenever an interrupt occurs (assuming it is not masked), the CPU simply goes to that address and begins executing code from there. If there is only one interrupt-generating device in the system, this code would be the handler for that device. If there are multiple devices, this code would have to be a generic interrupt handler that would query all the devices in priority order to determine the source of the interrupt, then branch to the appropriate service routine for that device. This basic, no-frills scheme is known as a *nonvectored* interrupt.

The obvious advantage of nonvectored interrupts is the simplicity of the hardware. Also, where multiple interrupt sources exist, any desired priority scheme can be established by writing the handler to query device status in the desired order. However, having to do this puts more of a burden on the software, and modifying system software to change device priorities is not something that should be lightly done. This software-intensive approach also significantly increases the latency between an interrupt request and the time the device is actually serviced (especially for the lower priority devices). In addition, having a fixed address for the location of the interrupt service routine places significant limitations on the design of the memory system. (One must make sure that memory exists at that address, that it can be loaded with the appropriate software, and that no other conflicting use is made of that area of memory.) Because of their several drawbacks, nonvectored interrupts are seldom seen in modern computer systems.

If the hardware devices in the system are capable enough, the use of *vectored* interrupts is a much more flexible and desirable alternative. In a system with vectored interrupts, the CPU responds to each interrupt request (of sufficient priority to be recognized) with an *interrupt acknowledge sequence.* During this special acknowledge cycle (Figure 5.6 shows a typical sequence of events), the processor indicates that it is responding to an interrupt request at a particular priority level and then waits for the interrupting device to identify itself by placing its device number on the system bus.

Upon receiving this device number, the CPU uses it to index into an *interrupt vector table* (or, more generically, an *exception vector table*) in memory. In some systems, this table begins at a fixed address; in others, it can be placed at any desired location (which is identified by the contents of a *vector table base register* in the CPU). Figure 5.7 shows an overview of the process. The entries in the vector table are generally not the interrupt service routines themselves, but the starting addresses of the service routines (which themselves can be located anywhere in memory). Thus, the value obtained from the table is simply loaded into the machine's program counter (after the previous contents of the PC, status flags, etc., have been saved), and execution of the service routine begins from that location.

The main advantage of the vectored interrupt scheme is that the correct handler for any device can be quickly and easily located by a simple table lookup process with no additional software overhead required.

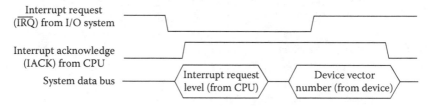

*Figure 5.6* Interrupt acknowledge sequence in a system with vectored interrupts.

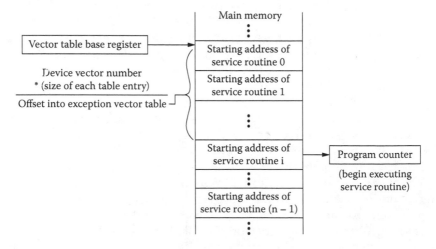

*Figure 5.7* Typical interrupt vectoring process.

Thus, the latency to respond to any type of interrupt of any priority level is essentially identical. Also, more flexibility is maintained because the interrupt service routines may reside anywhere in memory. New ones may be added or existing ones changed at any time by simply modifying the contents of the vector table. (Of course, this should be a privileged operation so that user programs cannot corrupt the interrupt system.) The only disadvantages are the slightly longer time required to determine the starting location of the service routine (because it is looked up instead of hardwired) and the increased complexity of the peripheral devices (which must be able to recognize the appropriate interrupt acknowledge sequence and output their device number in response). However, the extra cycle or two required to generate the address of the service routine is typically more than compensated for by the simplification of the handler software, and the additional logic needed in the peripheral device interfaces is generally not prohibitive.

If "dumb" devices are used in the system, a variation on the vectored interrupt scheme known as *autovectoring* can be used to preserve the advantages just discussed. In a system with *autovectored* interrupts, a device that is not capable of providing a vector number via the bus simply requests a given priority level interrupt while activating another signal to trigger the autovectoring process. When this occurs (or possibly when a given period of time elapses without a vector number being supplied) the CPU automatically internally generates a device number based on the interrupt priority level that was requested. (See Figure 5.8 for an example of an autovectoring sequence.) This CPU-generated device number is then used to index into the exception vector table as usual.

Vectored and autovectored interrupts can be, and often are, used in the same system. Devices that are capable of supplying their own vector numbers can make use of the full capabilities of vectored interrupts, and simpler devices can revert to the autovectoring scheme. This hybrid approach provides the system designer with maximum flexibility when a wide variety of different peripheral devices are needed.

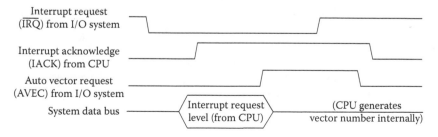

*Figure 5.8* Autovectored interrupt acknowledge sequence.

## 5.1.2   Software-related exceptions

The second major source of exceptions (in addition to hardware interrupts) is software running on the CPU itself. Various conditions can arise
from executing machine instructions that require at least the temporary
suspension of the program's execution; because of this, these conditions
are often best dealt with in a manner very similar to interrupts. Although
there are quite a variety of synchronous, software-related events that may
require attention, and although the number and details of these events
may vary significantly from one system to another, we can divide them
for discussion purposes into the general categories of traps, faults and
errors, and aborts.

*Traps*, also sometimes known as *software interrupts*, are the most common and least serious type of software-related exceptions. Traps are generally caused deliberately (either conditionally or unconditionally) by
programs in order to access operating system services (often for input or
output, but also for memory allocation and deallocation, file access, etc.).
In modern operating systems with virtual machine environments, user
programs run at a lower privilege level than system routines and are
generally not allowed to directly access hardware I/O devices because of
potential conflicts with other programs in memory. Thus, a program must
make a system call to request the OS to perform I/O (or other services
involving shared resources) for it. Traps provide a convenient mechanism
for this.

Almost any instruction set architecture one can think of has some
type of trapping instruction. A simple embedded microcontroller, such as
the Motorola 68HC12, has only one with the mnemonic SWI (for software
interrupt). The ubiquitous Intel x86 architecture defines an INT (interrupt) instruction with an op code followed by an 8-bit vector number
for a total of 256 possible software interrupts, each with its own exception vector. (In a given system, some of the vectors will be used for hardware interrupts, so not all 256 INT instructions are typically available
for use.) The Motorola 68000 implements 16 unconditional trap instructions, TRAP #0 through TRAP #15. The 68000 also has a conditional trap
instruction, TRAPV, which causes a trap only if the overflow flag in the
processor's condition code register (CCR) is set. (The 68020 and subsequent CPUs in that family defined additional TRAPcc instructions that
tested other conditions to decide whether or not to cause a trap.) Even
a reduced instruction set computer (RISC) architecture, such as SPARC,
defines a set of 128 trap instructions (half the total number of exceptions).
In each of these architectures, the process for handling traps, including
the vectoring/table lookup mechanism for locating the handlers, is essentially identical to that employed for hardware interrupts. The only difference is that instead of a device placing a vector number on the bus, the

vector number is generated from the op code of the trapping instruction (or a constant stored immediately following it); thus, the process is more akin to autovectoring.

The automatic vectoring process is important because it means user programs wanting to access operating system services via traps do not need to know where the system routine for a given service is located in memory. (If system services were accessed as subroutines, a user program wanting to call one would have to know or be able to calculate its address in order to access it.) The user program only needs to know which trap number to use for the service, plus any parameters or information that must be set up in registers or memory locations (such as the stack) to communicate with the trap handler. The exception vectoring and return process takes care of getting to the code for the service and getting back to the user program. This is a very significant advantage because the exact location of a given service routine is likely to vary from one system to another.

Although traps are synchronous events occurring within the CPU, they are otherwise similar to interrupts generated by external devices. In both situations, because the exception processing is due to a more or less routine situation rather than a catastrophic problem, control can generally be returned to the next instruction in the previously running procedure with no problems. In fact, because they are synchronous events resulting from program activity, traps are a bit easier to deal with than interrupts. Because the occurrence of traps is predictable, it may not be necessary for a trap handler to preserve the complete CPU state (as must be done to ensure transparency of hardware interrupts). Registers, flags, memory locations, etc., may be modified in order to return information to the trapping program, provided of course that any modified locations are documented.

*Errors and faults* are more serious than traps because they arise due to something the program has done wrong, for example, dividing by zero or trying to access a memory location that is not available (in other words, causing a page fault or segment fault in a virtual memory system). Other error conditions could include trying to execute an undefined op code, trying to execute a privileged instruction in user mode, overflowing the stack, overstepping array bounds, etc. Whatever it is, the error or faulting condition must be corrected (if possible) or at least adequately accounted for, before the offending program can be resumed. Depending on the particular error, it may not even be possible to complete the faulting instruction as is normally done with interrupts and traps. In order to service the faulting condition, the instruction being executed may have to be "rolled back" and later restarted from scratch, or stopped in progress and then continued—either of which is more difficult than handling an exception on an instruction boundary. Typically, when an error or fault occurs, processor state information is recorded in a predetermined area (such as the system stack). This information is then used to help the error or fault handler determine the cause of the problem and properly resume execution when the erroneous condition is corrected (for example, when a faulted page is loaded into physical main memory).

*Aborts* are the most critical software-related errors. They are generally unrecoverable, meaning that something catastrophic occurred while the program was running that makes it impossible to continue that program. Examples of aborts would include the 68000's "double bus error" (a hardware fault on the system bus that occurs during the handling of a previous bus error) and the x86's "double fault" (which arises when an exception occurs during the invocation of an exception handler, such as might occur if the exception vector table had been corrupted). In many cases, the problem is so serious that the CPU gets lost and it becomes impossible to tell the cause of the abort or which instruction was executing when it occurred. Because the program that was executing has been aborted, preservation of the processor state is not important for resumption of the program (although it could be worthwhile for diagnostic purposes). A well-written operating system running on a CPU in a protected, supervisor mode should be able to cleanly abort the program that crashed and continue running other programs (itself included). However, any user familiar with the "Blue Screen of Death" that is displayed by a certain family of operating systems knows that it is sometimes possible for an abort to crash the entire system and necessitate a reset.

## 5.2   Input and output device interfaces

Interfacing between the CPU and I/O devices is typically done via device interface registers that are mapped in an I/O address space. This space may be either common with, or separate from, the memory address space; we discuss the pros and cons of each approach in Section 5.3. Most I/O devices have associated control and status registers used to configure and monitor their operation along with one or more data input and output registers used to facilitate the actual transfer of information between the device and the CPU (or memory). Figure 5.9 shows a hypothetical device with four interface registers. Each of these device interface registers corresponding to an I/O *port* (or hardware gateway to a device) appears at a particular location in the address space, where it can be addressed by I/O instructions.

The principal reason I/O devices are interfaced via ports made up of interface registers is the difference in timing characteristics between the various types of devices and the CPU. The processor communicates with the memory devices and I/O ports over a high-speed bus that may be either *synchronous* (with a clock signal controlling the timing of all data transfers) or *asynchronous* (with a handshaking protocol.to coordinate transfers). In order for the main memory (and the fastest I/O devices)

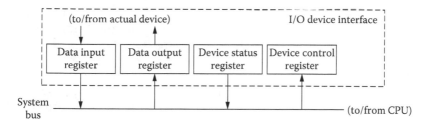

*Figure 5.9* Example of I/O device interface showing interface registers.

to perform well, this bus must be able to operate at the full bandwidth of memory. However, many I/O devices are slower—often considerably so—than memory devices. There also may be a delay between the time a device has data to transfer and the time the CPU is ready to receive it or vice versa. Because of these timing issues, there needs to be some sort of *buffering* or temporary storage between the bus and the device; a data register built into the device interface can serve this purpose. If several data items may accumulate in the interval over which the device is serviced, the interface may contain an enhanced buffer, such as a *first in, first out* (FIFO) memory. (Figure 5.10 shows an example of a device interface with FIFO buffering.) I/O ports composed of device interface registers provide the rest of the system with a standardized interface model that is essentially independent of the characteristics of any particular I/O device.

In addition to the basic functions of device monitoring and control and the buffering and timing of data transfers, I/O interfaces also sometimes perform additional duties specific to a particular type of device. For example, many interfaces provide *data conversion* between serial and parallel formats, between encoded and decoded representations, or even between digital and analog signals. Some may also perform *error detection*

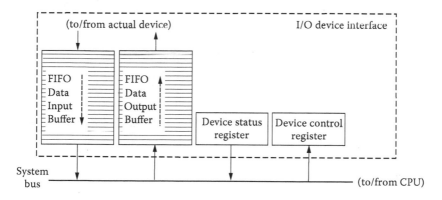

*Figure 5.10* Example I/O device interface with FIFO buffering.

and *error correction* on the data being transferred in order to enhance the reliability of communication. More sophisticated interfaces may serve as *device controllers*, translating commands from the CPU into specific signals needed by the hardware device. A classic example of this is a disk controller, which takes track/sector seek and read/write commands from the processor and derives motor drive signals that position the read/write head for accessing the desired data on the disk and the electrical signals needed by the head to read or write the data.

Given the typical I/O interface model presented in this section, how does the system go about performing transfers of data into or out of the machine? Several approaches, each with their own advantages and disadvantages, are possible. The CPU can directly perform input or output operations, either under dedicated program control or on demand (based on device-generated hardware interrupts). Alternatively, it is possible for the CPU to delegate the job of overseeing I/O operations to a special hardware controller or even to a second, independent processor dedicated to I/O operations. We examine each of these I/O methodologies in the next several sections.

## 5.3   *Program-controlled I/O*

The simplest, but most time-consuming (from the standpoint of the CPU), approach for performing I/O operations is to have the CPU monitor the status of each I/O device and perform the data transfer itself by executing the appropriate instructions when a given device is ready. The CPU determines whether each device is ready by periodically *polling* it (querying the appropriate bits in its status register). Because this monitoring and control of I/O devices is done by a program written for that purpose, it is referred to as *program-controlled* I/O. (Because the monitoring is done via status register polling, this method is also sometimes known as *polled* I/O.) For any given device, the code making up its part of the "polling loop" might look something like this:

| | | |
|------|------|------|
| MOVE | DEV1_STATUS_REG, R1 | ; get device status |
| AND | DEV1_STATUS_MASK, R1 | ; check to see if ready bit set |
| BZ | SKIPDEV1 | ; if not set, device not ready |
| MOVE | DEV1_DATA_REG, R2 | ; if device ready, read data |
| MOVE | DEV1_BUF_PTR, R1 | ; get pointer to device buffer |
| MOVE | R2, [R1] | ; copy data to buffer |
| INC | R1 | ; increment pointer |
| MOVE | R1, DEV1_BUF_PTR | ; store updated pointer |
| MOVE | DEV2_STATUS_REG, R1 | ; check next device |
| SKIPDEV1: | | |

This code checks the appropriate bit of an input device's status register to see if it has an item of data to transfer. If it does, the CPU reads the data from the input data register and stores it in a buffer in memory; if not, the CPU moves on and polls the next device. (A similar approach could be employed to check whether an output device is ready to receive data and, if so, to transfer the data to its output data register.) It would be possible to modify the code such that the status of a given device would be checked repeatedly until it was ready. This would ensure prompt servicing of that one device, but no other program activity (including polling of other devices) could occur until that device became ready and transferred its data. This would not be a desirable approach in any but the very simplest of embedded systems, if anywhere.

The advantage of program-controlled I/O is its obvious simplicity with respect to hardware. Unlike the other methods we examine, no devices or signals are required for data transfer other than the CPU, the I/O device's interface registers, and the bus that connects them. This lack of hardware complexity is attractive because it reduces implementation cost, but it is counterbalanced by increased complexity of the software and adverse implications for system performance. Code similar to the example shown above would have to be executed periodically for every device in the system—a task made more complex by the likelihood that some devices will need to transfer data more frequently than others. Any time spent polling I/O devices and transferring data is time not spent by the CPU performing other useful computations. If the polling loop is executed frequently, much time will be taken from other activities; if it is executed infrequently, I/O devices will experience long waits for service. (Data may even be lost if the device interfaces provide insufficient buffering capability.) Thus, program-controlled I/O is only useful in situations in which occasional transfers of small amounts of data are needed and latency of response is not very important, or where the system has very little to do other than perform I/O. This is unlikely to be the case in most general-purpose and high-performance computer systems.

In order to check I/O device status and perform input or output data transfers, the CPU must be able to read and write the device interface registers. These registers are assigned locations in an address space—either the same one occupied by the main memory devices or a separate space dedicated to I/O devices. The details of each of these approaches are discussed in the next sections.

## 5.3.1 Memory-mapped I/O

The simplest approach for identifying I/O port locations is to connect the device registers to the same bus used to communicate with main memory devices and decode I/O addresses in the same manner as memory addresses. (Figure 5.11 illustrates the concept.) When this technique, referred to as *memory-mapped* I/O, is used, there is only one CPU address space, which is used for both memory devices and I/O devices. Reads and writes of device interface registers are done with the same machine instructions (MOVE, LOAD, STORE, etc.) used to read and write memory locations.

The main advantage of memory-mapped I/O is its simplicity. The hardware is as basic as possible; only one bus and one address decoding circuit are required for a functional system. The instruction set architecture is also kept simple because data transfer instructions do double duty, and it is not necessary to have additional op codes for I/O instructions. This scheme is also flexible from a programming standpoint because any and all addressing modes that can be used for memory access can also

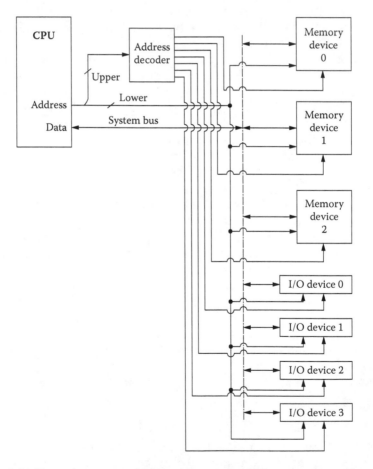

*Figure 5.11* Example system with memory-mapped I/O.

be brought to bear on I/O operations. These advantages led to the use of memory-mapped I/O in a wide variety of successful CISC and RISC architectures including the 680x0 and SPARC families.

On the downside, in a system with memory-mapped I/O, any addresses that are used for I/O device interface registers are unavailable for memory devices. In other words, I/O ports create "holes" in the memory address space, leaving less room for programs and data. (If there is an I/O device at address 400, there cannot also be a memory location with address 400.) Depending on how the I/O addresses are chosen, not all of the available physical memory space may be contiguous. This is admittedly less of a problem than it used to be because the majority of modern systems have very large virtual address spaces, and pages or segments that are not physically adjacent can be made to appear so in the virtual address space. Still, the addressing of I/O devices must be taken into account in the system design because all virtual references must ultimately be resolved to physical devices. Also, writing and examining machine-level code (for example, I/O drivers) can be somewhat more confusing, as it is difficult to tell the I/O operations from memory accesses.

## 5.3.2 Separate I/O

In contrast to systems using memory-mapped I/O, some architectures define a second address space to be occupied exclusively by I/O device interfaces. In this approach, known as *separate* I/O (sometimes referred to as *isolated* I/O), the CPU has a set of instructions dedicated to reading and writing I/O ports. The most popular example of this approach is the Intel x86 architecture, which defines IN and OUT instructions that are used to transfer data between CPU registers and I/O devices.

Separate I/O means that memory accesses and I/O transfers are *logically* separate (they exist in separate address spaces). They may or may not be physically separate in the sense of having distinct buses for I/O and memory operations as shown in Figure 5.12. Having separate buses would allow the CPU to access an I/O device at the same time it is reading or writing main memory but would also increase system complexity and cost. As we shall see later, it may be advisable in some cases to allow a direct connection between an I/O device and memory. This would not be possible in the dual-bus configuration unless there is a way to make a bridging connection between the two separate buses.

The other, more common, option for implementing separate I/O is to use one physical bus for both memory and I/O devices, but to provide separate decoding logic for each address space. Figure 5.13 shows how a control signal output from the CPU could be used to enable either the memory address decoder or the I/O address decoder, but not both at the same time. The state of this signal would depend on whether the processor

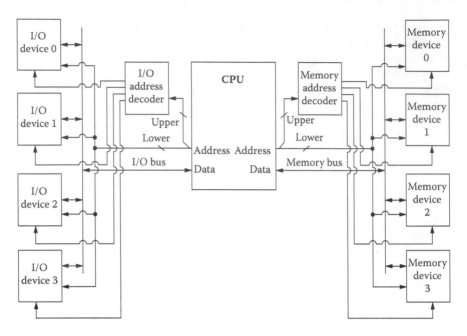

**Figure 5.12** System with separate buses for I/O devices and memory.

is currently executing an I/O instruction or accessing memory (either for instruction fetch or reading/writing data). Given this type of decoding scheme, it is possible for an I/O port numbered 400 to coexist with a similarly addressed memory location without causing a conflict, even when the same bus is used by the CPU to communicate with both devices. The bus is effectively *multiplexed* between the two types of operations.

The advantages and disadvantages of separate I/O are more or less the converse of those for memory-mapped I/O. Separate I/O eliminates the problem of holes in the physical address space that occurs when I/O ports are mapped as memory locations. All of the memory address space is available for decoding memory devices. It also makes it easy to distinguish I/O operations from memory accesses in machine-level programs (perhaps a nontrivial advantage given that not all code is well documented in the real world). Hardware cost and complexity may be increased, however; at the very least, some additional decoding logic is required. (If a completely separate bus is used for I/O, the cost increase may be considerable.) Defining input and output instructions uses up more op codes and complicates the instruction set architecture somewhat. Typically (this is certainly true of the Intel example), the addressing range and addressing modes available to the input and output instructions are more limited than those used with memory data transfer instructions. Although this takes some flexibility away from the

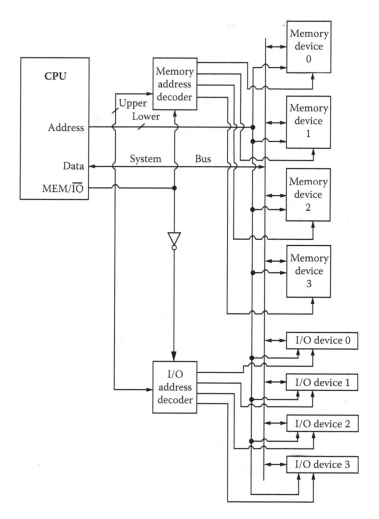

*Figure 5.13* Single-bus system with separate I/O address space.

assembly programmer (and the compiler), one could argue that most memory addressing modes are not really needed for I/O and that a typical system has far fewer I/O devices than memory locations. It is also worth noting that, although it is not feasible to implement separate I/O in a system designed to support memory-mapped I/O (because the needed instructions are not part of the ISA), one can always choose to memory-map devices in a system that supports separate I/O (because every architecture has memory data transfer instructions). In the final analysis, the fact that both of these approaches have been used in commercially successful systems with good I/O performance tends to imply that neither has a huge advantage over the other and that the choice between them is largely a matter of architectural taste.

## 5.4   Interrupt-driven I/O

The two big disadvantages of program-controlled I/O (whether memory-mapped or separate) are the inconvenience of, and time consumed by, polling devices and the overhead of the CPU having to execute instructions to transfer the data. In many cases, the first of these drawbacks is the more serious; it can be addressed by adopting an *interrupt-driven* I/O strategy. Interrupt-driven I/O is similar to program-controlled I/O in that the CPU still executes code to perform transfers of data, but in this case, devices are not polled by software to determine their readiness. Instead, each device uses a hardware interrupt request line to notify the CPU when it wants to send or receive data (see Figure 5.14). The interrupt service routine contains the code to perform the transfer(s).

The use of interrupts complicates the system hardware (and the CPU itself) somewhat, but interrupts are standard equipment on modern microprocessors, so unless a lot of external hardware devices (such as interrupt controller chips or priority encoders) are required, the incremental cost of using interrupts is minimal. Even an Intel x86-based PC with a single CPU interrupt request line (INTR) needs only two inexpensive 8259A interrupt controller chips cascaded together (see Figure 5.15) to collect and prioritize the interrupts for all its I/O devices. In modern machines, this functionality is usually incorporated into the motherboard chipset to further reduce cost and parts count.

Any slight additional hardware cost incurred by using interrupts for I/O devices is generally much more than offset by the simplification of system software (no polling loops), which also implies less overhead work for the CPU and therefore increased performance on non-I/O tasks.

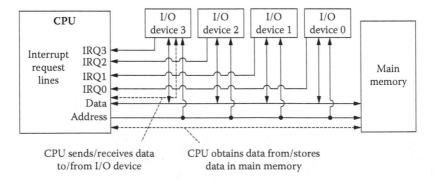

*Figure 5.14*  Interrupt-driven I/O concept.

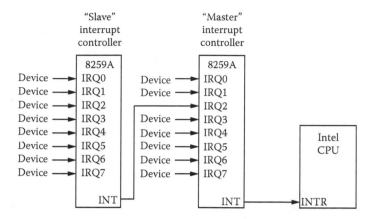

**Figure 5.15** Typical PC interrupt hardware configuration.

I/O code can be very efficient because each device can have its own interrupt handler tailored to its unique characteristics. (For best performance, there should be enough interrupt request lines that each device can have one dedicated to it; otherwise, service routines will have to be shared, and the processor will still have to perform some polling to distinguish between interrupt sources.) Executing the handlers still takes some time away from the CPU doing other things (such as running the operating system and user programs), but the elimination of polling means that no more processor time is spent on I/O than absolutely necessary. Also, devices typically are serviced with greatly reduced latency as compared to a polled implementation.

Interrupt-driven I/O is used to some extent in just about all types of computer systems, from embedded microcontrollers to graphics workstations to supercomputers. It is particularly suitable for general-purpose systems in which there are a variety of devices that require data transfers, especially when these transfers are of varying sizes and occur at more or less random times. The only situations in which interrupt-driven I/O is not a very good choice are those in which transfers involve large blocks of data or must be done at very high speeds. The following sections explore more suitable methods of handling I/O operations in such circumstances.

## 5.5   Direct memory access

In a system using either program-controlled or interrupt-driven I/O, the obvious middleman—and potential bottleneck—is the CPU itself. Unless input data are to be used immediately in a computation or unless results just produced are to be sent directly to an output device (both rare occurrences), data sent to or from an I/O device will need to be buffered for a time in main memory. Typically, the CPU stores data to be output in memory until the destination device is ready to accept it, and conversely reads data from a given input device when it is available and stores it in memory for later processing. Even a casual examination of this process will detect an inefficiency: In order to transfer data between main memory and an I/O device, the data goes into the CPU and then right back out of the CPU. If there were some way to eliminate the middleman and transfer the data directly between the device and memory, one would expect that efficiency of operations could be improved and the CPU would have more time to devote to computation. Fortunately, such a streamlined method for performing I/O operations does exist; it is known as *direct memory access* (DMA).

The reason program-controlled and interrupt-driven I/O are carried out by the CPU is that in a simple computer system such as the one illustrated in Figure 5.16, the CPU is the only device "smart" enough to oversee the data transfers. It is the only component capable of providing the address, read/write control, and timing signals necessary to operate the system bus and thus transfer data into or out of memory locations and device interface registers. In other words, it is the only *bus master* in the system. (A bus master is a device that is capable of "driving" the bus, or initiating data transfers.) Memory chips and most I/O devices are *bus slaves*; they never initiate a read or write operation on their own, but only respond to requests made by the current bus master. Thus, in this system, all data transferred from or to I/O devices must pass through the CPU.

In order to expedite I/O by performing direct transfers of data between two slave devices (such as main memory and an I/O port), there must be another device in the system, in addition to the CPU, that is capable of becoming the bus master and generating the control signals necessary to coordinate the transfer. Such a device is referred to as a *direct memory access controller* (DMA controller, or just DMAC). It can be added to the basic system as shown in Figure 5.17. It is also possible for individual devices or groups of devices to have their own DMACs in a configuration such as the one shown in Figure 5.18.

A typical DMAC is not a programmable processor like the system CPU, but a hardware state machine capable of carrying out certain

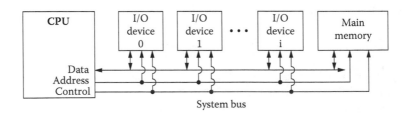

*Figure 5.16* Simple computer system without DMA capability.

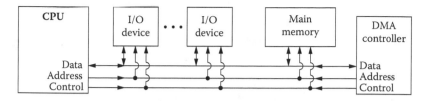

*Figure 5.17* Simple computer system with DMAC.

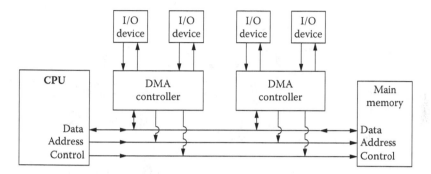

*Figure 5.18* Computer system with multiple DMACs.

operations when commanded by the CPU. Specifically, the DMAC is capable of generating the address and control and timing signals necessary to activate memory and I/O devices and transfer data over the system bus. By simultaneously activating the signals that cause a given device to place data on the bus and memory to accept data from the bus (or vice versa), the DMAC can cause a direct input or output data transfer to occur. Meanwhile, because the CPU is not occupied with the data transfer, it can be doing something else (presumably, useful computational work).

Because the DMAC is not a programmable device, it is not capable of making decisions about when and how I/O transfers should be done. Instead, it must be initialized by the CPU with the parameters for any given operation. (The CPU would do this when it became aware of a device's readiness to transfer data by polling it or receiving an interrupt.) The DMAC contains registers (see Figure 5.19 for a typical example) that can be loaded by the CPU with the number of bytes or words to transfer, the starting location of the I/O buffer in memory, the nature of the operation (input or output), and the address of the I/O device. Once these parameters are set up, the DMAC is capable of carrying out the data transfer autonomously. Upon completion of the operation, it notifies the CPU by generating an interrupt request or setting a status bit that can be polled by software. The simple DMAC shown in Figure 5.19 has only a single *channel*, meaning it can work with only one I/O device at a time. Integrated

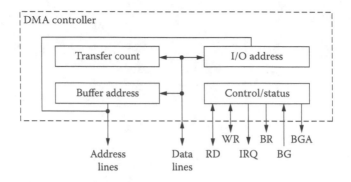

*Figure 5.19* Block diagram of typical DMAC.

circuit DMACs often have multiple channels and thus more than one set of control and status registers.

To the extent possible, the I/O operation using DMA proceeds simultaneously with normal code execution by the CPU. It is likely that the CPU will eventually want to read or write memory at the same time an I/O transfer is taking place. Unless main memory is *multiported* (meaning it has a second hardware interface so that more than one device can read or write it at a time) and a separate bus is provided for I/O, one or the other will have to wait because only one device can be bus master at a time. The decision as to which device will be bus master at any given time is made by *bus arbitration* logic, which may be implemented in a separate circuit or built into the CPU chip. Typically, any device wanting to become bus master asserts a hardware *bus request* signal to the arbitration logic (shown as BR in Figure 5.19), which responds with a *bus grant* (BG) signal when it is ready for that device to become bus master. The device acknowledges the grant, assumes control of the bus, and then relinquishes control of the bus when it is done. The DMAC shown in Figure 5.19 would do this by activating and then deactivating its *bus grant acknowledge* (BGA) signal; see Figure 5.20 for a timing diagram illustrating the sequence of events.

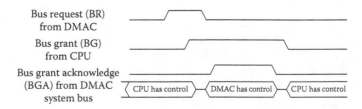

*Figure 5.20* Bus arbitration sequence for DMAC becoming bus master.

Usually, priority for bus mastership is given to the DMAC because it is easier for the CPU (as compared to an I/O device) to wait a few cycles and then resume operations. For example, once a disk controller starts reading data from a disk sector into memory, the transfers must keep up with the data streaming off the disk, or the controller will have to wait for the disk to make a complete rotation before continuing. Also, the CPU normally has an on-chip cache so that it can often continue executing code for some time without needing to read or write main memory, whereas all the I/O transfers involve main memory. (If the DMAC has an internal buffer, it is possible to transfer data directly from one I/O device to another without using main memory, but this mode is rarely if ever needed in most systems.) Depending on the characteristics (particularly speed) of the particular I/O device involved and the relative priority of the transfer compared to CPU activity, DMA may be done in *burst mode* or *cycle stealing mode*. In burst mode (as depicted in Figure 5.20), the DMAC assumes exclusive ownership of the bus for the duration of the I/O operation; in cycle stealing mode, transfers of I/O data are interwoven (time-multiplexed) with CPU bus activity. If DMA does not use the full bandwidth of the bus or the CPU needs relatively few bus cycles, the I/O operation may be done "for free" or nearly so, with little or no impact on processor performance.

Direct memory access is the most complex way of handling I/O we have studied so far; in particular, it requires more hardware than the other approaches. (After all, the CPU still needs all the circuitry and connections to be able to poll devices and receive interrupts to be aware of the need for a transfer to occur, but the DMAC is an additional device that adds its own cost and complexity to the system.) There is also the increased software overhead of having to set up the DMA channel parameters for each transfer. Because of this, DMA is usually less efficient than I/O performed directly by the CPU for transfers of small amounts of data.

However, large block transfers to or from a relatively small number of devices are easily, efficiently, and quickly handled via DMA. Its real power is seen when dealing with devices such as high-speed network communication interfaces, disk drive controllers, printers, and so forth. An entire disk sector, printer buffer, etc., can be conveniently transferred in (as far as the CPU is concerned) a single operation. These large blocks of data are normally transferred in or out at least twice as quickly as if the CPU performed them directly because each byte or word of data is only moved once (directly to or from memory) rather than twice (into and then out of the CPU). In practice, the speed of DMA may be several times that of other I/O modes because DMA also eliminates the need for the CPU fetching and executing instructions to perform each data transfer. This considerable speed advantage in dealing with block I/O transfers has made DMA a standard feature, not just in high-performance systems, but in general-purpose machines as well.

## 5.6   Input/output processors

Direct memory access is an effective way of eliminating much of the burden of I/O operations from a system's CPU. However, because the DMAC is not a programmable device, all of the decision making and initial setup must still be done by the CPU before it turns the data transfer process over to the DMAC. If the I/O requirements of a system are very demanding, it may be worthwhile to offload the work from the main CPU to an *input/output processor* (I/O processor or simply IOP). I/O processors are also sometimes called *peripheral processors* or *front-end processors*.

An I/O processor is simply an additional, independent, programmable processor that oversees I/O; it is able (among other things) to perform data transfers in DMA mode. Although the I/O processor could be another general-purpose processor of the same type as the system CPU, it is typically a special-purpose device with more limited capabilities, optimized for controlling I/O devices rather than normal computing tasks. Thus, although a system that uses one or more I/O processors is technically a parallel or *multiprocessor* system, it is considered a *heterogeneous* multiprocessor rather than the more typical *homogeneous* multiprocessor (such as the ones we will discuss in the next chapter) in which all CPUs are of the same type and are used to do more or less the same type of work. An I/O processor typically controls several devices, making it more powerful than a DMAC, which is typically restricted to one device per DMA channel. Because I/O processors are programmable, they are flexible and can provide a common interface to a wide variety of I/O devices. Of course, this power and flexibility come at a greater cost than is typical of nonprogrammable DMACs.

I/O processors as a separate system component date back to the IBM mainframes of the late 1950s and 1960s. IBM referred to its I/O processors as *channel processors*. These devices were very simple, von Neumann–type processors. The channel processors had their own register sets and program counters but shared main memory with the system CPU. Their programs were made up of *channel commands* with a completely different instruction set architecture from the main CPU. (This instruction set was very simple and optimized for I/O.) Channel processors could not execute machine instructions, nor could the CPU execute channel commands. The CPU and the channel processor communicated by writing and reading information in a shared *communication area* in memory. The I/O channel concept worked well for the types of applications typically run on mainframe computers, and IBM used this approach in many successful systems over a number of years.

A somewhat different approach was adopted by Control Data Corporation (CDC), whose large systems competed with IBM's high-end machines for many years. Starting in the 1960s, CDC's machines used I/O processors known as *peripheral processing units* (PPUs). These were not just autonomous processors sharing main memory like IBM's channels; they were simple computers, complete with their own memory that was separate from system main memory, dedicated to I/O operations. The PPUs had more complete instruction sets than were typical of channel processors; they were architecturally similar to a main CPU but with unnecessary features, such as floating-point arithmetic, omitted. Besides control of I/O devices, the PPUs could perform other operations, such as data buffering, error checking, character translation, and data formatting and type conversion. Although the name has not always been the same, this type of independent I/O processor with its own local memory has been used to good effect in many other systems over the years.

## 5.7    Real-world I/O example:
## the universal serial bus

Now that we have studied the basic techniques used to interface input and output devices to a system's CPU and memory, it is a good time to reinforce some of those concepts by looking at a practical example. Without a doubt, the most commonly used I/O interface in modern general-purpose computing systems is the Universal Serial Bus, or USB. This flexible interface can be used with a wide variety of peripheral devices, including keyboards, mice, printers, external hard drives, digital cameras, GPS units, and many others. USB was introduced in the mid-1990s as a way to standardize not only communication with, but also distribution of electrical power to the myriad devices that might be connected to a computer.

Before the introduction of USB, computer users had to deal with a number of different interface standards, each with its own connectors, cables, data transfer speeds, and peculiarities. The most commonly used interfaces for home and small business systems before 1996 were RS-232 serial ports and parallel printer ports (originally developed by Centronics and later standardized as IEEE 1284). These interfaces did not provide for sending power to a connected device, so the vast majority of peripherals had to have their own power supplies, adding cost, cords, and inconvenience to a typical system setup. The serial ports were painfully slow at transferring data and the parallel ports, although somewhat faster, initially allowed only unidirectional communication (although later versions allowed data transfer in both directions). Because of the (usually 8-bit) parallel data transfer path and the large, cumbersome connectors, the data cables were unwieldy and prone to being damaged. USB addressed all these issues with much more compact connectors and a slim, 4-wire shielded cable that delivered +5 volt power, ground, and a differential serial data connection (versions 1.0, 1.1, and 2.0). Versions 3.0 (introduced in 2008) and 3.1 (2013) have more signals and thus more complex cables, but support much higher data transfer speeds. Here we will concentrate mainly on the older, but still commonly used, versions (up through USB 2.0); the newer versions retain compatibility for use with less expensive and older devices.

A system that uses USB has a host machine (desktop or laptop computer, game console, set-top box, etc.) and some number of connected peripheral devices. The standard allows for up to 127 devices to be connected at a time, although practical considerations may limit the number to considerably less than that. Most laptops come with two or three USB connectors, and desktop machines may have 4, 6, or more. However, the USB standard allows for easy expandability to more ports by connecting a USB hub(s) to an existing USB port. Unpowered hubs allow data communication with multiple devices, but do not provide any additional charging capability beyond that of the single port they connect to (+5V at a maximum 500 mA of current, split among all devices connected to the hub), so they are only suitable for devices such as printers and scanners that have their own power supplies. Powered hubs provide multiple connection sockets, each with the ability to supply full +5V, 500 mA (2.5 W) power to an attached device, so they can be used with any USB-enabled peripheral. By cascading a sufficient number of hubs to create what is known as a *tiered star* topology (see Figure 5.21) with up to six levels,

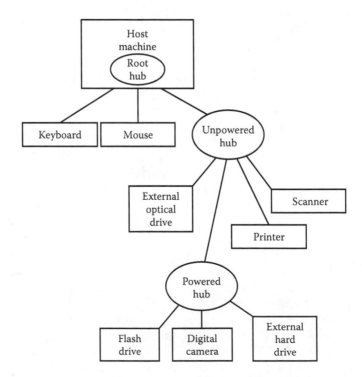

*Figure 5.21* Typical USB connection topology with multiple hubs.

a user could potentially build up to the maximum of 127 devices in a system.

There are two different classes of connectors, known as Type A and Type B, that are used in systems with USB. Type A connectors are intended for connecting to the host machine, which will have some number of the familiar-looking rectangular Type A sockets. The host end of a USB cable will thus normally have a Type A plug that fits into this socket. (This is the end of the cable from which power will be supplied if the connected peripheral requires it.) The opposite end of the cable will either be captive (as is usually the case with keyboards and mice) or will have one of several sizes of Type B plugs; the specific one chosen will depend on the socket present on the device one wishes to connect. The most common "B" connectors are standard, mini, and micro; frequently used cable end connector shapes are depicted in Figure 5.22. Note that connectors built to work with USB versions 1 and 2 are usually black (or the same color as the cable itself), while blue is normally used to identify connectors that support the higher-speed version 3 standard. The cables themselves can be up to 5 meters long; because there can be up to six levels of connections separated by hubs, a connected device can be up to 30 meters from the host.

What is the reason for the limitation on the number of ports, how are specific connected devices identified, and how does data communication between a device and the host actually take place? To answer these questions, we must look at some details of the USB architecture. First of all, we must recall that the "S" in USB stands for "serial." This means that all data, regardless of the word size used by the host computer or the peripheral, are transmitted one bit at a time over a pair of wires (Data+ and Data−, or more simply, D+ and D−). Multibit data values are transmitted least significant bit first. The data signal is *differential*, which means that both D+ and D− are driven with (usually) complementary electrical signals (as opposed to some other *single-ended* interfaces in which there is only one driven data signal, which is referenced to ground potential or 0 volts). The differential data transmission, along with the use of a twisted pair of wires inside of a shielded cable, helps USB resist data transmission errors due to noise and outside interference.

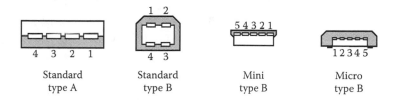

*Figure 5.22* Common USB cable connector shapes (not to scale).

D+ and D– are not always in opposite states. When no device is connected to a given port (known as the *detached* condition), both of them are pulled down by the host to a low voltage state. When a device is *attached*, it will pull one of the data signal lines (usually D+, although *Low Speed* devices use D–) to a higher voltage. Anytime no data are being transmitted, the lines will remain in this *idle* state. During the actual transmission of data, the D+ and D– lines can be in one of three states: J, K, and SE0 (*single-ended zero*). Voltage-wise, J is the same as the idle state, while K has the two lines in the opposite polarity. SE0 means that both lines are at a low voltage. The fourth possible condition, SE1 (*single-ended one*, or both lines at a high voltage) is illegal and should never occur unless there is a hardware malfunction.

Interestingly, the data bits being transmitted are not represented directly by the J and K states, but rather by the transitions between them. A binary 0 is transmitted by toggling from the J state to the K state or vice versa at the start of a new transmission time interval, and a binary 1 is transmitted by leaving the lines in their previous state through the next time interval. (To prevent loss of synchronization due to lack of signal transitions because of a long sequence of ones in the data, a technique known as *bit stuffing* is used; anytime six consecutive 1 bits have occurred, the sender artificially inserts a 0, which is discarded by the receiver.) This method of data transmission is known as *non-return-to-zero, inverted* (NRZI) encoding; an example is shown in Figure 5.23.

The time interval at which the D+ and D– lines can toggle (or not toggle when binary 1 is being transmitted) is governed by the applicable bit transfer rate. Low-speed transfers (typically used for slow human interface devices such as keyboards and mice) occur at 1.5 megabits per second, so the toggling interval in that case is 1/(1.5 MHz) = 666.7 nanoseconds. The signaling rates for the faster transfer modes are summarized in Table 5.1.

Data transfers do not normally achieve the maximum possible speed based on the signaling rate because there is always some communications

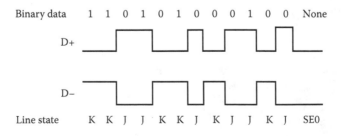

**Figure 5.23** NRZI-encoded data transmission showing line states and encoded binary values.

*Table 5.1* Data signaling rates supported by USB

| Transfer type | USB versions | Transfer rate (Mbit/s) | Bit transmission interval (ns) |
|---|---|---|---|
| Low Speed | 1.0 and above | 1.5 | 666.7 |
| Full Speed | 1.0 and above | 12 | 83.3 |
| High Speed | 2.0 and above | 480 | 2.083 |
| SuperSpeed | 3.0 and above | 5000 | 0.2 |
| SuperSpeed+ | 3.1 | 10,000 | 0.1 |

overhead, the amount and nature of which will depend on the type of transfer being done. There are three types of data transfers in a USB system; the type used by a particular device will depend on its characteristics. *Interrupt* transfers are typically used for human interface devices, such as keyboards and mice. These devices transfer only a small amount of data at a time (e.g., tracking information and button/key presses), but it is important for this information to be promptly brought to the attention of the host so the user interface can respond with minimum latency. *Isochronous* transfers are used for continuous, real-time streaming of data, such as audio or video over a USB connection. These data streams are guaranteed some fixed transfer bandwidth by the host, although this will not usually be the maximum supported bandwidth (as there will typically be other devices in the system that also need to transfer data); no error checking is performed on the data. Finally, devices such as printers, hard disks, and flash drives that transfer large amounts of data, but only sporadically (rather than continuously) usually make use of USB's *bulk* data transfer mode. Bulk transfers are error-checked to ensure accuracy of data transmission and can use any or all of the total communications bandwidth that remains after isochronous and interrupt transfers are taken care of. However, neither the amount of bandwidth nor the latency of any particular bulk transfer can be guaranteed.

The reason that overall data transfer bandwidth must be split among the various devices in the system is that only one data transfer, from the host to a specific peripheral or vice versa, can occur at a time. If the peripheral device is not connected directly to the host, the intervening hubs act as repeaters, forwarding data packets between their "upstream" (toward the host) and "downstream" (toward the peripheral) sides as appropriate.

Because the host communicates with only one specific device at a time, there obviously needs to be a mechanism for distinguishing among peripheral devices by assigning them unique identifiers (addresses). This is done during the process of *enumeration*. When a USB host first powers up (and again when new devices are plugged in), the host detects connected devices by sensing that they have pulled up the voltage on D+ or D–.

The host sends a *USB reset* (SE0 for at least 10 milliseconds) to initialize the newly discovered device to its default state. In this state, the device will initially respond to address 0. (The host only resets one device at a time, so there will never be duplicate devices with address 0.) The host then sends standardized commands to query the device for its *descriptor* and, subsequently, other configuration information. Among the information included in a device descriptor is its maximum packet size and its device class, which is identified by an 8-bit code. Some of the common classes are communications devices (for example Ethernet and Wi-Fi adapters and modems, class $02_{16}$), human interface devices (e.g., keyboards and mice, class $03_{16}$), printers ($07_{16}$), mass storage devices (flash drives, card readers, etc., $08_{16}$), USB hubs ($09_{16}$), and wireless controllers (e.g., Bluetooth, class $E0_{16}$). During the configuration process, the host transmits a Set Address command to the device; from this point on, it will respond to that (non-zero) assigned address instead of address 0. The limitation to 127 devices in a system comes from the fact that the address field in USB token packets (which are used for device setup as well as initiating data transfers) is 7 bits long, which gives $2^7 = 128$ combinations. Because address 0 is reserved for the special case of device initialization, 127 addresses remain to uniquely identify devices that have been configured by the host.

As part of the initialization process, the host queries each device regarding what type of data transfers (interrupt, isochronous, or bulk) it wants to perform and how much data transfer bandwidth it needs. Interrupt and isochronous devices can be allocated up to 90% of the total bandwidth of a Full Speed bus or up to 80% on a High Speed bus; the remainder is reserved for control packets and bulk transfers. Once the maximum percentage of available bandwidth has been reserved, the host will refuse to recognize any additional interrupt or isochronous devices. (For this reason, the 127-device limitation is rarely an issue; the overall USB system tends to run out of data transfer bandwidth before it runs out of addresses.) Bandwidth on Low Speed and Full Speed buses is allocated within *frames* that last 1 millisecond each; a High Speed connection uses *microframes* of 125 microseconds each. A given isochronous or interrupt device will be given up to a certain portion of each frame (or microframe) for data transmission and then must wait until the next (micro)frame to transfer additional data.

Within each frame, the individual communications between the host and each specific device are known as *transactions*. There are three types of transactions: SETUP, IN, and OUT. SETUP transactions are used to send control packets to devices, for example during the initial configuration process. IN transactions are used for transmission of data from a peripheral to the host, and OUT transactions are used to send data from the host to a peripheral.

Each USB transaction consists of a sequence of (normally) three packets: a *token* packet, a *data* packet, and a *handshake* packet. From a signaling standpoint, all three packet types begin with a "sync" pattern (to synchronize the receiver's clock signal to the transmitted bits) followed by a Packet Identifier (PID) that identifies the packet type; all packets end with an end of packet (EOP) sequence consisting of SE0 for two bit times and then the J state for one bit time. They differ in what comes between the PID and EOP.

Token packets act as message headers; they include the device address, a 4-bit endpoint field, and a 5-bit cyclic redundancy check (CRC) field for error detection. One physical device can have up to 16 different *endpoints*, which are individual sources or destinations for data and may not necessarily all transfer data the same way. For example, USB-connected stereo speakers might have two isochronous endpoints for transmission of left/right channel audio, plus an interrupt endpoint for communicating control settings, such as volume, balance, etc. Data packets include up to 1,024 bytes of data (depending on bus speed and transfer type) and a 16-bit CRC field. Handshake packets are sent by the device receiving information (the peripheral for SETUP and OUT transactions, the host for IN transactions) and contain only a PID. The four possible handshake PID values are ACK (acknowledging receipt of an error-free data transmission); NAK (negative acknowledge, meaning that an error occurred, the device is temporarily unable to transfer data, or has no data to transmit); STALL (device needs host intervention); and (High Speed only) NYET, which means "no response yet" from the receiving device. It is worth noting that isochronous data transfers do not use handshake packets as no error checking is done.

One other interesting point is that despite their name, interrupt transfers are not actually initiated by sending an interrupt to the host CPU. All USB data transfers, whether upstream or downstream, are initiated by the host's built-in (root) USB hub, not the peripheral. "Interrupt" transfers actually occur when the root hub polls the peripheral device for status, which happens at least at some minimum rate established when the device is first configured. The peripheral responds either by transferring data, or by sending a NAK packet if it has no data to transmit. The only USB-related interrupts to the host CPU will come from the root hub, which effectively acts as an intermediary between the CPU and all USB-connected devices. These interrupts can be used to trigger data transfers performed by the CPU under program control or directly to/from memory using DMA, depending on the volume of data being sent or received.

There are many more details involved in the operation of an input/output system based on USB, details that are not necessary for an end user to make use of its speed, flexibility, and ability to provide power to attached devices but would need to be understood when developing an operating system or writing drivers for devices to be connected via USB. These details will continue to change as new versions of the USB standard are introduced in the future. Although further elaboration is beyond the scope of this book, the reader should now have sufficient knowledge to be able to understand the role of USB in the I/O system of a modern computer and to pursue further study of the subject if the need should arise.

## 5.8   Chapter wrap-up

It has been said of human relationships that "love makes the world go 'round." Love, of course, is based on interaction with another person, and anyone who has been in such a relationship knows that communication is the key to success. So it is with computer systems as well. A system may have the fastest CPU on the market; huge caches; tons of main memory; and big, fast disks for secondary storage, but if it cannot effectively communicate with the user (and other devices as needed) it will not get any love in terms of success in the marketplace. Simply put, good I/O performance is essential to all successful computer systems.

In very simple systems, it may be possible for the CPU to directly oversee all I/O operations without taking much away from its performance of other tasks. It can do this by directly reading and writing I/O ports, which may be located in the memory address space or in a separate I/O address space. As systems become more complicated and more I/O devices must be supported, it becomes helpful—even necessary—to add complexity to the hardware in the form of an interrupt handling mechanism and support for DMA. Hardware interrupts generated by I/O devices requiring service are one specific example of a more general class of events known as exceptions. Trapping instructions (software interrupts), another exception type, are also important to I/O operations as they allow user programs to readily access I/O routines that must execute at a system privilege level. Handling all types of exceptions effectively and efficiently (including saving CPU state information, vectoring to and returning from the handler, and restoring the saved state) is a complex and important task for modern computer systems and one that must be done well to achieve good performance from the I/O system.

Although I/O performed by the CPU under program control or in response to interrupts is often sufficient for devices that transfer small amounts of data on a sporadic basis, it is not very efficient (nor is it a good use of CPU cycles) to handle block I/O transfers this way. A better approach uses the technique of direct memory access to copy data between memory and an I/O device in a single operation, bypassing the CPU. Not only can the transfer proceed more rapidly, but the CPU is freed up to perform other computations while the I/O operation proceeds. Almost all modern computers are capable of using DMA to send and receive large blocks of data quickly and efficiently. Some systems even go beyond the typical, "dumb" DMAC and employ dedicated, programmable I/O processors to offload work from the main system processor. Such systems, at the cost of significant extra hardware, maintain the advantages of DMA while adding flexibility and the power to control more I/O devices.

All I/O systems, from the simplest to the most complex (including those based on the versatile USB standard), are based on the concepts covered in this chapter. By studying and understanding these concepts, the reader will be better prepared to evaluate I/O techniques in existing systems and those of the future. Although the specific solutions change over time, moving bits from one place to another is a problem that never goes away.

## REVIEW QUESTIONS

1. What do we mean when we say that interrupts must be processed transparently? What does this involve and why is it necessary?

2. Some processors, before servicing an interrupt, automatically save all register contents. Others automatically save only a limited amount of information. In the second case, how can we be sure that all critical data are saved and restored? What are the advantages and disadvantages of each of these approaches?

3. Explain the function of a watchdog timer. Why do embedded control processors usually need this type of mechanism?

4. How are vectored and autovectored interrupts similar and how are they different? Can they be used in the same system? Why or why not? What are their advantages and disadvantages compared with nonvectored interrupts?

5. Given the need for user programs to access operating system services, why are traps a better solution than conventional subprogram call instructions?

6. Compare and contrast program-controlled I/O, interrupt-driven I/O, and DMA-based I/O. What are the advantages and disadvantages of each? Describe scenarios that would favor each approach over the others.

7. Systems with separate I/O have a second address space for I/O devices as opposed to memory and also a separate category of instructions for doing I/O operations as opposed to memory data transfers. What are the advantages and disadvantages of this method of handling I/O? Name and describe an alternative strategy and discuss how it exhibits a different set of pros and cons.

8. Given that many systems have a single bus that can be controlled by only one bus master at a time (and thus the CPU cannot use the bus for other activities during I/O transfers), explain how a system that uses DMA for I/O can outperform one in which all I/O is done by the CPU.

9. Compare and contrast the channel processors used in IBM mainframes with the PPUs used in CDC systems.

10. Name and describe the three types of data transfers that can be used for USB-connected input/output devices. Give an example of where each would be appropriate and explain why.

11. Suppose a USB data packet being transmitted at Full Speed consists of a sync pattern that lasts 8 bit times, an 8-bit PID, 1024 bytes of data, and 16 bits of CRC information. What is the minimum time required for the complete packet to be transmitted? Why might it take longer?

12. Fill in the blanks below with the most appropriate term or concept discussed in this chapter:

_____ A synchronous or asynchronous event that occurs, requiring the attention of the CPU to take some action.

_____ A special program that is run in order to service a device, take care of some error condition, or respond to an unusual event.

_____ When an interrupt is accepted by a typical CPU, critical processor status information is usually saved here.

_____ The highest priority interrupt in a system, one that will never be ignored by the CPU.

_____ A signal that causes the CPU to reinitialize itself and/ or its peripherals so that the system starts from a known state.

_____ The process of identifying the source of an interrupt and locating the service routine associated with it.

_____ When this occurs, the device in question places a number on the bus that is read by the processor in order to determine which handler should be executed.

_____ Another name for a software interrupt, this is a synchronous event occurring inside the CPU because of program activity.

_____ On some systems, the "blue screen of death" can result from this type of software-related exception.

_____ These are mapped in a system's I/O address space; they allow data and/or control information to be transferred between the system bus and an I/O device.

_____ A technique that features a single, common address space for both I/O devices and main memory.

_____ Any device that is capable of initiating transfers of data over the system bus by providing the necessary address, control, and/or timing signals.

_____ A hardware device that is capable of carrying out I/O activities after being initialized with certain parameters by the CPU.

_____ A method of handling I/O in which the DMAC takes over exclusive control of the system bus and performs an entire block transfer in one operation.

_____ An independent, programmable processor that is used in some systems to offload input and output activities from the main CPU.

_____ The type of connector that is used for the host end of a USB cable.

_____ A device that allows expansion of a USB system to include additional ports; during data transmission, it acts as a repeater, forwarding packets between its upstream and downstream connections as needed.

_____ The encoding scheme that is used to serially transmit bits over a USB connection, in which state transitions represent bits that are 0, and the lack of a transition represents a 1.

_____ A type of USB packet that is used to set up a device or start an IN or OUT data transmission.

## chapter six

# Parallel and high-performance systems

In Chapter 4, we explored a number of techniques, including pipelining, that can be used to make a single processor perform better. As we discovered, pipelining has its limits in terms of improving performance. Instruction execution can only be divided into so many steps, and operations such as memory accesses and control transfers (as well as dependencies between instructions) can cause delays. Superpipelined, superscalar, and very long instruction word (VLIW) designs are implementation and architectural approaches that have been used to overcome (to some degree) the difficulties inherent in extracting performance from a single processor, but each of these approaches has its own costs and limitations. Ultimately, given the level of implementation technology available at any point in time, designers can make a CPU execute instructions only so fast and no faster. If this is not fast enough for our purposes—if we cannot get the performance we need from a system with a single CPU—the remaining, obvious alternative is to use multiple processors to increase performance. Machines with multiple processing units are commonly known as *parallel processing* systems, although a more appropriate term might be *concurrent* or *cooperative* processing.

It should come as no surprise to the reader that there have been, and still are, many types of high-performance computer systems, most of which are parallel to some extent. The need for high-performance computing hardware is common across many types of applications, each of which has different characteristics that favor some approaches over others. Some algorithms are more easily parallelized than others, and the nature of the inherent parallelism may be quite different from one program to another. Certain applications, such as computational fluid dynamics (CFD) codes, may be able to take advantage of *massively parallel* systems with thousands of floating-point processors. Others, for example, game tree searching, may only be able to efficiently use a small number of central processing units (CPUs), and the operations required of each one may be quite different than those required of the CFD machine. Thus, a wide variety of systems ranging from two to tens of thousands of processors have been built and have found some degree of success.

Parallel systems are not only distinguished by the number and type of processors they contain, but also by the way in which these processors are connected. Any task carried out by multiple processors will require some communication; multiple CPUs with no interconnection between them can only be considered as separate machines, not as a parallel system. Indeed, the communications networks used to interconnect parallel systems are a key component—often the most important aspect—of the overall system design. Of course, the amount of communication required varies widely across applications, so a number of different techniques can be used to implement it. Thus, a large portion of this chapter (after the first section on the classification of high-performance and parallel systems and their characteristics) will be devoted to the important topic of interconnection networks for parallel systems.

## 6.1   Types of computer systems: Flynn's taxonomy

A variety of computer architectures, including parallel ones, have existed for many years—much longer than many modern computer professionals realize. The first parallel systems, including the Univac LARC, Burroughs D825, and IBM Sage, appeared in the late 1950s (scarcely 10 years after the first programmable computers were built). Within a few years, several types of architectures that are still in use today emerged. In 1966, Michael Flynn published an article in which he defined four general classifications of computer systems, three of which were parallel in some sense of the word. Five decades later, Flynn's classifications are still widely used to describe the computers of today. With the exception of a few specialized architectures (some of which are discussed in Chapter 7) that were developed after Flynn's work, almost all practical systems still fall into one of his four categories:

1. *Single Instruction Stream, Single Data Stream (SISD)* machines have a single CPU that executes one instruction on one operand (or pair of operands, such as those for a typical scalar arithmetic or logical operation) at a time. Nonparallel machines built around a single processor with a von Neumann (Princeton) or modified Harvard architecture, like most of the ones we have studied so far, fall into this category. Although pipelined and superscalar machines internally process multiple instructions simultaneously, they appear to process one instruction at a time and thus are still generally considered to be SISD architectures. This type of system with a single processor (which executes a single stream, or thread, of machine instructions) is also known as a *uniprocessor* system.

2. *Single Instruction Stream, Multiple Data Stream (SIMD)* machines execute a single, sequential program, but each instruction specifies

an operation to be carried out in parallel on multiple operands (or sets of operands). Typically, these multiple operands are elements of an array of values; thus, SIMD machines are often known as *array processors*. The SIMD architecture generally implies a machine with a single control unit but many datapaths (processing elements), although a pipelined *vector processor* is very similar from a software point of view. This architectural classification is discussed in more detail in Section 6.1.1.

3. *Multiple Instruction Stream, Single Data Stream (MISD)* machines were described by Flynn, presumably for symmetry with the remaining categories, but they did not exist at the time of his writing— nor have they proliferated since. An MISD machine would be one that performed several different operations (or executed multiple sequences of instructions) on the same stream of operands. It is generally agreed that no commercially available machine is a true MISD system. However, one could argue that some *vector processors*, particularly those with *vector chaining*, and some special-purpose architectures, such as *artificial neural networks* and *dataflow machines* (covered in Chapter 7), resemble, at least to some extent, MISD architectures as described by Flynn.

4. *Multiple Instruction Stream, Multiple Data Stream (MIMD)* machines simultaneously execute different programs (instruction streams) on different sets of operands. Such a system is composed of more than one SISD-type (von Neumann or Harvard architecture) CPU connected in such a fashion so as to facilitate communication between the processors. This is the most common type of parallel system in existence today. Depending on the nature of inter-processor communication, MIMD systems may be known as *multiprocessor* systems (Section 6.1.3) or *clusters* (a.k.a. multicomputer systems, Section 6.1.4); hybrids of the two classifications are also possible.

Figure 6.1 breaks down the various classifications of computer systems using Flynn's taxonomy as a guide (although some other types of systems are shown as well). Note that all the systems except SISD machines are parallel in some sense of the word. The characteristics of SISD systems are discussed in some detail in the first five chapters of this book. The few systems with MISD-like characteristics are pointed out as appropriate in the discussion that follows. The vast majority of parallel systems in use today are built around an MIMD or SIMD architecture (or a hybrid of the two). Thus, the rest of this chapter deals primarily with those types of systems. The following sections describe in more detail several common categories of high-performance computer architectures that are introduced above.

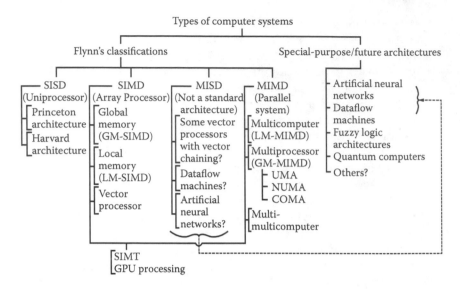

**Figure 6.1** Classifications of computer architectures.

### 6.1.1    Vector and array processors

The first types of parallel architectures we shall examine are those in which individual machine language instructions can operate on multiple-element data structures (vectors or arrays) instead of being limited to scalar operands (individual operands or pairs of operands). In terms of computational mathematics, an *array* is just an orderly arrangement of numerical values in one or more dimensions. A *vector* is a one-dimensional array; a *matrix* is a two-dimensional (rectangular) array consisting of (horizontal) rows and (vertical) columns. Each row or column of a matrix can be considered to be (and processed as) a vector. When we are working with arrays, it is common for the same mathematical operation to be carried out on all of the individual elements, so it makes sense to define an instruction set architecture that specifies these parallel operations. For example, in a typical general-purpose scalar processor, there is typically an addition instruction of the form ADD A, B, C, which means "add value A to value B and store the result in location C." In a vector or array processor, the same type of machine instruction would mean "add every element of A to the corresponding element of B, placing the results in the corresponding elements of C."

Of course, we could perform this same operation on a general-purpose machine by placing the scalar ADD instruction inside a program loop, using indirect or indexed addressing to work our way from one element to the next. The advantage of a vector or array processor is that either pipelining or spatial parallelism (or both) is used to speed up the performance of the calculations on multiple array elements, eliminating the need to do repetitive processing in a loop. It is common for vectors and matrices to be used extensively in scientific computing applications; thus, machines intended for use in such applications have often been designed and built to optimize this type of computation by adopting the Single Instruction Stream, Multiple Data Stream (SIMD) approach as categorized by Flynn. Array processors use the classical, parallel SIMD architecture, while vector processors accomplish much the same result using deeply pipelined arithmetic units.

The most commonly cited examples of vector processing systems are the Cray-1 and its successors, including the Cray X-MP and Y-MP vector supercomputers, although similar machines were also built by other companies, such as NEC, Fujitsu, and Hitachi. As we mentioned earlier, their computational instructions operate not on individual (scalar) values, but on vectors (one-dimensional arrays of values). MUL V1, V2, V3, for example, would multiply each element of vector V1 times the corresponding element of V2 to produce a product vector V3. *Vector reduction* operations (such as dot product) that produce a scalar result from a vector(s) may also be implemented as single machine instructions. These types of operations are usually accomplished by passing the vector elements, one after the other, through a (usually quite deeply) pipelined arithmetic/logic unit; Figure 6.2 depicts this process.

Because the vectors being processed are typically made up of a large number of elements (the length may be fixed, or specified by some means such as a vector length register) and because the same operations are performed on each element of the vector, the heavily pipelined approach used in vector processors may yield a large speedup compared with performing the same computations on a scalar processor. (To be able to supply data quickly enough to keep the pipeline full, the vector operands are kept in low-order interleaved memory or, in some machines, special vector registers inside the processor.) Vector processing also significantly decreases the effort involved in fetching and decoding instructions because it is not required that a machine instruction be executed for each computation. Instead, efficiency is enhanced as the fetching, decoding, and execution of just one instruction causes a large number of similar arithmetic or logical operations to be performed.

Although the vector approach can be quite useful in carrying out scientific calculations, it is of little use in general-purpose machines. This lack of general application makes vector processing systems more expensive as the (usually high) development cost is spread over fewer machines. High cost and low generality are significant reasons why vector processors have mostly disappeared from the supercomputing arena, having largely been replaced by massively parallel systems using CPUs and/or GPUs (see Section 6.1.2).

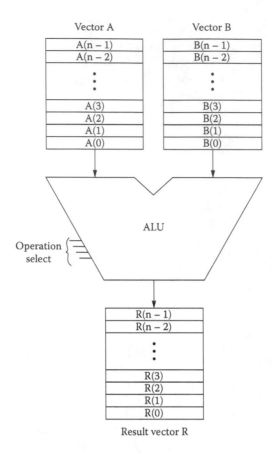

*Figure 6.2* Vector addition example.

*Array processors* also carry out the operation specified by a single machine instruction on a number of data elements. (Depending on the specific architecture, the operands may be structured as arrays of one, two, or more dimensions.) However, instead of doing this with a single (or just a few) deeply pipelined execution units that overlap computations in time (temporal parallelism), array processors typically make use of spatial parallelism: they employ many simple processing elements (PEs), each of which operates on an individual element of an operand array. If these processing elements are pipelined at all, they typically have only a small number of stages. Figure 6.3 shows the structure of a typical SIMD array processor; note the single control unit that coordinates the operations of all the PEs.

Because they share a single control unit, the PEs of an array processor operate in lockstep; each one performs the same computation at the same

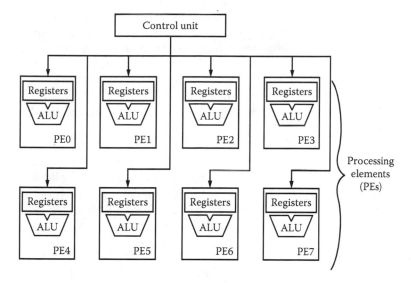

*Figure 6.3* Array processor (SIMD machine) example.

time, but on different elements of the operand array(s). In some cases, it may be possible to set or clear bits in a mask register to enable some PEs while disabling others; but if during a given cycle a PE is performing any computation at all, it is performing the same one as the other PEs. Figure 6.4 illustrates a situation in which six of the eight PEs in our example array processor will perform a given computation while the other two remain idle.

Like the MIMD machines to be considered below, SIMD computers may be constructed with either shared (globally accessible) or distributed (local) memory systems; each approach has its own set of pros and cons. Because most high-performance array processors have used the massively parallel approach and because of the difficulty of sharing memory among

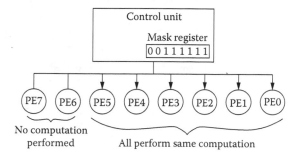

*Figure 6.4* Array processor with some processing elements masked.

a large number of processing elements, the distributed memory approach was more prevalent in such machines. Such an array processor, in which each PE has its own *local memory*, may be designated as a LM-SIMD system; a simple example is depicted in Figure 6.5. The alternative structure, in which PEs share access to a *global memory* system, may be referred to as a GM-SIMD architecture (see Figure 6.6).

Although SIMD machines are designed to greatly accelerate performance on computations using arrays of data, like pipelined vector processors they provide little or no speed advantage for general-purpose, scalar computations. For this reason, array processors were, and are, almost never built as standalone systems; rather, they typically serve as coprocessors (or "array accelerators") to a conventional CPU that serves as the front-end, or main, processor for the overall system. Historically, most array processors of the 1970s, 1980s, and early 1990s were built for government agencies and large corporations and were used primarily for scientific applications.

One of the earliest array processors was the ILLIAC IV, a research machine originally developed by Daniel Slotnick at the University of Illinois and later moved to NASA Ames Research Center. The ILLIAC was a fairly *coarse-grained* SIMD system; it had only 64 processing elements, each capable of performing 64-bit arithmetic operations at 13 MHz.

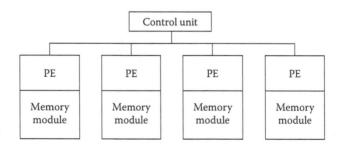

**Figure 6.5** Array processor with distributed memory (LM-SIMD architecture).

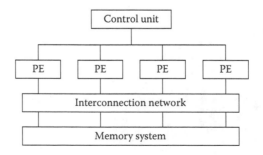

**Figure 6.6** Array processor with shared memory (GM-SIMD architecture).

(The original design called for 256 PEs, but the hardware proved to be so expensive that only one quarter of the array was ever built.) Each processing element had its own local memory; the single control unit could access the memory of any PE. The front-end processor for the ILLIAC IV was originally a Burroughs B6700, which was later replaced by a DEC PDP-10.

Perhaps as a reaction to the high cost (more than $30 million) and delays (several years) involved in getting the ILLIAC IV to work, most of the other high-performance array processors that were developed in the following decades adopted the *massively parallel* approach: Each used many more processing elements than ILLIAC, but each PE was much simpler (often just a 1-bit processor). This approach, also known as *fine-grained* SIMD, became more attractive as very large-scale integration (VLSI) technology matured and allowed the integration of a large number of simple processing elements in a small physical space. The ICL Distributed Array Processor (DAP), the Goodyear Aerospace Massively Parallel Processor (MPP), and the Thinking Machines Connection Machine (CM-1 and CM-2) systems were all composed of large numbers (4K, 16K, and 64K, respectively) of 1-bit processing elements. (The CM-2 did augment its huge array of 1-bit PEs with 2K 64-bit floating-point processors.) Eventually, implementation technology developed to the point at which some complexity could be restored to the processing elements. The MasPar MP-1 and MP-2 systems, introduced a few years after the Connection Machine, returned to a parallel datapath design. Each of these systems could contain up to 16K processing elements; those used in the MP-1 were 4 bits wide, and the PEs in the MP-2 were 32 bits wide. Each PE also contained floating-point hardware.

All the SIMD systems just mentioned were high-performance machines directed at the scientific processing market. Like vector processors, these large-scale array processors did not have a place in the general computing arena, so only a relatively few copies of each system were built and sold. However, SIMD technology has more recently found its way into personal computers and other less expensive systems—particularly those used for multimedia applications, signal and image processing, and high-speed graphics. The most ubiquitous example of (very coarse-grained) SIMD in today's computers is in the *multimedia extensions* that have been added to the original Intel x86 architecture. Intel's MMX enhancements, introduced in 1997, were executed by a simple array coprocessor integrated onto the chip with the CPU and floating-point unit. These new MMX instructions were capable of performing arithmetic operations on up to eight integer values at once. Advanced Micro Devices (AMD), manufacturer of x86-compatible processors, responded the following year by introducing its *3DNow!* instruction set architecture, which was a superset of MMX, including floating-point operations. Further SIMD operations were later added by Intel with streaming SIMD extensions (SSE), SSE2,

SSE3, SSSE3, and SSE4. AMD countered with Enhanced *3DNow!* and later included full SSE support in processors supporting Professional *3DNow!* before announcing in 2010 that it was discontinuing *3DNow!* support in favor of Intel compatibility. Newer processors from both Intel and AMD support advanced vector extensions (AVX). These architectural enhancements, along with others, such as SPARC's Visual Instruction Set (VIS); the MIPS MDMX, MIPS-3D, and MXU multimedia instructions; the AltiVec extensions to the PowerPC architecture; and the NEON advanced SIMD extensions to the ARM Cortex-A8 processors, all make use of parallel hardware to perform the same operation on multiple data values simultaneously and thus embody the SIMD concept as they enhance performance on the multimedia and graphics applications being run more and more on home and business systems.

In addition to general-purpose microprocessors like the ones just mentioned, some specialized processors used for graphics, digital signal processing, and other applications have also made use of SIMD architecture. One of the first array processors built for graphics applications was Pixar's Channel Processor (Chap), developed around 1984 for use in pixel processing operations, such as rotations, transformations, image blending, and edge filtering. Chap, which performed only integer operations, was later followed by Floating-Point Array Processor (Flap), which was used for three-dimensional transformations, clipping, shading, etc. Subsequently, the Pixel-Planes 5 and PixelFlow graphics machines developed at the University of North Carolina also used SIMD arrays for rendering, and ClearSpeed Technology (formerly PixelFusion) adapted its Fuzion 150 array processor for use in high-speed network packet processing applications. Analog Devices' Hammerhead SHARC family of DSP chips, among others, used SIMD to support processing of audio streams.

Although pipelined vector processors have waned in popularity in recent years, it appears that array processors based on the SIMD architecture are not going to follow suit. Although large-scale, massively parallel SIMD supercomputing machines are mostly defunct, smaller-scale systems are increasingly making use of moderate, coarse-grained array parallelism. In summary, SIMD array processors have succeeded to a greater extent than pipelined vector machines in recent years because their spatially parallel architecture scales better as chip feature sizes shrink and because they have been found to have a wider range of useful applications.

## 6.1.2   GPU computing

In recent years (late fifth-generation and sixth-generation systems as we defined them in Section 1.2), the use of SIMD in the form of traditional vector and/or array processors has been largely supplanted—particularly in high-performance systems used for scientific computing—by the use of

Graphics Processing Units (GPUs). Although GPUs were originally developed (as their name implies) as specialized processors for image rendering, as their flexibility and performance have increased dramatically and prices have dropped, GPUs have been found to be excellent choices to supplement conventional CPUs in handling a variety of demanding computing jobs. The same type of hardware parallelism that makes GPUs good at processing image data has also been found by researchers to be extremely efficient in other applications that have large amounts of inherent data parallelism. In the early days of general-purpose computing on graphics processing units (GPGPU), it was difficult to develop software for these systems; programmers had to work directly with graphical operations available in graphics application programming interfaces (APIs), such as DirectX and OpenGL, to write code to carry out nongraphical tasks. The development of proprietary programming languages, such as NVIDIA's Compute Unified Device Architecture (CUDA, introduced in 2007), and cross-platform languages, such as OpenCL (2008), has allowed programmers to more easily apply the power of GPUs to many of these nonvisual, but computationally intensive, problems.

How can we classify systems based on modern GPU architectures? They are related to, but clearly not the same as, more conventional array processors that fall under the SIMD category of machines. It is generally agreed that GPUs, such as the AMD/ATI FireStream and NVIDIA's Fermi and Kepler (as well as GPU-like processors, including the STI Cell Broadband Engine and Intel's Xeon Phi), do not directly correspond to any of Flynn's original four categories, so other terms have been proposed. NVIDIA, the leading provider of GPU hardware, refers to its GPU programming model in Flynn-like fashion as SIMT, or "Single Instruction Multiple Thread." Others refer to the type of computing done by GPUs as "stream processing." Both of these terms appear to be catching on among computing researchers, but what do they mean?

One way to explain SIMT is as a combination of multithreaded processing, as described previously in Section 4.5.4, with SIMD array processing. From a hardware standpoint, a GPU is comprised of several (e.g., 15 in NVIDIA's Kepler) *streaming multiprocessors*. (NVIDIA uses the acronym SMX for these, presumably to avoid confusion with the long-established use of "SMP" to abbreviate *symmetric multiprocessor*—a concept to be explained in Section 6.1.3.) Each streaming multiprocessor, in turn, contains a large number of individual cores or *stream processors* (SPs), each of which can normally execute one instruction from one thread during each clock cycle. Subsets of the stream processors are grouped together much like the processing elements (PEs) of a SIMD array, such that they perform common operations on data from multiple threads at the same time. Each of the stream processors has its own register set, computational hardware, and program counter; but other resources, such as L1 instruction and data cache, context storage memory, and instruction fetch/decode/dispatch ("front end") logic, are shared among multiple SPs to make more efficient use of resources. Figure 6.7 shows a simplified block diagram of a generic GPU; Figure 6.8 gives a more detailed view of a small group of SPs. The overall GPU can contain hundreds or even thousands of SPs (in the case of Kepler, each of the 15 SMX units contains 12 groups of 16 SPs each, for a total of 192 SPs per SMX, or 2880 overall). This allows for massively data-parallel computation without the entire collection of processing elements having to be in lockstep as they would in a massively parallel, purely SIMD machine.

Not only are the architecture and physical implementation of a GPU rather different from those of an array processor, but the programming model used to exploit the parallelism is different as well. SIMD array processors employ a single-threaded programming model, in which the instructions in the executing thread are simultaneously applied to multiple elements of an array of data. Conversely, as the SIMT acronym implies, efficient use of a graphics processor requires the simultaneous execution of a large number of threads. This multithreading is very fine-grained, and threads are managed directly by the hardware rather than

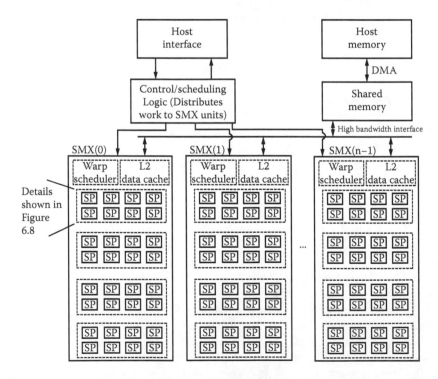

*Figure 6.7* Typical graphics processing unit (GPU) as used to do stream processing.

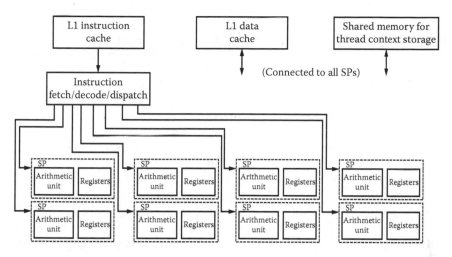

*Figure 6.8* One group of stream processor (SP) cores within a GPU's streaming multiprocessor.

by the operating system (as would be typical on a general-purpose CPU). Individual threads operate on scalar data, but they are normally grouped together for operation on parallel data. A batch of (usually 32–64) threads that run together on one of a GPU's streaming multiprocessors is called a "warp" (NVIDIA) or "wavefront" (AMD/ATI); for simplicity, we will adopt NVIDIA's term. The threads in a warp execute the same program code, or *kernel*, on different data. (In graphics terminology, a kernel corresponds to a shader.)

A key distinction between SIMT and SIMD is that the threads within a warp do not have to be in lockstep; although they are running the same code, individual threads can take different execution paths through that code. For example, an IF structure in the code could evaluate to true for some threads and false for others, causing divergent execution. Although arbitrary control flow is allowed, execution of a warp is more efficient when the thread behavior is convergent rather than divergent. (In other words, independent branching is supported, but if branches go different ways, the hardware has to compute both paths even though any given thread only takes one path.) Thus, for best performance, the compiler should optimize code by grouping together similar threads as much as possible.

Threads can be used to schedule operations in various ways depending on the needs of a particular application. For example, when a scalar processor would execute a loop to process multiple data elements, in SIMT an individual thread could be used for each data element, or each thread could loop to process multiple elements (but less than the total number). For reduce-type operations (e.g., summing the elements of an array), one could start with a thread for each pair of elements, then pass those sums to half of the original threads for another addition, etc., until a single thread ultimately accumulated the overall sum.

Because a modern GPU contains a large number of stream processing cores, it can (depending on manufacturer and model) run several to several dozen warps at the same time. For example, running on the latest NVIDIA GPUs, CUDA prefers to run thousands of threads at a time to make maximum use of hardware resources. (In the CUDA environment, up to 32 warps of 32 threads each can be grouped into a *block*, and multiple blocks can make up a *grid*. All threads that belong to the same grid will be executing the same kernel function.) One of the big advantages of this extreme multithreading approach is the fact that it helps the system mask memory latency (which can be a significant performance issue in single-threaded computing). When a cache miss occurs within any of the threads of a warp, the hardware scheduler puts that warp on a waiting list and schedules another warp to run. Given enough available threads and massively parallel hardware, a large number of computations will be in progress at any given time; and although (due to memory latency and/or other factors) each thread may take many cycles to complete, the overall rate of computation is very high. Thus, GPU processing works well for applications in which overall computational throughput is more important than the latency of any individual computation.

The use of GPUs to do nongraphical processing in high-performance systems has rapidly increased during the sixth generation of computing. The first two GPU-enhanced systems (using NVIDIA Tesla C2050 GPUs) cracked the Top 500 supercomputer list in June 2010; by November 2012, 62 systems on the list were using GPU accelerators/coprocessors. Three years later, the proportion of top-performing systems with GPU acceleration had risen to 21% (104/500). All signs point to a continuation of this trend for the foreseeable future.

### 6.1.3   *Multiprocessor systems*

*Multiprocessor* systems fall under Flynn's category of MIMD machines. They differ from array processors in that the several CPUs execute completely different, usually unrelated, instructions at the same time rather than being in lockstep with each other. Specifically, multiprocessors are MIMD systems with two or more independent (though usually similar, that is, *homogeneous*) CPUs that communicate via writing and reading shared main memory locations. Because they share memory, the CPUs in a multiprocessor system are said to be "tightly coupled." Such an architecture in which at least some—usually all—of main memory is globally accessible may be referred to as a GM-MIMD system. A block diagram of a typical multiprocessor system is shown in Figure 6.9.

Multiprocessors may have as few as two or as many as hundreds of CPUs sharing the same memory and input/output (I/O) systems. (Most massively parallel MIMD systems with thousands of CPUs are *multi- computers* in which memory and I/O are not globally shared.) As with most parallel machines, the key feature of any multiprocessor architec- ture is the interconnection network used to tie the system together—in this case, to connect the processors to the shared main memory mod- ules. (A *multicore* CPU is a single-chip multiprocessor with several inde- pendent computational cores, and the interconnections between them, all fabricated on the same semiconductor chip.) These interprocessor networks vary considerably in terms of construction, cost, and speed; we examine them in more detail in Sections 6.2 through 6.4. Although the cost of interconnection hardware may be considerable, multiproces- sor systems offer a simple programming model without the message- passing overhead inherent in parallel systems that do not have shared memory. It is relatively easy to port existing code from a uniproces- sor environment and achieve a reasonable increase in performance for many applications.

The classical multiprocessor system architecture, which has been in use at least since the early 1970s, has shared memory modules that are *equally* accessible by all processors. Given that no contention exists (in other words, a given module is not already in the process of being read or written by another CPU), any processor can access any memory loca- tion in essentially the same amount of time; this attribute is known as the *uniform memory access* (UMA) property. This type of system, in which a number of similar (usually identical) processors share equal access to a common memory system, is known as a *symmetric multiprocessor* (SMP) system. SMPs have a single address space that is accessible to all CPUs. All CPUs are managed by a single operating system, and only one copy of any given program is needed.

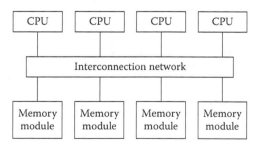

*Figure 6.9* Typical multiprocessor system (GM-MIMD architecture).

As far as a user is concerned, a symmetric multiprocessor looks, and is programmed, much like a uniprocessor; the difference is that multiple processes can run in a truly concurrent (rather than time-sliced) fashion. The performance of each CPU's cache is important, just as it is in a uni-processor system, but the placement of code and data in main memory is not particularly critical. The main advantages of SMP, therefore, are sim-plicity and ease of programming. One of the principal disadvantages of an SMP architecture is that (because of the characteristics of the intercon-nection networks that are used to connect the CPUs to memory) it does not scale well to a large number of processors. As the number of processors increases, a single bus becomes a bottleneck, and performance suffers. If a multiported memory system or a higher-performance network such as a crossbar switch is used to overcome this limitation, its complexity and cost can become prohibitive for large numbers of CPUs. Thus, a typical symmetric multiprocessor system has 16 or fewer CPUs, with the practical limit being a few dozen at the very most.

A more recently developed approach that is primarily used in larger multiprocessor systems is the *nonuniform memory access* (NUMA) archi-tecture. In this type of system, there is still a global address space that is accessible to all CPUs, but any given memory module is more local to one processor (or one small group of processors) than to others that are physically farther away. A given CPU has a direct physical connection (as it would in an SMP system) to certain memory modules, but only an indirect (and considerably slower) connection to others. (Figure 6.10 illus-trates a typical system based on this architectural plan.) Thus, although all shared memory locations are accessible to all CPUs, some of them will

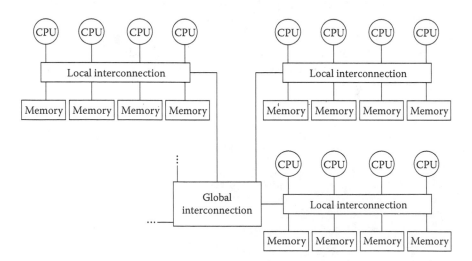

*Figure 6.10* Multiprocessor with distributed shared memory (NUMA architecture).

be able to access a given location more quickly than others. One can consider such a NUMA system to be an example of *distributed shared memory* (DSM). As such, the NUMA architecture represents a middle ground between shared memory SMP systems with only a few processors and the (often much larger) distributed memory multicomputers to be discussed below.

The performance of a multiprocessor with a NUMA architecture depends substantially on how information (code and data) needed by each processor is mapped in memory. Although a program written and compiled for an SMP architecture will run without modification on a cache coherent NUMA (CC-NUMA) machine, it will be unlikely to achieve the performance gain one might expect given the number of processors. This is largely because if a processor frequently needs to access memory locations that are physically remote, it will spend much of its time waiting for information. To maximize performance, the programmer and/or the compiler must be "NUMA aware," tuning the program such that as much as possible the information needed by each CPU is available in its directly connected, local memory. This process requires some effort and is more successful for some programs than others. A CC-NUMA system will also generally incur a higher hardware cost due to a more complex cache coherence scheme (to be discussed below) compared to an SMP system. If hardware is not used to ensure cache coherence, the programming effort (and performance trade-offs) required to do so in software will increase cost in a different way.

On the plus side, the NUMA architecture scales well to a large number of processors, although the differential access time between local and remote memory access does tend to increase as the system grows larger. Not only may a given system be built with many (in some cases, hundreds or thousands of) CPUs, but most NUMA-based machines are designed to be highly configurable, such that a user may start out with a smaller system and add extra processor and memory boards as needed. This is possible because while the interconnection structure within local groups of processors is like that of an SMP and can only handle a limited number of components, the global interconnection network (although slower) is generally much more flexible in structure. Because of their power and flexibility, systems with this type of architecture became increasingly popular during the fifth and sixth generations of computing. Notable historical examples include the Data General/EMC Numaline and AViiON systems, the Sequent (later IBM) NUMA-Q, the Hewlett-Packard/Convex Exemplar line of servers, and the Silicon Graphics Origin 2000 and 3000 supercomputers. Most servers based on modern Intel (Core i7 and Xeon) and AMD (Opteron) x86-64 family processors are based on NUMA architectures.

A few experimental multiprocessors, such as the Data Diffusion Machine (multiple test systems have been built), the Kendall Square

Research KSR-1, and Sun Microsystems' Wildfire prototype, have used a refinement of NUMA known as a *cache-only memory architecture* (COMA). (Most of these systems were actually based on a simplified implementation of this architecture known as *simple COMA* [S-COMA].) In such a system, the entire main memory address space is treated as a cache. Portions of the cache are local to each CPU or small group of CPUs. Another name for the local subsets of cache is "attraction memories." This name comes from the fact that in a COMA-based machine, all addresses represent tags rather than actual, physical locations. Items are not tied down to a fixed location in memory, although they appear that way to the programmer (who views the system as though it were a large SMP). Instead, blocks of memory (from the size of pages down to individual cache lines, depending on the implementation) can be migrated—and even replicated—dynamically so that they are nearer to where they are most needed. By using a given part of memory, a processor "attracts" that block of memory to its local cache (and local DRAM, which acts like just another level of cache). A directory is used to make the current overall cache mapping available to all processors so that each can find the information it needs. Although the COMA approach requires even more hardware support than the traditional multiprocessor cache coherence methods we are about to study and has yet to be adopted in commercial machines, it shows some potential for making larger multiprocessors behave more like SMPs and thus perform well without the software having to be tuned to a particular hardware configuration.

*Cache coherence* in multiprocessors is an important issue that we have briefly mentioned without giving a formal definition or explanation. In any multiprocessor system, processes running on the several CPUs share information as necessary by writing and reading memory locations. As long as access to the shared main memory is done in proper sequence—and as long as main memory is the only level involved—this approach is fairly straightforward. However, systems with main memory only (in other words, without cache) would be much slower than cache-based systems. This would directly contradict the main purpose (speed) for building a parallel system in the first place. Thus, cache is essentially universal in multiprocessor systems. However, the existence of multiple caches in a system, which can possibly contain the same information, introduces complications that must be dealt with in one way or another.

Recall our discussion of cache memories in Section 2.4. We noted that cache operation is fairly straightforward as long as only memory read operations are considered. Writes, however, introduce complications because sooner or later main memory must be updated such that its contents are consistent with those of the cache. We studied two write policies, write-through and write-back, commonly used in cache design. One potential problem with a write-back policy is that for some amount of time, main memory may contain *stale data*; that is, the cache may contain a newly computed value for an item while main memory contains a previous value that is no longer correct. This situation is not corrected until the line containing the updated item is displaced from cache and written back to memory. In the meantime, if any other operation (such as a DMA transfer from memory to an output device) were to read the contents of that memory location, the incorrect, old value would be used.

One way to solve this problem in a uniprocessor system (although it may cost a little in performance) is to simply adopt a write-through policy such that every time an item in cache is updated, the same value is immediately written to main memory. This makes sure that the contents of cache and main memory are always consistent with each other. Main memory will never contain invalid or "stale" data. Now consider the addition of a second CPU, with its own cache, to the system as shown in Figure 6.11. What effects may result from the existence of multiple caches in a system?

As Figure 6.11 shows, a potential problem arises anytime the same line from main memory is loaded into both caches. It is possible that either processor may write to any given memory location in the line. A write-through policy guarantees that main memory will always be updated to contain the information written by either of the processors, but it does nothing to ensure that the other CPU's cache will be updated with the new value. If the location in question remains in the second processor's cache long enough to be subsequently referenced by a program (resulting in a cache hit), the data obtained will be the old, incorrect value instead of the new one. The two caches are inconsistent with each other and thus present an *incoherent* view of the contents of main memory. (The problem only

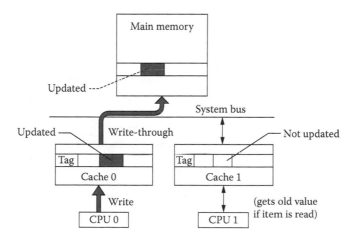

*Figure 6.11* Cache coherence problem in a multiprocessor system.

gets more complex as additional CPUs, each with its own cache, are added to the system.) This is not a desirable situation, as it can result in incorrect operation of the software running on the parallel system. Maintaining *coherence* of stored information (while keeping complexity and cost to reasonable levels) is a major design issue in multiprocessor systems. Let us examine some of the strategies commonly used to address this problem.

The simplest scenario as far as cache coherence is concerned is a small, symmetric multiprocessor in which all CPUs use the same interconnection hardware (often a shared bus as in Figure 6.12) to access main memory. If a write-through policy is used, every time a value is updated in any cache in the system, the new data value and the corresponding address will appear on this common network or bus. By monitoring, or "snooping on," system bus activity, each cache controller can detect whether any item it is currently storing has been written to by another processor and can take appropriate action if this is the case. A cache coherence protocol that uses this approach is called (with apologies due to *Peanuts* creator Charles Schulz) a *snoopy* protocol.

What is the appropriate action to take when a cache controller detects another processor's write to a location it has cached? One of two approaches will suffice to keep another CPU from using the stale data currently in its cache. Either its duplicate copy of the cache entry must be *invalidated* (for example, by clearing the valid bit in its cache tag) such that the next reference to it will force a miss and the correct value will be obtained from main memory at that time, or it must be *updated* by loading the changed value from main memory. (Depending on timing constraints, it may be possible in some systems to copy the item from the bus as it is written to

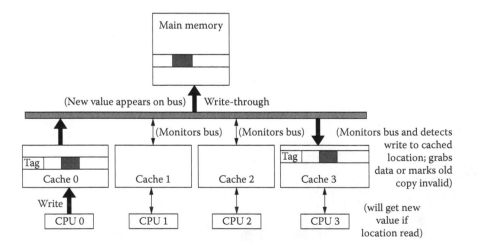

*Figure 6.12* Symmetric multiprocessor with shared bus interconnection.

main memory.) Not surprisingly, these two approaches are referred to, respectively, as *write-invalidate snoopy cache* and *write-update snoopy cache* protocols. Their operation is illustrated in Figures 6.13 and 6.14.

Each of these two approaches has certain advantages and disadvantages. The write-update protocol makes sure that all caches that hold copies of a particular item have a valid copy. For data that are heavily shared (frequently used by more than one CPU), this approach works well because it tends to keep hit ratios high for the shared locations. It is especially useful in situations in which read and write accesses tend to alternate. However, write-update can significantly increase overhead on the system bus, especially considering that in many systems an entire line must be transferred to or from cache even if only one byte or word is affected. This additional bus activity required for updates may reduce the bandwidth available to other processors and interfere with other activities such as I/O.

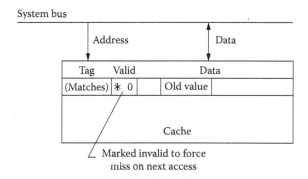

*Figure 6.13* Maintaining cache coherence using the write-invalidate approach.

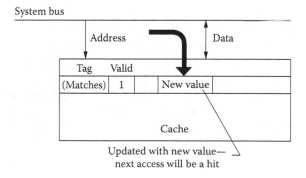

*Figure 6.14* Maintaining cache coherence using the write-update approach.

If data are only lightly shared (seldom referenced except by one particular CPU), much of the time and effort spent updating all the other caches on each write may be wasted because the line in question is likely to be evicted from most of the caches before it is used again. In this case, the write-invalidate protocol may be preferred even though it will sometimes cause a miss on shared data for which write-update would have resulted in a hit. (The overall hit ratio for the cache containing the invalidated line may increase because it may be replaced with another line that turns out to be hit more frequently.) Write-invalidate does not perform as well as write-update when reads and writes alternate, but has the advantage when several writes to the same line occur in succession because it avoids the multiple update transactions. A line only needs to be invalidated once no matter how many times it is written.

System designers cannot precisely predict the behavior of the many programs that will ultimately be run on a given machine, but they can study (either live or in simulation) the effect of different protocols on benchmarks or other representative code. In practice, most commercial SMP systems use some form of the write-invalidate protocol (several implementation variants exist) because it works better with the larger line sizes that are common in today's cache designs and because it saves bandwidth on the system bus. As the number of processors in an SMP increases (as has been occurring in recent years), system bus bandwidth is at more and more of a premium; this has given a fairly decisive edge to write-invalidate protocols in most applications.

The above discussion assumed a write-through policy for simplicity, but many caches are designed with a write-back policy (which can also help to reduce traffic on the system bus). Although a snoopy cache as just described depends on the cache controllers monitoring line write-throughs to main memory, the same coherence effect can be achieved with a write-back policy if the capability exists for broadcasting information between caches without a main memory write having to occur on the system bus. If this capability includes a way to transfer data, then copies of the data in other caches can be updated that way without main memory being written. If (as is more common) no provision is made for direct data transfer among the caches, only the write-invalidate protocol can be used. (This is yet another reason for its popularity over write-update.) For the write-invalidate approach to work with write-back caches, it is necessary for a given cache to claim exclusive ownership of an item the first time its CPU writes to it. Subsequent writes to the same line (by the owning processor only) can be done at cache speed with a single write-back to main memory occurring later. However, all other caches must not only mark their copies of this line as invalid, but must block access to the corresponding main memory locations because they are invalid too. If another CPU subsequently needs to perform a read or write of an invalid location, it must ask the first processor to relinquish ownership and perform the write-back operation before the memory access can proceed.

Snoopy cache protocols work well for small SMP systems with only a single bus to be snooped, but they become much more complex and expensive for systems with multiple buses or other complex interconnection structures. There is simply too much hardware to snoop. In particular, it is significantly more difficult to achieve coherence in a NUMA architecture than in an SMP. The snoopy cache approaches described above are inappropriate for NUMA systems in which different SMP-like processor groups have no way to snoop on, or otherwise be directly informed of, each other's internal transactions. Instead, *directory-based* protocols are normally used to implement cache-coherent NUMA machines.

A *directory* is simply a hardware repository for information concerning which processors have cached a particular (cache line–sized) block of shared memory. The directory maintains a large data structure containing information about each memory block in the system, including which caches have copies of it, whether or not it is "dirty," and so on. (Figure 6.15 shows an example of a system with 16 CPUs and $2^{24}$ memory blocks.) Because a *full-map directory* like the one shown contains information about the presence of every block in the system in each cache, its size is essentially proportional to the number of processors times the number of memory blocks. If this proves to be too large to implement, the directory may be constructed with fewer entries (which means only a limited number of blocks from main memory can be cached at once) or with fewer bits per entry (which means only a limited number of caches may contain a given

| Block number | Cached? | Exclusive? | Shared? | Presence bits | | | | | | | | | | | | | | | |
|---|---|---|---|---|---|---|---|---|---|---|---|---|---|---|---|---|---|---|---|
| | | | | 0 | 1 | 2 | 3 | 4 | 5 | 6 | 7 | 8 | 9 | 10 | 11 | 12 | 13 | 14 | 15 |
| 0 | 0 | 0 | 0 | 0 | 0 | 0 | 0 | 0 | 0 | 0 | 0 | 0 | 0 | 0 | 0 | 0 | 0 | 0 | 0 |
| 1 | 1 | 0 | 1 | 1 | 1 | 1 | 1 | 0 | 0 | 0 | 0 | 1 | 1 | 0 | 0 | 1 | 1 | 0 | 0 |
| 2 | 1 | 1 | 0 | 0 | 0 | 0 | 0 | 0 | 0 | 1 | 0 | 0 | 0 | 0 | 0 | 0 | 0 | 0 | 0 |
| 3 | 0 | 0 | 0 | 0 | 0 | 0 | 0 | 0 | 0 | 0 | 0 | 0 | 0 | 0 | 0 | 0 | 0 | 0 | 0 |
| . . . | . . . | . . . | . . . | | | | | | | | | | | | | | | | |
| $2^{24}-2$ | 0 | 0 | 0 | 0 | 0 | 0 | 0 | 0 | 0 | 0 | 0 | 0 | 0 | 0 | 0 | 0 | 0 | 0 | 0 |
| $2^{24}-1$ | 1 | 0 | 1 | 0 | 0 | 1 | 1 | 0 | 1 | 0 | 1 | 0 | 0 | 0 | 0 | 0 | 1 | 0 | 1 |

*Figure 6.15* Full-map directory example.

block). The directory approach works with any interconnection structure (not just a single bus) because all communications related to cache coherence are point-to-point (between a given cache and the directory), and no broadcasting or snooping is required.

What sort of events require communication between a cache and the directory? Fortunately, in most applications, the majority of accesses are memory reads resulting in cache hits. Because read hits use only the local cache and do not affect the work of any other processor, they require no interaction with the directory. However, a read (or write) that results in a miss means the cache needs to load a new block from main memory (usually evicting a previously stored line in order to do so). This requires notifying the directory of the line being displaced (so its directory information can be updated to reflect this action) and requesting the new information to be loaded from main memory with the directory noting that transaction as well. Also, write hits to shared, previously "clean" cache locations must be communicated to the directory because they directly affect the consistency of data. Exactly how and when this is done depends on the specific directory protocol being used.

There are many different implementations of directory-based cache coherence protocols, some relatively simple and some more complex. The details vary from system to system and, for the most part, are beyond the scope of this book. We shall content ourselves with describing a simple scheme in which each memory block is always in one of three states: *uncached*, *shared*, or *exclusive*. The uncached state means exactly what it says: The block exists only in main memory and is not currently loaded in any processor's cache. The shared state means the block is present in one or more caches in addition to main memory and all copies are up to date. (The block will remain in this state until it is displaced from all caches or until a processor tries to write to it.) When a write occurs, the block enters the exclusive state. Only one cache (the one local to the CPU that performed the write) has a valid copy of the block; it is said to be the owner of the block and can read or write it at will. Before granting ownership, the directory sends a message to any other caches that contain the block, telling them to invalidate their copies. If a processor other than the owner later wants to access a block that is in the exclusive state, it has to stall while the directory tells the owning cache to write it back to main memory; the block then returns to the shared state (if the access is a read) until it is written again (and thus returns to the exclusive state) or displaced from all caches (returning to the uncached state).

Figure 6.16 is a state diagram that depicts the possible transitions between the three states of this simple directory-based protocol. Note that the logic behind the states and transitions is very similar to that used by a typical write-invalidate snoopy protocol. The main difference is that in this case the state transitions are driven by messages sent from a cache

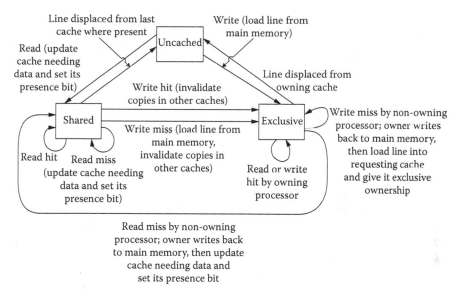

*Figure 6.16* State diagram for simple directory-based cache coherence protocol.

to the directory (or vice versa) rather than bus transactions observed by snooping.

The original, and simplest, form of the directory-based approach involved the use of a *centralized directory* that kept track of all memory blocks in the system. However, this approach does not scale well because a large number of CPUs (and therefore caches) may saturate a single directory with requests. The alternative to a centralized directory scheme, which works better for larger systems (especially CC-NUMA machines) is a *distributed directory* protocol. In such an implementation, a number of smaller directories (typically one for each local group of processors) exist with each one being responsible for keeping track of the caches that store the portion of the memory address space most local to it. For example, if a CC-NUMA system had 32 processors divided into four local groups of eight processors each, the directory might be split into four parts with each part keeping track of coherence information for one fourth of the blocks. Any cache needing to communicate with a directory would send a message to one of the four based on which part of the address space it was referencing. Distributed directory protocols scale well to almost any size multiprocessor system, although cost and communication delays may become limiting factors in very large systems.

An additional aspect of cache coherence that pertains to both snoopy and directory-based protocols is *serialization*, which means ensuring the proper behavior of multiple write operations. (So far, we have only considered the effects of individual writes to memory.) When more than one CPU updates a given memory location at or about the same time, the updates must be *serialized* (forced to be sequential) so that they are seen in the same order by all processors. If this was not done and multiple write operations took effect in different sequences from the standpoint of different CPUs, incorrect operation could result. In a single-bus SMP, writes (as well as reads) are automatically serialized by the bus because only one main memory operation can occur at a time, but in DSM multi-processors with more complex interconnection schemes, it is possible for multiple write operations to be initiated at once. The directory (or any other coherence scheme used by the system) must ensure that ownership of a data item is granted exclusively and that all processors see the effect of grants of ownership (and the subsequent memory writes) in the same order. Subsequent reads or writes of the same item must be delayed until the first write is committed to memory.

Finally, we note that not all multiprocessors solve the cache coherence problem by using hardware. Another approach that has been tried in a few systems (mostly larger ones in which hardware cache coherence schemes are costly) is a *non-cache-coherent* (NCC) architecture. (The Cray T3D and T3E systems are prominent examples of NCC-NUMA architecture.) In such a system, consistency of data among the various caches is not enforced by hardware. Instead, software is responsible for ensuring the integrity of data shared among multiple processors. One simple way to do this is to divide data into two types: cacheable and noncacheable. Data that are read-only or used only by one CPU are considered cacheable; read/write data that may be used by more than one processor are marked as noncacheable. Because all shared read/write data are stored in main memory and never in cache, the cache coherence problem is solved by simply avoiding it.

The problem with such a technique, known as a *static coherence check* (which is usually performed by the compiler with support from the operating system) is that it is an extremely conservative approach. It works, but it may impair performance significantly if there is a large amount of data that are shared only occasionally. Designating all shared variables as noncacheable means many potential cache hits may be sacrificed in order to avoid a few potential problems. In many cases, there may be only certain specific times when both read and write accesses can occur; the rest of the time, the data in question may effectively be read-only (or used by only one CPU) and thus could be cacheable. If the compiler (or the programmer) is capable of performing a more sophisticated analysis of data dependencies, it may be possible to generate appropriate cacheable/noncacheable intervals for shared data and thus improve performance in an NCC system. This type of advanced compiler technology is an area of ongoing research in computer science. For now, however, the difficulty of achieving good performance with software coherence approaches is non-trivial, which is another way of saying that compilers are not very good at it yet and it is a real pain to do manually. This lack of easy programmability explains why very few NCC-NUMA systems exist.

*Synchronization and mutual exclusion* are also important issues in multiprocessor systems. In a system using the shared memory model, processes, which may be running on the same or different CPUs, generally communicate by sharing access to individual variables or data sets in memory. However, the programmer must be careful that accesses to shared data are done in the correct order for the results to be predictable. As a simple example, assume that two processes (A and B) both want to increment a shared variable X that has initial value 0. Each process executes the same sequence of three instructions as follows:

| | | |
|---|---|---|
| $I_1$: | LOAD | X, R1 |
| $I_2$: | ADD | 1, R1 |
| $I_3$: | STORE | R1, X |

What will be the final value of X after this code is executed by both processes? Ideally, X would have been incremented twice (once by process A and once by process B) and thus would have the final value 2. This will be the case if either process executes all three instructions in order before the other begins executing the same sequence. However, absent explicit synchronization between the two processes, it is possible that X could be incremented only once. For example, it is possible that the order of execution could be $I_1$ (A), $I_2$ (A), $I_1$ (B), $I_2$ (B), $I_3$ (B), $I_3$ (A). In this case, it is easy to see that the final value of X would be 1—an erroneous result.

How can we ensure that operations on shared data will be done correctly in a parallel system? The obvious solution is to somehow ensure that when a process begins executing the crucial statements that affect the shared item (instructions $I_1$, $I_2$, and $I_3$ in the above example) it completes this entire *critical section* of code before any other process is allowed to begin executing it. Only one process at a time will have access to the critical section and thus the shared resource (generally an I/O device or memory location—in this case, the variable X). In other words, the processes that share this resource can be synchronized with respect to the resource using *mutual exclusion*.

How can the system enforce mutually exclusive access to shared resources? Remember, in a multiprocessor, multiple CPUs execute instructions at the same time; each is unaware of what instructions the others are processing, so absent some sort of "locking" mechanism, it is possible that more than one could be in a critical section at the same time. Even in a uniprocessor system, a process could be interrupted while in a critical section, and then another process could enter the critical section before the first process is resumed. Clearly some mechanism (software instructions supported by the underlying hardware) needs to be in place to restrict access to critical sections of code. Such a mechanism could be placed in front of each critical section to effect the necessary synchronization.

The classical mechanism for regulating access to shared resources uses a special shared variable called a *semaphore*. The simplest type of semaphore is a binary variable that takes on the value true or false (set or clear). Before entering a critical section of code, a process tests the associated semaphore; if it is clear, the process sets it and enters the critical section. If it is already set, the process knows another process has already entered the critical section and waits for a period of time, then retests the semaphore and either enters the critical section or waits some more as appropriate. After exiting the critical section, a process must release (clear) the semaphore to allow access by other processes. More general types of semaphores (for example, counting semaphores) may be used to regulate access to multiple units of a shared resource, but the basic idea behind them is similar.

The alert reader has probably already spotted a potential flaw in the solution just described. The semaphore itself is a shared resource just like the one being accessed inside the critical section. If we are using the semaphore to enforce mutually exclusive access to some other shared resource, how can we know access to the semaphore itself is mutually exclusive? Would it not be possible for multiple processes to read the semaphore at nearly the same time and find it clear, then perform redundant set operations and proceed together into the critical section, thus defeating the mutual exclusion mechanism?

The answer to the above question would be "yes," and the semaphore scheme would fail, unless the system had a means of ensuring, in hardware, that each access to the semaphore was *atomic* (complete and indivisible). In other words, its contents must be read and written in one uninterruptible memory cycle. The trick to accomplishing this is to provide a hardware lock that prevents any other device from accessing memory while the semaphore is read, tested, modified if necessary, and written back. A memory cycle in which all of these operations take place without the possibility of interruption is called a *read-modify-write* (RMW) cycle.

Machine instructions that implement indivisible RMW cycles are known as *mutual exclusion primitives*. Almost all modern microprocessors, because they are designed to be usable in multitasking and multiprocessor systems, provide one or more such instruction. One common example is a *test and set* instruction; in one operation, it reads the current value of a memory location (whether clear or set) and sets it to

true. Another equivalent instruction (but with more general functionality) is a *swap* instruction. It exchanges the contents of a register and a
memory location (which could be a semaphore) in a single indivisible
instruction.

In bus-based uniprocessor or multiprocessor systems, a hardware signal (called LOCK or something similar) inhibits other accesses to memory
while it is asserted. When a mutual exclusion primitive is executed, the
lock signal stays active until the semaphore value has been checked and
the updated value is committed to memory. Interrupts are also temporarily disabled while the lock signal is active, thus ensuring that the RMW
cycle is atomic. Figure 6.17 shows a typical sequence of events for an RMW
cycle.

Mutual exclusion is more difficult to implement in larger multiprocessor systems based on the NUMA or COMA models due to the lack of
a single interconnection structure that can be locked for a given atomic
operation to take place. Because in a system with DSM memory it is possible for multiple memory operations to be in progress at the same time,
the management of mutual exclusion primitives must be integrated with
and enforced by the cache coherence scheme (which is usually directory-
based). When a lock is requested for exclusive access to a memory location,
that location must be put into the exclusive state under the ownership of
the processor that is going to test and modify it. It cannot enter any of the
other states until the lock is released by the owning processor and the
updated value has been committed to memory. Other processors that may
be waiting to access the synchronization variable must be queued by the
directory. Because of the directory overhead required, atomic operations
are more costly in performance in DSM systems than they are in SMP
systems.

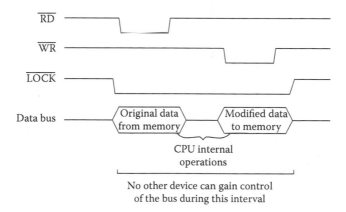

*Figure 6.17* Timing diagram showing a RMW sequence.

### 6.1.4   Multicomputer systems

*Multicomputer* systems (also called *clusters*) are parallel systems that, like multiprocessors, come under the MIMD classification in Flynn's taxonomy of computer systems. The difference—and it is a significant one—is that multicomputers do not directly share memory. As the name implies, each part of a multicomputer system is a complete computer in its own right with its own processor, memory, and I/O devices. (Unlike multiprocessors, it is not even necessary that the individual computers within a cluster run the same operating system as long as they understand the same communications protocol.) The memory of each computer is local to it and is not directly accessible by any of the others. Thus, the memory access model used in multicomputers may be referred to as no remote memory access (NORMA)—the next logical step beyond the UMA and NUMA models typical of multiprocessors. Because of the lack of shared main memory, multicomputers may be referred to as loosely coupled systems or, architecturally speaking, LM-MIMD systems. A typical multicomputer is illustrated in Figure 6.18.

Multicomputer systems may be constructed from just a few computers or as many as tens of thousands. The largest MIMD systems tend to be multicomputers because of the difficulties (noted previously) with sharing main memory among large numbers of processors. Beyond a few hundred processors, the difficulties in constructing a multiprocessor system tend to outweigh the advantages. Thus, most massively parallel MIMD supercomputing systems are multicomputers.

Multicomputer systems date back to the early to mid-1980s. The Caltech Cosmic Cube is generally acknowledged to have been the first multicomputer, and it significantly influenced many of the others that came after it. Other historical examples of multicomputers include Mississippi State

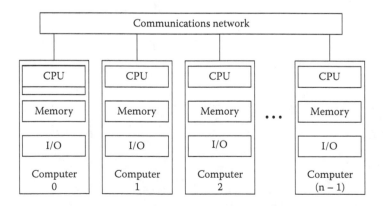

*Figure 6.18* Typical multicomputer system (LM-MIMD architecture).

University's Mapped Array Differential Equation Machine (MADEM), nCUBE Company's nCUBE, the Intel iPSC and Paragon, the Ametek 2010, and the Parsys SuperNode 1000. During the fifth generation, the Beowulf class of cluster computers (introduced by Donald Becker and Thomas Sterling in 1994) quickly became the most widely used type of multicomputer. Beowulf clusters are parallel machines constructed of commodity computers (often inexpensive PCs) and inexpensive network hardware; to further minimize cost, they typically run the open-source Linux operating system. Although each individual computer in such a cluster may exhibit only moderate performance, the low cost of each machine (and the scalability of the connection mechanism) means that a highly parallel, high-performance system can be constructed for much less than the cost of a conventional supercomputer.

As is the case with other parallel architectures, the key feature of any multicomputer system is the communications network used to tie the computers together (see Sections 6.2 through 6.4). One major advantage of multicomputers (besides the low cost of the individual machines, which do not have to be internally parallel) is the relative simplicity and lower cost of the interconnections. Because memory does not have to be shared, the types of networks used to connect the individual computers can be much more flexible, and in many cases less expensive, than those used in multiprocessors. Frequently, the communications network may be implemented by something as simple and cheap as off-the-shelf 10/100/1000 Mb/s Ethernet, which is used in many inexpensive cluster machines. Maximizing performance in an environment with a heavier interprocessor communications load may require higher performance networks such as 10 Gigabit Ethernet, Myrinet (the latest version of which supports full-duplex communication at 10 Gb/s in each direction), or Infiniband (with some versions running at up to 40 Gb/s); these are still usually less expensive than the specialized, proprietary interconnections used in some multiprocessors. It is interesting to note that by November 2015, 237 of the world's top 500 high-performance supercomputers used Infiniband networks, 119 used 10 Gigabit Ethernet, 62 used Gigabit Ethernet, and the remaining 82 used custom or proprietary networks.

The main disadvantage of multicomputers is that the simple, global shared memory model common to uniprocessors and multiprocessors is not feasible. Because there is no way to directly address the memory belonging to other processors, interprocess communication must be done via a message-passing programming model instead of the shared memory model. In other words, rather than communicating *implicitly* (via accessing shared memory locations), processes running on a multicomputer must communicate *explicitly* (via sending and receiving messages that are either routed locally or over the network, depending on the locations of the communicating processes).

The message-passing programming model for multicomputers is generally implemented in high-level language programs by using standardized Message Passing Interface (MPI) or Parallel Virtual Machine (PVM) functions. These message-passing models are more complex and are not generally considered to be as intuitive to programmers as the shared memory model; the different programming paradigm also makes it more difficult to port existing code from a uniprocessor environment to a multicomputer than from a uniprocessor to a multiprocessor. There can also be a negative effect on performance as message-passing always incurs some overhead. Of course, this is counterbalanced somewhat by the flexibility in constructing the communications network and the additional advantage that parallel algorithms are naturally synchronized via messaging rather than requiring artificial synchronization primitives such as those needed in multiprocessor systems. The relative desirability of the two approaches from a programmer's point of view can be summed up by the observation that it is relatively easy to implement message-passing on a machine with shared memory, but difficult to simulate the existence of shared memory on a message-passing multicomputer.

In general, because network performance (which is usually slower than the interconnection between CPUs and memories in a multiprocessor) and messaging overhead tend to be limiting factors, multicomputers work best for relatively coarse-grained applications with limited communications required between processes. The length of the messages is not usually as important as the frequency with which they are sent; too many messages will bog down the network, and processors will stall waiting for data. Multicomputer applications that transmit a few larger messages generally perform better than those that need to send many small ones. The best case is that in which data are transferred between computers only intermittently with fairly long periods of computation in between.

It is possible to construct a large-scale parallel system that combines some of the attributes of multiprocessors and multicomputers. For example, one can construct a multicomputer system in which each of its computing nodes is a multiprocessor. Such a hybrid system may be considered to occupy a middle ground, architecturally speaking, between a typical multicomputer system (NORMA) and a NUMA-based multiprocessor system and is sometimes referred to as a *multi-multicomputer*. Because memory can be shared between CPUs in a local group but not between those that are physically remote, a message-passing protocol must still be supported as in other multicomputers. Early examples of multi-multicomputer systems included the IBM SP (Scalable POWERparallel) series machines and the Cluster of Multiprocessor Systems (COMPS), a joint venture (launched in 1996) between the University of California at Berkeley, Sun Microsystems, and the National Energy Research Scientific Computing Center (NERSC). Each local computing node in these highly parallel systems was a symmetric multiprocessor containing four to eight CPUs. Given the fact that virtually all modern high-performance micro-processors are multicore (and, therefore, constitute small SMPs on a chip), the latest sixth-generation high-performance computing clusters all technically fall under this hybrid classification.

## 6.2   Interconnection networks for parallel systems

From the discussion of parallel systems in the previous section, the reader may have correctly concluded that one of the most important keys to performance is the speed of communications among the processors. Regardless of how fast a given CPU may be, it can accomplish no useful work if it is starved for data. In a uniprocessor system, access to the needed data depends only on the memory system or, in some cases, an I/O device interface. In a parallel system, it is unavoidable that one processor will need data computed by another. Getting those data from their source to their intended destination, reliably and as quickly as possible, is the job of the parallel system's interconnection network.

### 6.2.1   Purposes of interconnection networks

In the previous section, we noted the two principal purposes for interconnection networks in parallel systems. The first was to connect CPUs or PEs (and possibly other bus masters, such as I/O processors or DMA controllers) to shared main memory modules as is typical of multiprocessor systems and a few array processors. The second was to enable the passing of messages that are communicated from one processor to another in systems without shared memory.

Some interconnection networks serve additional purposes. In many cases, networks are used to pass control signals among processors or processing elements. This use of a network is particularly notable in SIMD systems, but MIMD systems may also require certain control signals to be available to all CPUs. For example, signals required to implement the system's cache control strategy or coherence policy may need to be connected to all the processors. Likewise, other hardware control lines such as bus locks, arbitration signals for access to shared resources, etc., may be distributed via the network.

One other purpose of some interconnection networks is to reorganize and distribute data while transferring values from one CPU or PE to another. Architectures intended for specialized applications, such as processing fast Fourier transforms (for which data are shuffled in a predetermined pattern after each iteration of the computations), may be constructed with network connections that optimize the routing of data for the performance of that application. However, because networks with more generic topologies have improved considerably in performance and because a specialized network is unlikely to improve (and may even hurt) performance on applications for which it is not intended, this type of network is rarely encountered in modern systems.

## 6.2.2   Interconnection network terms and concepts

In order to understand the operation of interconnection networks in parallel systems, we must first define some terms and concepts that we will use to describe and compare the many types of networks that are and have been used in such systems. First, a *node* is any component (a processor, memory, switch, complete computer, etc.) that is connected to other components by a network. (Another way to say this is that each node is a junction point in the network.) In an array processor, each processing element (or PE–memory combination if main memory is local to the PEs) might constitute a node. In a symmetric multiprocessor, each CPU and each shared memory module would be a node. In a multicomputer, each individual computer is considered a node.

### 6.2.2.1   Master and slave nodes

A *master node* is one that is capable of initiating a communication over the network. *Slave nodes* respond to communications when requested by a master. In a multiprocessor, the CPUs are master nodes and the memories are slave nodes. In a multicomputer, all nodes are generally created equal, and a given node may act as the master for some communications and as a slave for others. The details of array processor construction vary considerably, but in general, the common control unit is responsible for coordinating transfers of data between nodes as necessary and thus could be considered the master node, and the PEs and memories are generally slaves.

### 6.2.2.2   Circuit switching versus packet switching

At different times, a given node in a parallel system will need to send or receive data to or from various other nodes. In order to do so, it must be able to select the desired node with which to communicate and configure the state of the network to enable the communication. Configuring the network for a communication to take place is done in two general ways. A *circuit-switched* network is one in which physical hardware connections (generally implemented via multiplexers and demultiplexers or their equivalents) are changed in order to establish a communications link between desired nodes. This pathway for information and the connections that comprise it remain in place for the duration of the communication (see Figure 6.19 for an example). When it becomes time for different nodes to communicate, some or all of the previously existing connections

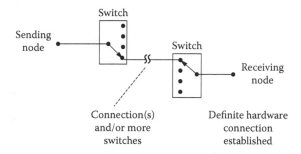

*Figure 6.19* Circuit-switched network.

are broken, and others are made to facilitate the new pathway for data to travel. In this sense, a circuit-switched interconnection for a parallel computing system is directly analogous to the landline telephone network. Circuit-switched networks are an example of solving a problem (in this case, communication) using a hardware-oriented approach. Circuit-switched networks can be advantageous when the number of nodes is not too large, particularly if certain connection patterns are needed frequently, and we want to be able to optimize the speed of transfers between certain sets of nodes. Circuit switching is particularly applicable when there is a need to transfer relatively large blocks of data between nodes.

The alternative to circuit switching is known as *packet switching*. A packet-switched network is one in which the same physical connections exist at all times. The network is connected to all the nodes all the time, and no physical connections are added or deleted during normal operation. In order for one node to communicate with a specific other node, data must be routed over the proper subset of the existing connections to arrive at its destination. In other words, a *virtual connection* must be established between the sending and receiving node. Because specific node-to-node communications are made virtually rather than by switching physical connections, packet-switched networking is considered to be a more software-oriented approach than circuit-switched networking (although hardware is definitely needed to support the transmission of data). The routing hardware and software are local in nature; each node has its own copy.

The terminology used for this type of network comes from the fact that each message sent out over the network is formatted into one or more units called *packets*. Each packet is a group of bits formatted in a certain predetermined way (a hypothetical example can be seen in Figure 6.20). A given packet is sent out from the originating (sending) node into the network (which may be physically implemented in any number of ways) and must find its way to the destination (receiving) node, which then must recognize it and take it off the network. Often a packet might take multiple possible

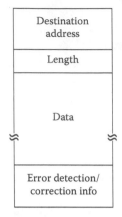

*Figure 6.20* Typical message packet format.

paths through the network, as shown in Figure 6.21. Because the packets appear to find their way through a blind maze of network connections in order to arrive (apparently by magic) at their correct destinations, they are sometimes referred to as "seeing-eye packets." Of course, this does not really occur by either magic or blind luck. Rather, the hardware and software present at each node in the network must have enough intelligence to recognize packets meant for that node and to route packets meant for other nodes so that they are forwarded ever closer to their destinations. The packets do not really find their own way to the destination—they are helped along by the intermediate nodes—but they appear to get there by themselves because no physical connections are changed to facilitate their passage.

In order for intermediate nodes to route packets and for destination nodes to recognize packets meant for them to receive, each message packet sent must include some type of routing information in addition to the data

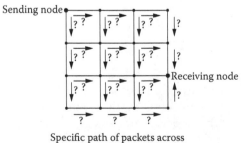

*Figure 6.21* Routing of messages in packet-switched network.

being transmitted (plus, perhaps, information about what the data represent or how they are to be processed). The originating node must provide the address of the destination node on either an absolute or relative basis; this part of the packet will be examined by each node that receives it in order to determine how to perform the routing.

Unlike circuit-switched interconnections, packet-switched networks tend to favor the sending of short to medium-length messages; they are less efficient for transfers of large blocks of data. In many cases, the length of the packets sent over the network is fixed at some relatively small size; messages that contain more data than will fit into one packet must be divided up into multiple packets and then reassembled by the destination node. In other networks, the length of the packets may be variable (generally, up to some maximum size). If this is the case, then one of the first items in each packet (either just before or just after the addressing information) is a length specifier defining the size of the packet.

Our discussion has treated networks as being either purely circuit switched or purely packet switched. This is usually, but not always, the case. A few systems have adopted an approach known as *integrated switching*, which connects the nodes using both circuit and packet switching. The idea is to take advantage of the capabilities of each approach by moving large amounts of data over circuit-switched paths, while smaller chunks of information are transmitted via packet switching.

As parallel computer systems have become larger and more complex, there has been a trend toward creating networks that use a hybrid approach in which some node-to-node connections use circuit switching and others are made via packet switching. This is particularly true of large multiprocessors that employ the NUMA and COMA architectural models. Shared-memory systems with many processors often have a number of fast, usually circuit-switched, networks (such as a crossbar, multistage network or bus) that tie together local groups of a few nodes. These smaller subnetworks are then tied together by a global, "wider area" network, which often uses packet switching over a ring, mesh, torus, or hypercube interconnect. Systems that have used this hybrid, hierarchical networking approach include the Cray T3D, Stanford DASH and FLASH, HP/Convex Exemplar, and Kendall Square Research KSR-1.

### 6.2.2.3   Static and dynamic networks

As far as we are concerned, these terms are virtual synonyms for packet-switched and circuit-switched networks. In a *dynamic* network, the connections between nodes are reconfigurable, while the pattern of connections in a *static* network remains fixed at all times. Thus, in general, networks that use circuit switching may be termed "dynamic." (A bus may be considered a special case; its basic connection topology is static, but various bus masters can take control of it at different times dynamically and use it to communicate with various slave devices, and in that sense, it is circuit-switched.) Although it would be possible to route packets over a dynamically configured network, in practice, packet-switched networks are almost always static in their connection topology. Therefore, unless otherwise specified, we will consider dynamic networks as being circuit-switched and static networks as being packet-switched.

### 6.2.2.4   Centralized control versus distributed control

In any network, whether static or dynamic, the physical or virtual connections must be initiated and terminated in response to some sort of control signals. These signals may be generated by a single network controller, in which case the network is said to use *centralized control*. Alternatively, it may be possible for several controllers, up to and including each individual node in the network, to allocate and deallocate communications channels. This arrangement, in which there is no single centralized network controller, is known as *distributed control*.

Packet-switched networks are an example of distributed control. There is no centralized controller for the communications network as a whole; rather, the control strategy is distributed over all the nodes. Each node in the network only needs to know how to route messages that come to it, and it normally handles these routings independently, without consulting any other node. Because this type of network uses distributed control, it has few if any limitations on physical size and can be local area or wide area in nature. The Internet, which stretches to all parts of the globe, can be considered the ultimate example of a packet-switched wide area network with distributed control.

Circuit-switched networks usually are characterized by centralized control. Processors (or other master nodes) request service (in the form of connections) from a single network controller that allocates communications links and generates the hardware signals necessary to set them up. In the case of conflicts, the central network controller arbitrates between conflicting requests according to some sort of priority scheme, which only it (and not the nodes) needs to know. The network controller could be an independent processor running a control program, but for speed reasons, it is almost always implemented in hardware (much like a cache controller or DMA controller). Because of the need to quickly send control information from the network controller to the switching hardware in order to make and break connections, it is difficult and expensive to use a centralized control strategy in networks that cover large physical areas. Instead, centralized control is typically used in small, localized networks.

### 6.2.2.5   Synchronous timing versus asynchronous timing

Another way to classify the interconnection networks used in parallel systems is by the nature of the timing by which data transfers are controlled.

In any network, timing signals are necessary to coordinate the transfer of data between a sending node and a receiving node. Data being transferred must be maintained valid by the sender long enough for the receiver to acquire them. In a *synchronous* network, this problem is solved by synchronizing all transfers to a given edge (or, in some cases, both edges) of a common, centrally generated clock signal. The sender is responsible for having data ready for transfer by this time, and the receiver uses the clock signal (or some fixed reference to it) to capture the data. This approach works well for localized interconnection networks, such as a bus or small circuit-switched network, in which all attached components are physically close together and operate at roughly the same speed. However, in situations in which some nodes are much faster than others, when there are many nodes, and/or when nodes are spread over a considerable physical area, synchronous networks are at a disadvantage. If components are of disparate speeds, a synchronous network must operate at the speed of the slowest component (or there must be a means of inserting "wait states" to allow slower devices to respond to communication requests). If there are many nodes or they are widely spread out, it is difficult to distribute a common clock signal to all of them while keeping it in phase so that transfers are truly synchronized.

The alternative is to use an *asynchronous* network in which there is no global timing signal (clock) available. In such a network, the timing of any one transfer of data is unrelated to the timing of any other transfers. Instead, each transfer of data between nodes is regulated by *handshaking* signals exchanged between the two components. Each indicates to the other when it is ready to send or receive data. A simple handshaking protocol is illustrated in Figure 6.22. It is worth noting that message-passing networks, which are usually packet switched, are generally asynchronous due to their distributed control, while circuit-switched networks are more likely to be synchronous (although this is not always the case).

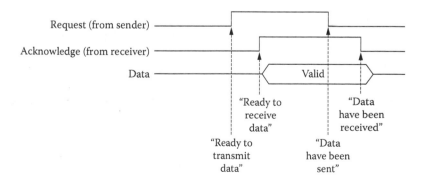

*Figure 6.22* Example handshaking protocol for asynchronous data transfer.

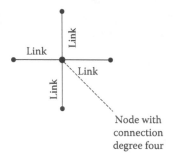

**Figure 6.23** Connection degree of a node.

### 6.2.2.6   Node connection degree

Node connection degree refers to the number of physical communications links connected to a given node. For example, a node with four connections to other nodes (like the one shown in Figure 6.23) is said to be of degree four. In some networks, all nodes are of the same degree; in other networks, some nodes have more connectivity than others. Generally speaking, nodes with high connection degree are more expensive to build, as each additional connection to another node takes up a certain amount of space, dissipates more power, requires more cables to be routed in and out, and so on. However, all else being equal, networks with higher-degree nodes tend to exhibit better communications performance (because of having shorter communication distance; see next section) and may be more tolerant of the failure of some links.

### 6.2.2.7   Communication distance and diameter

Communication distance and diameter are important parameters for interconnection networks. The *communication distance* between two nodes is the number of communication links that must be traversed for data to be transferred between them. In other words, the distance is the number of network "hops" an item must make to get from one node to the other. If there is more than one possible path, the distance is considered to be the length of the shortest path. In Figure 6.24, the distance between nodes A and B is 2, and the distance between nodes B and C is 3. In some networks, the distance between any arbitrarily chosen nodes is a constant; more often, the distance varies depending on which nodes are communicating. One factor that is often of interest (because it can affect network performance) is the average communication distance between nodes. Another important parameter, the *communication diameter* of a network, is the maximum distance between any two nodes. In Figure 6.24, nodes A and C are as widely separated as possible. The distance between them is 4, so the diameter of the network is 4.

In general, the time and overhead required for communication increase as the number of hops required for a data transfer increases. Not only is

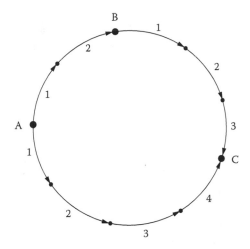

**Figure 6.24** Communication distance and diameter in a network.

there a basic propagation delay associated with the physical length of the communications path and the number and type of components between the sending and receiving nodes, but the longer the path, the greater the chance of contention for resources along the way. If two or more communications are trying to use the same hardware, some must wait while another goes through. Because contention can occur at any node along a path, the longer the distance of the communication, the greater the chance of contention. Therefore, a smaller average or maximum distance between nodes is better—all else being equal. The average distance between nodes thus gives some indication of relative communications performance when technically similar networks are being compared. A network with smaller diameter means that the worst-case communication time is likely to be smaller as well.

### 6.2.2.8 *Cost, performance, expandability, and fault tolerance*

The cost of a network, like the cost of all other computer hardware, is ultimately measured in dollars (or whatever the local unit of currency might be). However, many things factor into that bottom line cost figure. In general, the cost of a network depends on its size and speed; larger, faster networks will inevitably cost more to build than smaller, slower networks. The relationships, however, are not always (or even usually) linear, as we shall see in our examinations of the different types of networks in the following sections of this chapter.

Each link between nodes costs a certain amount to build, not only in terms of dollars spent constructing hardware, but in other ways, such as increased power dissipation, maintenance costs, and (sometimes most importantly of all) physical space taken up by the communications hardware on each end and the cables or other media used to make the connection. If a wireless network is used (uncommon in today's high-performance computers, but perhaps more common in the future), then each link costs a certain amount of electromagnetic spectrum. (As economists tell us, anything that makes use of scarce resources, whatever form those resources may take, incurs a cost.)

The cost of the individual links, however they are implemented, is normally related to their bandwidth (see below; all else being equal, higher-capacity links will cost more). The total number of links depends on the number of nodes and their connection degree (which is governed by the topology of the network). The total cost of a network is roughly proportional to the number of links that compose it, although economies of scale are applicable in some situations, while in others increases in size beyond a certain point may increase cost more rapidly.

The performance of a network depends on many factors and can be measured in different ways. One very important factor is the raw information *bandwidth* of each link. (This is defined in the same sense as it was in Section 1.6 for memory and I/O systems: the number of bits or bytes of information transferred per unit of time. Common units of bandwidth are megabits, megabytes, gigabits, or gigabytes per second.) More bandwidth per link contributes to higher overall communications performance. In conjunction with the number of links, the bandwidth per link determines the overall peak bandwidth of the network. (This is an important performance parameter, but, like peak MIPS with respect to a CPU, it does not tell us all we need to know because it fails to take topology or realistic application behavior into account.) We might also note that in networks (as in memory and I/O systems) the peak available bandwidth is not always used.

Other factors are just as important as link bandwidth in determining overall network performance. Node connection degree (which, in conjunction with the number of nodes, determines the number of links) also plays a major role; a node connected via multiple links has access to more communications bandwidth than one with a single link of the same speed. Another consideration that is significant in some types of networks is communications overhead. Besides the actual data transferred, the network must (in many cases) also carry additional information, such as packet headers, error detection and correction information, etc. This overhead, although it may be necessary for network operation, takes up part of the information-carrying capacity of the network. Any bandwidth consumed by overhead takes away from the gross bandwidth of the network and leaves less net bandwidth available for moving data.

Although peak and effective bandwidth (per node or overall) are important performance measures, the one that is most significant in many situations is simply the time taken to deliver an item of data from its source to its destination. This amount of time is defined as the *network latency*. In some networks, this value is a constant; in many others it is not. Therefore, measures of network performance may include the minimum, average, and maximum times required to send data over the network.

Latency may depend on several factors, most notably the speed of the individual communications links (which is related to bandwidth) and the number of links that must be traversed. It also may depend on other factors, including the probability of delays due to congestion or contention in the network. In some cases, the network latency is a known, constant value; but in most networks, although there may be a minimum latency that can be achieved under best-case conditions, the maximum latency is considerably longer. (It is not even necessarily true that network latency is bounded; for example, any network using Transmission Control Protocol/Internet Protocol [TCP/IP] as the transport mechanism does not guarantee that message packets will be delivered at all.) Perhaps the best measure for evaluating the performance of the network is the *average latency* of communications, but one must be careful to measure or estimate that value using realistic system conditions. It is likely to vary considerably from one application to another based on the degree of parallelism inherent in the algorithm, the level of network traffic, and other factors. Thus, as in the case of evaluating CPU performance, we should be careful to choose the most applicable network benchmark possible.

*Expandability* is the consideration of how easy or difficult, how costly or inexpensive, it is to add more nodes to an existing network. If the hardware required for a node is cheap and the structure of the network is fairly flexible, then it is easy to expand the machine's capabilities. This is one of the most desirable characteristics of a parallel system. A situation often arises in which we, the users, didn't get the performance gain we hoped for by building or buying a system with $n$ nodes, but we think that if we add $m$ more, for a total of $n + m$ nodes, we can achieve the desired performance goal. The question then becomes, "what (above and beyond what the additional node processors, memories, etc., cost) will it cost us to add the necessary connections to the network?"

In some types of networks, adding a node costs a fixed amount, or, to borrow the "big O" notation used in algorithm analysis, the cost function is $O(n)$, where $n$ is the number of nodes. In other words, a 16-node configuration costs twice as much as an 8-node configuration but half as much as a 32-node configuration. For other types of networks, the cost can vary in other ways, for example, logarithmically [$O(n \log n)$] or even quadratically [$O(n^2)$]. In this last case, a network with 16 nodes would cost four times as much as one with 8 nodes, and a network with 32 nodes would cost 16 times as much as an 8-node network. All else being equal, such a network is not preferable if there is any chance the system will need to be expanded after its initial acquisition.

*Fault tolerance* in interconnection networks, as in most other systems, generally ties in with the idea of redundancy. If there is only one way to connect a node to a given other node and the hardware fails somewhere along that path, then further communication between those nodes is impossible and, most likely, the system will not continue to function. (Alternatively, one may be able to stop using some of the nodes and function in a degraded mode with only a subset of the nodes operating.) If the network has some redundancy (multiple paths from each node to other nodes) built in, then a single hardware fault will not cripple the system. It will be able to keep operating in the presence of a failed link or component with all communications able to get through, although perhaps a bit more slowly than normal, and all nodes will still be accessible.

Another benefit of a network with redundant connections, even in the absence of hardware faults, is that temporary contention or congestion in one part of the network can be avoided by choosing a different routing or connection that bypasses the affected node. This may improve performance by reducing the average time to transmit data over the network.

Both packet-switched and circuit-switched networks may have redundancy built in. In circuit-switched networks, the redundancy must be in the hardware (redundant switching elements and connections). In some packet-switched networks, however, additional redundancy can be achieved using the same set of hardware connections by adopting control and routing strategies that are flexible enough to choose alternate paths for packets when problems are encountered. This can have beneficial implications for both fault tolerance and performance.

### 6.2.3   Bus-based interconnections

The simplest and most widely used interconnection network for small systems, particularly uniprocessors and symmetric multiprocessors, is a *bus*. A bus, as depicted in Figure 6.25, is nothing more or less than a set of connections (e.g., wires or printed circuit traces) over which data, addresses, and control and timing signals pass from one attached component to another. Connections to a bus are normally made using tri-state

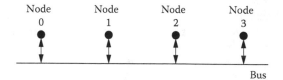

*Figure 6.25* Use of a bus to interconnect system nodes.

devices, which allow devices not involved in a particular data transfer to open-circuit their outputs and effectively isolate themselves from the bus temporarily. Meanwhile, the sending device enables its outputs onto the bus while the receiving device enables its inputs.

Control signals define the type and timing of transfers in a bus-based system. Buses may be either synchronous or asynchronous in their timing. Normally, each component connected to a bus is of degree one although it is possible for processors or memory modules to have multiple bus interfaces. The distance between any two components, and thus the diameter of the bus, is also one because the bus is not considered a node in and of itself but merely a link between nodes. (If the bus were considered a node, the distance between connected components would be two.) Because all the nodes have degree one and are connected to the same bus, any master node (CPU, DMA, IOP, etc.) can access any other node in essentially the same time unless there is a conflict for the use of the bus.

The advantage of a bus interconnection is its relative simplicity and expandability. Up to a point, it is easy to add additional master or slave nodes to the bus and expand the system. The one great limitation of a bus is that only one transaction (transmission of data) can take place at a time; in other words, it exhibits no concurrency in its operation. The data transfer bandwidth of the bus is fixed and does not scale with the addition of more components. Although in a multiprocessor system most memory accesses are satisfied by local caches and thus do not make use of the system bus, a certain fraction will require use of the bus to reach main memory. I/O transactions, including DMA, also use the bus. The more processors there are in the system, the more traffic will be seen on the bus. At some point, the needed bandwidth will exceed the capacity of the bus, and processors will spend significant time waiting to access memory. Thus, a single-bus interconnection is practically limited to a system with relatively few nodes.

To get better performance using the same general approach, designers can use a multiple-bus architecture, such as the dual-bus configuration shown in Figure 6.26; this would allow more pairs of nodes to communicate simultaneously, but at increased hardware cost. Scheduling and arbitrating multiple buses is also a complex issue, and cache coherence schemes (see Section 6.1.3) that depend on snooping a system bus are also

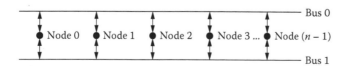

*Figure 6.26* Dual-bus interconnection.

more difficult when multiple buses must be monitored. In order to be able to keep up with the demand for data communications in a larger system (cluster or multiprocessor with many nodes), other types of networks that scale up better may be required.

When multiple buses are used in a system, it is common for them to have different widths, speeds, and other characteristics depending on the nature of the components they are used to connect. A very common example is the bus structure of a typical sixth-generation Intel- or AMD-based PC as shown in Figure 6.27. A particular CPU chip may have two, four, or more computing cores, each with its own private L1 (and often L2) caches. These cores share equal access to the lowest level (usually L3) of cache and to main memory, forming a symmetric multiprocessor on a chip. An internal or "back side" bus can be used to connect the cores and caches, and an external or "front side" bus connects to the rest of the system via the motherboard chipset. The chipset itself is normally divided into a *northbridge* (or memory controller hub), which operates at high speed to connect the CPU to DRAM and the GPU, and a *southbridge*

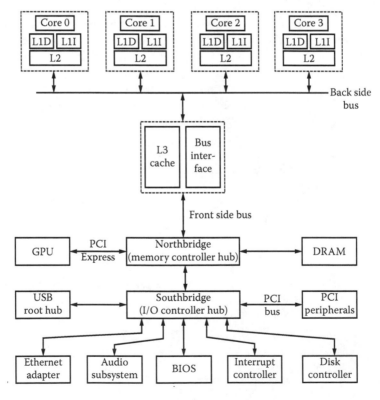

*Figure 6.27* Bus structure of a typical PC.

(or I/O controller hub), which operates at lower speed to transfer data to/from the USB root hub, disk controller, wired and/or wireless network adapters, etc.

It is worth noting that newer (since 2011) processors based on AMD's Accelerated Processing Unit and Intel's Sandy Bridge (and subsequent) microarchitectures have moved the northbridge logic onto the CPU chip itself, replacing the front side bus (FSB) with a HyperTransport (a.k.a. HT, created by AMD) or QuickPath Interconnect (QPI, found in Intel chips) connection. Besides allowing for faster access to RAM than is possible through the FSB and northbridge, HT and QPI can also be used to connect to additional CPU chips on a motherboard with multiple processor slots, forming a NUMA system.

## 6.3    Static interconnection networks

A static interconnection network, simply put, is one in which the hardware connections do not change over time. As noted in the previous section, a given connection topology is put into place and remains in place while the system is running regardless of which nodes need to communicate with each other. At various times, different nodes use the same physical connections to transfer data. Static networks are commonly found in clusters (multicomputers) in which message packets must be sent from one node to another. They may also be found in large (NUMA/COMA) multiprocessors, in which they can serve as the global, system-wide network that connects together the smaller, local networks. We examine some of the more common static network topologies in this section.

### 6.3.1    Linear and ring topologies

Two of the simplest static networks are *linear* and *ring* networks; an example of each is shown in Figure 6.28. The topology of each is fairly obvious from its name. A linear network, or linear array, is a structure in which each node is directly connected to its neighbors on either side. The two terminal (end) nodes have neighbors only on one side. Thus, all nodes are degree two except for those on the ends, which are of degree one. The ring network is identical to the linear network except that an additional link is

(a)                                            (b)

*Figure 6.28* (a) Linear and (b) ring networks.

added between the two end nodes. This makes all nodes identical in connectivity with a connection degree of two.

A linear network with $n$ nodes has a communication diameter of $n-1$, and a ring network of the same size has a diameter of $\lfloor n/2 \rfloor$ ($n/2$ rounded down to the next smaller integer). Thus, a ring of eight nodes (as shown in Figure 6.28b) has a diameter of 4, while one with seven nodes would have a diameter of 3. The above analysis presumes that all communication links are bidirectional, which is virtually always the case. A linear network with unidirectional links would not allow some nodes to send data to others. A ring network could be made to work with unidirectional links but would not perform well, nor would it exhibit the enhanced fault tolerance (due to the two possible routing paths between any two nodes) that is the other main advantage of a ring over a linear network. (Between any two nodes in a linear network, there is only one possible path for the routing of data, while there are two possible paths between arbitrary nodes in a ring network. Thus, if one link fails, the ring can still function in degraded mode as a linear array. All nodes will still be able to communicate, although performance is likely to suffer to some extent.)

Linear and ring networks are simple to construct and easily expandable (all one has to do is insert another node with two communications links anywhere in the structure). However, they are impractically slow for systems with large numbers of nodes. Communications performance can vary over a wide range because the distance between nodes may be as little as 1 or as great as the diameter (which can be very large for a network with many nodes). A linear network is not fault tolerant, and a ring network breaks down if more than one fault occurs.

### 6.3.2   Star networks

Another simple network with a static interconnection structure is the *star network* (see Figure 6.29). In a star network, one central node is designated as the *hub*, and all other nodes are directly connected to it. Each of these peripheral nodes is of degree one, and if there are $n$ peripheral nodes the

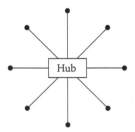

*Figure 6.29* Star network.

hub is of degree $n$. The diameter of the star network is 2 regardless of the number of nodes; it takes one hop across a network link for a piece of data to reach the hub and one more for it to reach its destination, assuming the destination is not the hub itself. This assumption is generally true. Because the hub sees much more communications traffic than any other node, it is common for it to be dedicated to communications rather than consisting of general-purpose computing hardware. The hub may include a dedicated communications processor, or it may simply be a nonprogrammable hardware switching device.

The main advantage of a star network is its small diameter for any number of nodes. Its main disadvantage is the large amount of traffic that must be handled by one node (the hub). As more nodes are added to the network, the amount of communications bandwidth required of the hub increases until at some point it is saturated with traffic. Adding more nodes than this will result in nodes becoming idle as they wait for data. If the advantages and disadvantages of a star network sound familiar, they should; they are essentially identical to those of a bus. If one considers the bus hardware itself to be a node, it corresponds directly to the hub of a star network. Both are potential bottlenecks that become more of a problem as more nodes are added to the system.

### 6.3.3    Tree and fat tree networks

A *tree network* is one in which the nodes are connected by links in the structure of a binary tree. Figure 6.30 shows a tree network with seven nodes. It is easy to see that the binary tree network illustrated here is not constant degree. Leaf nodes are of degree one, the root node has degree two, and the nodes at intermediate levels of the tree all have a connection degree of three, as they are connected to one node above them and two below.

The communication diameter of a binary tree network is obviously the distance between leaf nodes on opposite sides of the tree. In Figure 6.30, the diameter of the seven-node tree network is 4. Adding another level (for a total of 15 nodes) would give a diameter of 6. In general, the diameter is $2(h - 1)$, where $h$ is the height of the tree (the number of levels of nodes).

*Figure 6.30* Tree network.

*Figure 6.31* Fat tree network.

A tree network is not ideal if communication between physically remote nodes is frequent. It has a relatively large diameter compared to the number of nodes (although not as large a diameter as the ring or linear networks). Another disadvantage is that all communications traffic from one subtree (right or left) to the other must pass through a single node (the root node); that node and the others near it will tend to be the busiest in terms of data communications. Although this is not quite as serious a performance limitation as the hub node in a star network (through which all traffic must pass), it does imply that the root (and other nodes close to it) may tend to saturate if the tree is large or the volume of communication is heavy.

In such a case, if designers want to maintain the tree topology, a better solution is the use of a *fat tree* network (see Figure 6.31). It has the same overall structure as the regular binary tree, but the communications links at levels close to the root are duplicated or of higher bandwidth so that the increased traffic at those levels can be better handled. Lower levels, which handle mostly local communications, can be lower in bandwidth and thus can be built in a less costly fashion.

Tree and fat tree networks have few advantages for general computing, although the tree topology (like any specialized interconnection) may map well to a few specific applications. Although these structures do scale better to a larger number of nodes than the linear, ring, or star networks, they lack some of the advantages of the network topologies covered in the following sections. Thus, tree networks are rarely seen in large-scale parallel machines.

### 6.3.4  Nearest-neighbor mesh

One type of interconnection network that has been used in many medium- to large-scale parallel systems (both array processors and multicomputers)

**Figure 6.32** Two-dimensional nearest-neighbor mesh network.

is the *nearest-neighbor mesh*. It has most frequently been constructed in two dimensions as shown in Figure 6.32, although it is possible to extend the concept to three (or more) dimensions. (Note that a linear network as shown in Figure 6.28a is logically equivalent to a one-dimensional nearest-neighbor mesh.) Each interior node in a two-dimensional mesh interconnection is of connection degree four, with those on the edges being of degree three and the corner nodes, degree two. One can think of the nodes as equally spaced points on a flat map or graph with the communication links being lines of latitude and longitude connecting them in the east–west (x) and north–south (y) directions.

The higher degree of connectivity per node as compared to the networks we previously considered tends to keep the average communication distance (and the diameter or worst-case distance) smaller for the nearest-neighbor mesh. For example, consider a system with 16 nodes. If they are connected in linear fashion, the diameter is 15; if in a ring, 8; if they are connected in the mesh of Figure 6.32, the diameter is only 6. In general, the diameter of an $n$ by $n$ (also written as $n \times n$) square mesh (with $n^2$ nodes) is $(n - 1) + (n - 1) = 2(n - 1) = 2n - 2$. The price one pays for this decreased diameter (which tends to improve communication performance) is the price of building more links and making more connections per node. The routing of messages from one node to another is more complex as well, because multiple paths exist between any two nodes. However, this flexibility provides fault tolerance, as a nonfunctional or overburdened link may be routed around (provided the situation is recognized and the routing protocol allows for this).

## 6.3.5   Torus and Illiac networks

The one exception to the regular structure of the two-dimensional nearest-neighbor mesh just discussed is the reduced connectivity of the edge and corner nodes. It is often convenient to build all the nodes in a network the same way; if most of them are to have four communications interfaces, for simplicity's sake, the others might as well have that capability, too. In the nearest-neighbor mesh, some nodes do not make use of all four of their possible links. Two other networks, the *torus network* and the *Illiac network*, have all degree four nodes and thus make use of the full number of links per node. The 2-D torus network is simply a two-dimensional nearest-neighbor mesh with straightforward edge connections added to connect the nodes at opposite ends of each row and column. Looking at the nearest-neighbor mesh shown in Figure 6.32, it is clear that if links were added to connect the leftmost and rightmost nodes in each row, the result would be a cylindrical topology as shown in Figure 6.33a. By making the same type of connections between the nodes at the top and bottom of each column, one transforms its geometrical configuration into that of a torus (better known to most of us as the shape of a doughnut or bagel), as shown in Figure 6.33b.

Figure 6.34 shows a 16-node torus network depicted to emphasize its similarity to the nearest-neighbor mesh rather than its toroidal geometry.

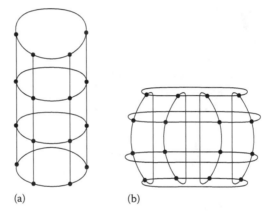

(a)                              (b)

***Figure 6.33*** Formation of (a) cylindrical and (b) toroidal networks from nearest-neighbor mesh.

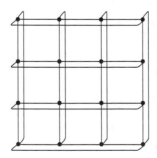

***Figure 6.34*** Torus network with 16 nodes.

As one might imagine, the characteristics, advantages, and disadvantages of this network are similar to those of the two-dimensional mesh. Because of the additional connectivity, its communication diameter is somewhat smaller than that of a mesh without the edge-to-edge connections. In general, the diameter of a torus with $n^2$ nodes (an $n \times n$ mesh plus edge connections) is $2 \lfloor n/2 \rfloor$ ($= n$, if $n$ is even). In the case of the 16-node network shown, the diameter is only 4 instead of the two-dimensional mesh's 6. The torus is also somewhat more fault-tolerant than the mesh because more alternative routings exist between nodes. Of course, the cost is slightly higher because of the additional links that must be put in place.

It is worth noting that, like its relative the nearest-neighbor mesh, the torus network is not necessarily limited to connecting in the usual two (x and y) dimensions. A specific example of a sixth-generation supercomputer family built with a three-dimensional torus network is the Cray XT series of machines (XT3, XT4, XT5, and XT6). The SeaStar (and later Gemini) routers used in these systems connect to their local AMD Opteron-based processing node over a HyperTransport link, and to six remote nodes (+x, +y, +z, −x, −y, and −z) over high-speed network links. By forming fully connected rings in all three dimensions, the linked routers become nodes in a 3-D torus network that allows efficient communication between tens of thousands of processors. Although we shall not attempt to depict such a network on the flat pages of this book, the reader can imagine the structure shown in Figure 6.34 with three more sets of 16 nodes behind the ones shown and with all four 16-node layers of the structure connected in the z dimension just as the nodes shown already are connected in the x and y dimensions.

Another network that is very similar in topology to the torus is the Illiac network—so named because it was used to interconnect the processing elements of the ILLIAC-IV array processor. The original Illiac network had 64 nodes; for simplicity, Figure 6.35 depicts a network with the same structure but only 16 nodes. Notice that the edge connections in one

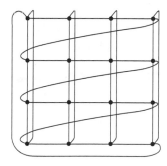

*Figure 6.35* 16-node Illiac network.

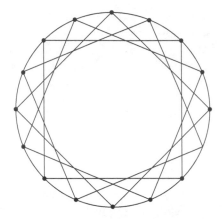

***Figure 6.36*** Alternative representation of Illiac network.

dimension (the columns in Figure 6.35) are identical to those in the torus network. In the other direction, the connections are a bit different. The rightmost node in each row is connected to the leftmost node in the next row, with the rightmost node of the last row being linked to the leftmost in the first row. As in the 2-D torus network, this results in all nodes being of connection degree four.

Figure 6.35 shows the Illiac network as a modified two-dimensional nearest-neighbor mesh. Alternatively, it can be viewed as a large ring network plus several smaller, internal ring networks that provide additional connectivity. The reader can trace the horizontal connections in Figure 6.35 from one row to the next and back to the first, confirming that they alone form a continuous ring among the 16 nodes. It is easy to see that each set of nodes in vertical alignment forms a ring of four nodes when the top-to-bottom edge connections are considered. Thus, an $n \times n$ Illiac network with $n^2$ nodes consists of $n$ rings of $n$ nodes each, interspersed within a larger ring that connects all $n^2$ nodes. This interpretation of the network's structure is emphasized in Figure 6.36.

The characteristics of an Illiac network are similar to those of a 2-D torus network. The connection degree of the nodes is the same (four), and the diameter is the same or slightly less (for $n^2$ nodes, the diameter is $n - 1$). Both the Illiac and torus networks exhibit somewhat better communications performance than the simple two-dimensional mesh at a slightly higher cost.

### 6.3.6   Hypercube networks

The nearest-neighbor mesh, torus, and Illiac networks have nodes of higher degree than the linear, ring, and star networks. Their resulting lower communication diameters make them more suitable for connecting larger numbers of nodes. However, for massively parallel systems in which logically remote nodes must communicate, even these networks may have unacceptably poor performance. One type of network that scales well to very large numbers of nodes is a *hypercube* or *binary n-cube* interconnection. Figure 6.37 shows an eight-node hypercube (three-dimensional hypercube or binary 3-cube) that has the familiar shape of a geometrical cube. Each node has a connection degree of three, and the diameter of the network is also 3. These numbers are not coincidental; a hypercube network of $n$ dimensions always has a diameter of $n$ and contains $2^n$ nodes, each of degree $n$.

Notice that if we number the nodes in binary, from 000 (0) to 111 (7), it takes 3 bits to uniquely identify each of the eight nodes. The numbering is normally done such that the binary addresses of neighboring nodes differ only in one bit position. One bit position corresponds to each dimension of the cube. Assigning the node addresses in this way simplifies the routing of messages between nodes.

It is interesting to note that hypercubes of dimension less than three degenerate to equivalence with other network topologies we have studied. For example, a binary 2-cube, with four nodes, is the same as a 2 × 2 square nearest-neighbor mesh (or a 4-node ring network). A binary 1-cube, with just two nodes connected by one communications link, equates to the simplest possible linear array. In the lower limit, a binary 0-cube (with a single node) would not be a parallel system at all; instead, it would represent a uniprocessor system with no network connectivity.

Hypercubes of more than three dimensions are not difficult to construct and are necessary if more than eight nodes are to be connected with this type of topology. They are, however, increasingly difficult to diagram on a flat sheet of paper. Figure 6.38 shows two alternative representations of a four-dimensional hypercube (binary 4-cube). The important thing to notice is that each node is connected to four others, one in each dimension

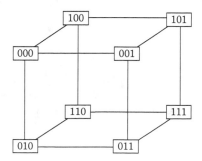

*Figure 6.37* Three-dimensional hypercube network.

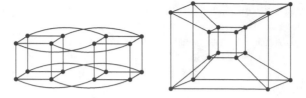

***Figure 6.38*** Four-dimensional hypercube network.

of the cube. (Binary node addresses would in this case be composed of 4 bits, and each of the four nodes adjacent to a given node would have an address that differs in one of the four bit positions.) By extension, one can easily infer (if not easily depict on a flat surface) the structure of a binary $n$-cube with five, six, or more dimensions. Hypercubes of up to 16 dimensions (up to 65,536 nodes) have been constructed and used in massively parallel systems.

Hypercube networks, by keeping the communication diameter small, can provide good performance for systems with large numbers of nodes. However, they do have certain limitations. First, the number of nodes must be an integer power of two to maintain the cube structure, so incrementally increasing performance by adding just a few nodes is not feasible as it is for some of the other network types. Also, the node connection degree increases with each new dimension of the network. This may cause difficulties with construction and result in a network that tends to be more expensive than others for the same number of nodes. When we increase the order of the cube by one, we only add one more connection per node but we double the number of nodes. Therefore, the total number of communication links increases dramatically with the dimensionality of the network. An $n$-cube with $2^n$ nodes requires $(n \times 2^n)/2 = n \times 2^{n-1}$ connections. That is only 12 connections for a binary 3-cube with 8 nodes, but the number grows to 24,576 connections for a 12-cube with 4096 nodes and 524,288 connections for a 16-cube with 65,536 nodes. In this as in so many other aspects of computer systems design, there is no such thing as a free lunch.

### 6.3.7   Routing in static networks

Because the network connections themselves are not changed in static networks, the logic used to decide which connections must be utilized to get data to the desired destination is crucial. We do not usually want to broadcast a particular item to all nodes, but rather to send it to a particular node where it is needed. The method of determining which connection, or sequence of connections, must be followed in order for a message to get to its destination in a packet-switched network is referred to as the routing protocol. In addition, the manner in which message packets are transferred from node to node along the chosen routing path has great significance with regard to performance, especially in larger networks. We shall examine some aspects of message routing over static networks in a bit more detail.

*Store and forward routing* is a data transfer approach that has been in use in networks (most notably the Internet) for many years. In a system using this type of routing, messages are sent as one or more packets (which are typically of a fixed size). When a message is sent from one node to another, the nodes may (if we are fortunate) be directly connected, but in most cases the packets will have to pass through one or more intermediate nodes on their way from the originating node to the destination. To facilitate this, each node has FIFO storage buffers the size of a packet. As a message packet traverses the network, at each intermediate node the entire packet is received and assembled in one of these buffers before being retransmitted to the next node in the direction of the destination (as determined by the particular routing protocol being used). Packets are treated as fundamental units of information; at any given time, a packet is either stored (completely buffered) in one node or in the process of being forwarded (transmitted) from one node to a neighboring node; thus, this method is referred to as store and forward routing. When store and forward routing is used, a packet is never dispersed over multiple nodes, regardless of how many hops must be traversed between the sending and receiving nodes. This approach simplifies the process of communication, but delays can be substantial when the sending and receiving nodes are widely separated.

*Wormhole routing* is an alternative strategy that was developed later to improve message-passing performance, especially in large networks with many nodes through which messages must pass. It was first introduced by William Dally and Charles Seitz and was first used commercially in the Ametek 2010 multicomputer, which featured a two-dimensional nearest neighbor mesh interconnection. Academic research on wormhole routing was done using Caltech's Mark III hypercube and Mississippi State's Mapped Array Differential Equation Machine (MADEM).

In this approach, each message packet is divided up into flow control digits, or *flits*, each typically the size of an individual data transfer (the channel width between nodes). When a message packet is transmitted from one node to the next, the flits are sent out one at a time. The receiving node checks to see if the message is intended for it and, if so, stores the flits in a receive buffer. If the message is passing through on the way to another node, the correct output channel is identified, and the first flit is sent on its way as soon as possible. Subsequent flits are sent after the first one as rapidly as possible, without waiting for the entire packet to be assembled in any intermediate node. If the destination node is several hops away from the sending node, the message will likely become spread out over several intermediate nodes and links, stretching across the network like a worm (hence the name of the technique).

Figure 6.39 shows how message packets might be formatted in a parallel machine connected by a two-dimensional mesh network. This example uses a *deterministic* routing, which means that any message sent from one node to another specific node always follows the same path. Specifically, in this case messages are always routed in the X direction (east–west) as far as they need to go, and only then in the Y direction (north–south). The first flit is always the X address, expressed in a relative format (+3 means three hops to the east; –2 means two hops to the west).

At each intermediate node, the absolute value of the X address is decremented by one so that it reaches zero at the node at which the message must change directions. At that point, the X address flit is stripped off, and the second flit (the Y address) becomes the leading flit. If it is found to be zero, the message is at the destination node and is pulled off the network as the remaining flits come in. If it is positive, the message is routed north, and if it is negative, the message is routed south. Again, at each node in the Y direction, the Y address flit is decremented toward zero, and when it reaches zero, the message has reached its destination. If the message packets are not all the same length, the third flit (behind the X and Y address flits) would contain the message length information, informing the routing hardware of how many additional flits are to follow so that each node knows when a message has passed through in its entirety. Once this happens, the sending and receiving *channels* (input or output pathways in particular directions) are freed up so that they can be allocated to other messages as necessary.

It is important to note that a node cannot reallocate the same input or output channel to another message until the current message is completely received or transmitted over that channel. Channels must be dedicated to one message at a time. Intermixing flits from more than one

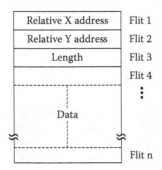

*Figure 6.39* Wormhole routing example.

message over the same channel does not work because there is no way to tell them apart. Address, length, and data flits all appear the same to the hardware; the only way the routers can tell which flit is which is by counting flits until the end of a message is reached. Intermixing flits from multiple messages would spoil the channel's count. (It is worth noting that some networks split the bandwidth of one physical channel into several *virtual channels* by providing multiple buffers and keeping track of the state of each virtual channel separately.)

The dedication of specific channels to a particular message means that often a message will have to temporarily stall in the network because some other message has already started using a channel that this message needs in order to make progress toward its destination. For example, in Figure 6.40, node A1 is trying to send a message to node A2, and node B1 is trying to send a message to node B2. Both messages have to pass through node C's router, exiting toward their destinations via that node's north output channel. Say the message from A1 arrives at C first and a connection is established between C's west input channel and its north output channel to service this message. Subsequently, the first flit of the message from B1 arrives at C via C's south input. The routing hardware at node C examines the address flit of this message and discovers that it needs to be sent out via the north output channel, but that is already allocated to the other message. Therefore, this message must wait until the last flit of the first message exits via C's north output.

In this type of situation, the flits of a stalled message may be temporarily left spread out over one or more nodes of the network, with further progress temporarily impossible. Because each intermediate node has an input and output channel reserved for the stalled message, performance may drop due to network congestion as messages block other messages. Performance can be improved if each node has FIFO buffers (like those

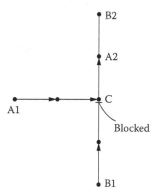

*Figure 6.40* Example of stall caused by contention for a channel.

used in systems with store and forward routing) associated with each of its input channels to store incoming flits as they arrive. This hybrid approach, known as *cut-through* (or *virtual cut-through*) routing, allows flits to proceed as far as possible toward their destination and thus helps keep stalled messages from blocking others.

Stalling may also be reduced and performance further improved if the system uses an *adaptive* (or *dynamic*) routing algorithm such that messages do not always have to follow the same path from point A to point B. In such a system, messages may be routed around areas of congestion where channels are already in use. For example, in Figure 6.41, if node A1 is communicating with node A2, the routing hardware might choose to send a message from B1 to B2 over a different path as shown, avoiding a delay.

The major caveat in designing an adaptive routing algorithm for a network using wormhole routing is the absence of *deadlock*, a situation in which messages permanently stall in the network because of mutual or cyclic blocking. (In other words, message A blocks B at one point, but somewhere else B blocks A; or A blocks B, which blocks C, which blocks A, or some such.) The simple east–west, then north–south, routing algorithm presented first has been shown to be deadlock-free, guaranteeing that all messages will eventually reach their destinations. Adaptive routing algorithms attempt to be as flexible as possible, but some path restrictions must be placed on routing in order to meet the deadlock-free criterion.

In a system with wormhole routing, communications are effectively pipelined. (For this reason, wormhole routing is sometimes referred to as "pipelined routing.") When the sending and receiving nodes are adjacent in the network, wormhole routing is no faster than store and forward routing. When they are far apart, however, the pipelining effect means that the message may arrive at its destination much sooner. Figure 6.42

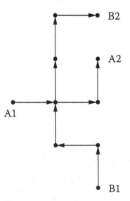

*Figure 6.41* Alternative routing in a system with wormhole routing.

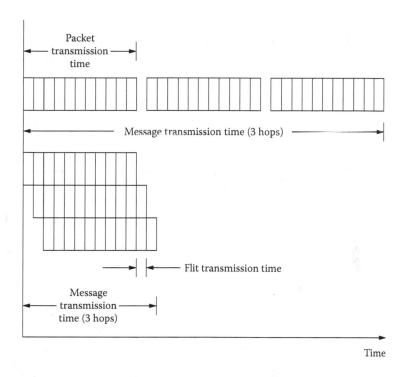

**Figure 6.42** Store and forward routing versus wormhole routing.

compares the two approaches based on the time taken to communicate between remote nodes.

When store and forward routing is used, a message packet works its way sequentially across the network, moving from one node along its path to the next. The packet does not even begin to arrive at the destination node until after it has been completely received and buffered by the next to last node in its path. In the absence of congestion-induced stalls, the total time required for communication is proportional to the number of hops between nodes.

When wormhole routing is used (assuming no stalls occur along the path), the message starts being received as soon as the first flit can traverse the network to the destination. The remaining flits follow the first as fast as the intermediate nodes can receive and send them, arriving right on the heels of the first flit. The overall time to send the message is dominated by the size of the message rather than the number of network hops; each additional node traversed adds only a minimal increment (in the best case, the time to send and receive an individual flit) to the overall communication time. Thus, the advantage of wormhole (or cut-through) routing over store and forward routing becomes greater as the size of the network increases.

In summary, static, packet-switched networks come in a variety of topologies and use a variety of routing techniques to get messages to their intended destinations. Each static network configuration has its own characteristics, advantages, and disadvantages. In general, increased performance and robustness can be obtained by utilizing more sophisticated routing techniques and by increasing the number of communications links relative to the number of nodes. (Of course, this also increases the cost of implementation.) Because packet-switched networks are commonly used in multicomputers/clusters, which are becoming more and more popular in high-performance computing, continued research, innovation, and improvement in this aspect of parallel systems design is likely.

## 6.4   Dynamic interconnection networks

Dynamic networks are those in which the physical connections within the network are changed in order to enable different nodes to communicate. Hardware switching elements make and break connections in order to establish and destroy pathways for data between particular nodes.

Some dynamic, circuit-switched networks have direct connections (single switches) between nodes; these are known as *single-stage* networks. These networks can be very complex and expensive to build and can be very costly for large numbers of nodes. If the network has a large number of nodes, it is more usual to have signals travel through several, simpler, cheaper switches (with a small number of inputs and outputs) to get from one node to another rather than one complex, expensive switch. If the same set of switches is used iteratively in order to transfer data from one node to another, it is called a recirculating network; if several sets (stages) of smaller switches are connected in sequence to take the place of a single, large switch, we refer to the network as a multistage interconnection network.

### 6.4.1   Crossbar switch

The highest-performance dynamic interconnection network is a single-stage network known as a *crossbar switch*, specifically a *full crossbar* (so termed to distinguish it from a staged network made up of smaller crossbar switching elements). An $m \times n$ crossbar switch is a hardware switching network with $m$ inputs and $n$ outputs (usually, but not always, $m = n$) in which a direct connection path can be made from any input to any output. The hardware is so duplicated that no connection interferes with other possible connections; in other words, making a connection from input $i$ to output $j$ does not prevent any input other than $i$ from being connected to any output other than $j$. At any time, several independent, concurrent connections may be established (as many as the smaller of $m$ or $n$), subject only to the restriction that a given input can only be connected to one output and vice versa.

The switching elements in a crossbar are often bidirectional, enabling data to be sent or received over the same connection. Thus, it might be more accurate in many situations to simply refer to nodes as being on one or the other side of the crossbar switch. Often, particularly in symmetric multiprocessors, the nodes on one side are master nodes (such as CPUs), and those on the other side are slave nodes (e.g., memory modules). An 8 × 8 full crossbar switch, which would allow up to eight simultaneous connections, is depicted in block diagram form in Figure 6.43.

It is worthwhile to take the time to consider how a crossbar switch might be implemented in hardware. Consider the simplest possible case of a 2 × 2 crossbar switching element. (We take this example not only for the sake of simplicity but because such elements are used as building blocks in other types of networks that we shall consider later.) The most basic 2 × 2 switch has only two states: *straight-through* connections (input 0 to output 0 and input 1 to output 1) and *crossover* connections (input 0 to output 1 and input 1 to output 0). A single control input is sufficient to select between these two states. This basic 2 × 2 crossbar is depicted in block diagram form in Figure 6.44.

Now consider what type of circuitry is required to implement such a switch. First, consider the simplest case, in which data transfer through

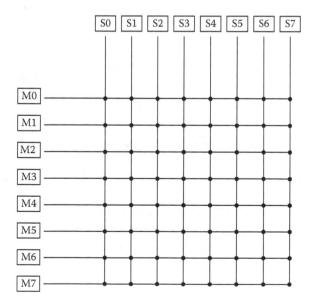

*Figure 6.43* 8 × 8 full crossbar switch.

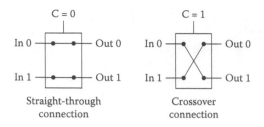

Straight-through
connection

Crossover
connection

*Figure 6.44* Block diagram of basic 2 × 2 switching element.

the switch is unidirectional (only from the designated input side to the output side). Considering that either of the two inputs must be able to be connected to a given output, the circuitry driving either output is equivalent to a 2 to 1 multiplexer. Thus, the complete circuit for the 2 × 2 switch (see Figure 6.45) effectively consists of two such multiplexers, with the sense of the control or select signal inverted such that the input that is not routed to one output is automatically routed to the other.

If the switching element needs to be bidirectional, such that data may be transferred from the designated output side to the input as well as from input to output, the logic gates making up the multiplexers must be replaced by bidirectional tri-state buffers, CMOS transmission gates, or some other circuit that allows signals to flow in either direction. Two possible implementations of this type of circuit are illustrated in Figure 6.46.

In some systems, it is important for a single node to be able to *broadcast* data (send them to all nodes on the other side of the crossbar). To facilitate this, we can construct the switch such that in addition to the straight-through and crossover configurations, it can be placed in the *upper broadcast* or *lower broadcast* modes. Upper broadcast simply means that input 0 is connected to both outputs while input 1 is open-circuited, and lower broadcast means the reverse. The four possible states of such a

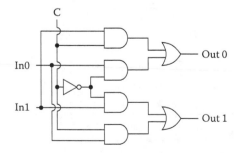

*Figure 6.45* Logic diagram for basic 2 × 2 switching element.

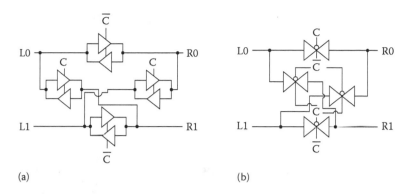

(a)                                                    (b)

*Figure 6.46* Logic diagrams for bidirectional 2 x 2 switching element. Bidirectional switch using (a) tri-state buffers and (b) CMOS transmission gates.

switching element (which requires two control signals instead of one) are shown in Figure 6.47.

This circuit may be implemented much like the one shown in Figure 6.45 (as a combination of two 2 to 1 multiplexers) except that some additional control logic is required to allow one input to be connected to both outputs. Figure 6.48 illustrates this approach. If broadcast capability is desired in a bidirectional switch, implementation is more complex than simply replacing logic gates with bidirectional buffers as we did in the circuits of Figure 6.46. Separate broadcast paths must be constructed for each side to avoid the situation in which one input connected to two outputs becomes two inputs connected to a single output.

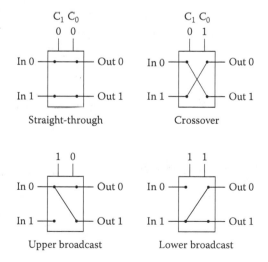

*Figure 6.47* Block diagram of 2 × 2 switching element with broadcast capability.

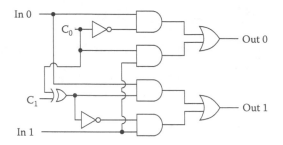

**Figure 6.48** Logic diagram for 2 × 2 switching element with broadcast capability.

These diagrams represent one bit of the smallest possible (2 × 2) cross-bar switch. A switch to be used for parallel data transfers would require this hardware to be replicated for each of the bits to be simultaneously transferred. The reader can easily imagine that increasing the capability of these small switches to 4 × 4, 8 × 8, or larger crossbars would make them significantly more complex and expensive to build.

Full crossbars are most commonly used to interconnect CPUs and memory modules in high-end symmetric multiprocessor systems. One side of the switching network is connected to the processors and the other to the memories. The main advantage of a crossbar interconnection is high performance. At any time, any idle input can be connected to any idle output with no waiting, regardless of connections already in place. Although CPUs in a bus-based multiprocessor system may frequently have to wait to access main memory because only one bus transaction can take place at a time, processors connected to memory modules via a crossbar only have to wait to access a module that is already in use by another CPU. Conversely, the main disadvantage of a crossbar is its high cost. Because of the complexity of controlling the switch and the replication of hardware components that is necessary to establish multiple connections simultaneously, the cost of a full crossbar is a quadratic function. In "Big O" notation, an $n \times n$ crossbar has an implementation cost of $O(n^2)$. Thus, it is very expensive to build a crossbar switch with large numbers of inputs and outputs.

## 6.4.2   *Recirculating networks*

The problem with a full crossbar switching network is that it is expensive, particularly for large numbers of inputs and outputs. To save on cost, it is possible to build a switching network from a number of smaller switching elements (often 2 × 2 switches like the ones we studied above) such that each node is directly connected only to a limited number of other nodes. Because not all inputs can be connected to all outputs given a single set (or stage) of switches like those shown in Figure 6.49, communication between a given source and most destinations must be accomplished by forwarding data through intermediate nodes. Because the data must often pass through this type of network more than once to reach their destination, it is referred to as a *recirculating network*. Because (like a crossbar) it is implemented with a single set of switches, a recirculating network is also known as a *single-stage network*.

The advantage of a recirculating network is its low cost compared to other networks, such as a crossbar switch. Interconnecting $n$ nodes requires only $n/2$ switch boxes (assuming they are $2 \times 2$)—fewer if the boxes have more inputs and outputs. Therefore, its cost function for $n$ nodes is $O(n)$. Its obvious disadvantage is reduced performance. Unless a communication is intended for one of the nodes directly connected to the originating node, it will have to be transmitted more than once; each additional pass through the network will introduce some overhead, slowing the process of communication.

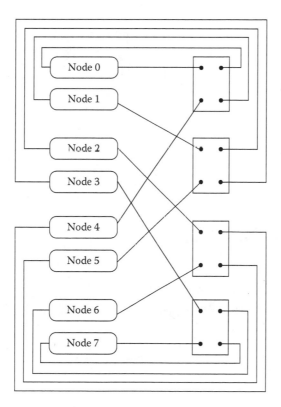

*Figure 6.49* Recirculating network with eight nodes.

### 6.4.3   *Multistage networks*

A *multistage network* (or *staged network*), like a recirculating network, is built from a number of smaller switch boxes rather than one large switch. However, to avoid the need to make multiple passes through the same set of switches, the multistage network adds additional sets of switches so that any input may be connected directly to any output. For the most usual case of $2 \times 2$ switch boxes making up an $n \times n$ network, it takes at least $\log_2 n$ stages of $n/2$ switches each to enable all possible input to output combinations. Figure 6.50 depicts a network designed in this manner. To connect any of eight inputs to any of eight outputs, this interconnection (known as an Omega network) uses three stages of four $2 \times 2$ switch boxes each.

   Although this is more expensive (in this case, by a factor of three) than the recirculating network discussed above, it generally performs better because all direct source-to-destination connections are possible. It is also much less expensive than an $8 \times 8$ full crossbar. Analysis shows that this $n \times n$ multistage network has a cost that scales as $O(n \log n)$, which is much less expensive than $O(n^2)$ as $n$ becomes large.

#### 6.4.3.1   *Blocking, nonblocking, and rearrangeable networks*
With a little analysis, one can confirm that a multistage network, such as the Omega network depicted in Figure 6.50, is capable of connecting any

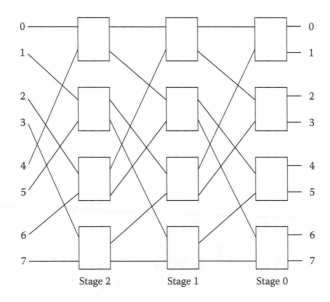

*Figure 6.50* Example multistage network (Omega network).

input to any chosen output given proper configuration of three switch boxes (one in the first stage, one in the second, and one in the third). The specific boxes that must be controlled in each stage depend on the particular nodes one is trying to connect. The ability to connect any given input to a chosen output is one of the characteristics we attributed to a full crossbar switch. Does it follow that the multistage network shown is equivalent to an 8 × 8 full crossbar? By examining the diagram more critically, we can see that this is not the case.

Look again at Figure 6.50. After studying the structure of the network, the reader should be able to convince himself or herself that there is a single possible path through this network between any specified input and output and that establishing this path requires specific control signals to be sent to a particular set of switching elements (one particular box in each of stages 0, 1, and 2). What may not be as obvious at first is that this first configuration, whatever it may be, precludes some other input-to-output connections from being simultaneously made.

Consider the situation depicted in Figure 6.51, in which a connection is made from node 6 on the left side of the network to node 2 on the right side of the network. This requires the three switch boxes in the only possible connection path to be set to the crossover, straight-through, and straight-through configurations, respectively. (Incidentally, if a crossover connection is enabled by a 1 on a box's control input and a straight-through

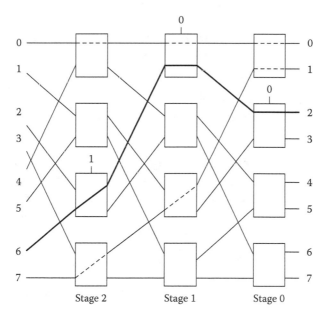

*Figure 6.51* Connections in Omega network.

connection is enabled by a 0, the proper control settings may be obtained by exclusive-ORing the binary addresses of the sending and receiving nodes. In this case, 110 XORed with 010 yields 100 as the proper value to set up the three switching elements.) Once these three boxes are configured in this way, it is easy to see that several (up to seven) other simultaneous connections are possible. For example, node 0 on the left could be connected to node 0 on the right, node 7 on the left could be connected to node 1 on the right, and so on.

However, it is not possible to make arbitrary additional connections across the network once a given connection is in place (as would be the case for a full 8 × 8 crossbar switch). For example, with an existing connection from node 6 to node 2, it is not possible to connect node 4 on the left to node 3 on the right. This is because two of the switch boxes (the top one in stage 1 and the second one from the top in stage 0) are common to the two desired connections. Connecting node 6 to node 2 requires both of these boxes to be in the straight-through configuration, while connecting node 4 to node 3 would require both of them to be in the crossover connection. If one of these connections is currently made, the other cannot be made until the first connection is no longer needed. Thus, one connection is said to block the other, and this network is termed a *blocking network*. A full crossbar, in which any connections involving distinct nodes are simultaneously possible, is known as a *nonblocking network*. Once again, we see a typical engineering trade-off of cost versus flexibility (and, ultimately, performance). A blocking network like the Omega network is less expensive and less flexible; because (at least some of the time) connections will have to wait on other connections, all else being equal, it will not enable the system to perform as well as a full crossbar (which offers maximum flexibility but at maximum cost).

There is another possible trade-off point between the blocking and nonblocking networks. By adding some additional switching hardware beyond that required to create a blocking network, it is possible to allow for multiple possible paths between nodes. Any path between nodes will still block some other connections, but if it is desired to make one of those blocked connections, a different (redundant) path between the original two nodes can be selected that will unblock the desired new connection. In other words, the original connection (or set of connections) in place can be rearranged to allow a new connection. Such a network is called a *rearrangeable network*. An example of a rearrangeable network, known as a Benes network, is shown in Figure 6.52.

The cost of a rearrangeable network such as this one is greater than the cost of a similar blocking network. In general, the topologies of rearrangeable networks are similar to those of blocking networks except for the addition of extra stages of switch boxes that serve to provide the path redundancy necessary to rearrange connections. The 8 × 8 Benes network

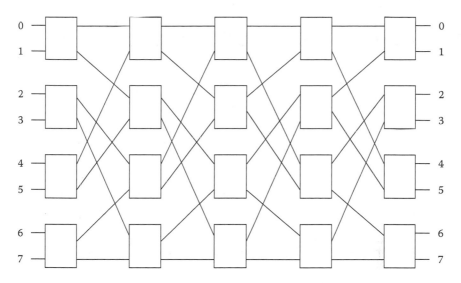

*Figure 6.52* Example rearrangeable network (Benes network).

shown is similar to the 8 × 8 Omega network discussed previously but has two extra sets of switch boxes. It looks like an Omega network folded over on itself such that the last two sets of connections between switches are a mirror image of the first two sets. By adding these two extra stages, we have ensured that there are four possible paths, rather than just one, between arbitrarily chosen nodes on the left and right sides. For example, Figure 6.53 shows the four possible paths that connect node 7 with node 0. Notice that each of these paths goes through a different switch box in the middle stage. This is true for any left–right node pair. Because there are two good routings through each switch box and because any input–output pair can be joined through any of the four central switch boxes, it should not be surprising that it is possible to simultaneously connect all eight nodes on one side to any combination of the eight nodes on the other side.

In general, if there are $n = 2^m$ inputs and outputs and the blocking (Omega) network has $m = \log_2 n$ stages, the rearrangeable (Benes) network with the same number of inputs and outputs and the same size switch boxes will require $2m - 1$ stages. Notice that the cost function of an $n \times n$ rearrangeable network like this one is still $O(n \log n)$ as in the case of the blocking network, but with a larger (by a factor of two) proportionality constant. All of this analysis assumes the use of 2 × 2 switch boxes. It is also possible to construct rearrangeable networks by increasing the size (number of inputs and outputs) of the individual switch boxes rather than the number of stages, but this approach is not as common; building more, smaller switches is usually cheaper than building fewer, but more complex, switches.

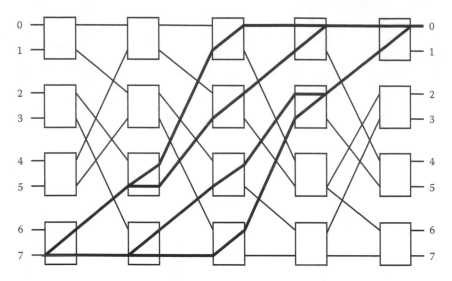

**Figure 6.53** 8 × 8 Benes network showing multiple possible connections between nodes.

The main feature of this type of network is that (similar to a full crossbar) given one or more (up to $n - 1$) connections already in place, it is always possible to make a connection between a currently idle node on the right to an idle node on the left. In some cases, making this new connection may not require any existing connections to be rerouted, but in other cases it will. (This is where the rearrangeable network differs from a nonblocking network.)

For example, assume that nodes 1, 3, and 5 on the left side of the 8 × 8 Benes network are currently connected to nodes 1, 6, and 7 on the right side as shown in Figure 6.52. Further suppose that we want to connect node 2 on the left to node 4 on the right. We could do this without rearranging any of the existing connections as shown in Figure 6.52.

Now assume that we want to add a connection from node 0 on the left to node 0 on the right. We find that the existing four connections do not permit this. However, it is possible to reroute existing connections—for example, the connection between nodes 2 and 4 as shown in Figure 6.52. This frees up a path so that our newly desired connection can be established. After this, it would still be possible to connect the three remaining nodes on the left to the three remaining nodes on the right in any combination; however, some additional rearrangement of the existing connections would likely be necessary.

Recirculating networks, blocking and rearrangeable multistage networks, and full crossbar switches are different ways of implementing a dynamic, circuit-switched interconnection. Each has its own characteristics and each represents a different trade-off point in the price–performance continuum. As these networks are most often found in small-to-medium multiprocessors, they are likely to continue in use for some time to come, but because parallel systems show a continuing trend of expanding the number of processors to increase performance, dynamic networks will likely often be used in conjunction with static networks in large systems.

## 6.5 Chapter wrap-up

The history of computing devices is the history of a quest for performance. The time of the human user is valuable, so he or she always wants a system to run the chosen applications faster. However, at any given time, existing technology only allows a single processor to perform computations at a certain maximum speed. That speed is not always sufficient to meet performance demands, and thus parallel systems were born.

Many types of parallel systems have been developed over the years. Except for a few unconventional architectures that we shall discuss in Chapter 7, most of them fall into one or the other of Flynn's two practical parallel architectural classifications: single instruction stream, multiple data stream (SIMD) machines, also known as array (or vector) processors, and multiple instruction stream, multiple data stream (MIMD) machines, which are variously known as multiprocessors or multicomputers/clusters depending on their structure. MIMD machines are by far the more common of the two basic parallel architectures, but SIMD machines have also played an important part in the history of high-performance computing and are unlikely to disappear in the foreseeable future. In recent years, high-performance systems have increasingly made use of graphics processing units (GPUs), which have some SIMD-like features but are programmed in a highly multithreaded fashion, giving rise to an environment that their most popular manufacturer (NVIDIA) refers to as single instruction, multiple thread (SIMT).

Perhaps the most significant factor that can be used to distinguish one parallel architecture from another is whether or not the processors (or processing elements in a SIMD) communicate by sharing memory or by passing messages back and forth. Sharing memory, as in multiprocessor systems, allows processors to interact more extensively using relatively large amounts of data. It also presents a programming model that is familiar to those who are accustomed to the uniprocessor (SISD) architecture. However, sharing memory among processors has its complications, especially when large numbers of processors are involved. Among the most significant of these complicating factors is the need to maintain coherence of data among main memory and the caches local to each processor.

(Other significant design issues for multiprocessors include serialization and synchronization via mutual exclusion primitives.) Depending on system size and topology, various approaches, including snoopy cache and directory-based protocols, may be employed to allow the system to realize the performance benefits of cache while maintaining the integrity of data. Alternatively, software-based approaches may be used to avoid the cache coherence problem (though generally at significant cost in performance and programming effort).

Some parallel systems avoid the problems inherent to sharing memory by limiting memory accesses to the local processor. Because processors cannot share data by reading each other's memories, they must communicate by passing data in messages sent over a communications network. A processor needing data may send a message to the node that has the data, and that node sends back a message containing the requested information. This approach not only avoids the problem of consistency between caches, but it provides a built-in synchronization mechanism in the form of sending and receiving messages. This approach works very well when the volume of data sharing (and thus communications) is light; if the amount of data to be shared is great, it may overwhelm the communications capacity of the system and cause it to perform poorly. As off-the-shelf technologies, such as Ethernet, have become faster, less expensive, and more widely available, the use of "clusters" (networks of commodity computers built around relatively inexpensive CPUs and GPUs) to run demanding applications has increased. This trend appears likely to continue for the foreseeable future.

Any parallel system, whether it is SIMD or MIMD and whether or not the processors share memory directly, must have some sort of interconnection network to allow nodes to communicate with one another. Interconnection networks generally fall into the categories of dynamic, or circuit-switched, networks and static, or packet-switched, networks. Within each category, networks may be classified as synchronous or asynchronous depending on their timing, and control of a network may be distributed or centralized. Network topology differences lead to distinctions between networks in terms of node connection degree, communication diameter, expandability, fault tolerance, etc. With increasing frequency, communications networks for larger systems are being constructed as hybrids of more than one basic type of network. However a network is constructed, if the parallel system is to perform up to expectations, the capabilities of the network must be matched to the number and speed of the various processors and the volume of traffic generated by the application of interest. As demand for high-performance computing continues to grow, parallel systems are likely to become more and more common, and the challenges for system designers, integrators, and administrators will no doubt continue as well.

## REVIEW QUESTIONS

1. Discuss at least three distinguishing factors that can be used to differentiate among parallel computer systems. Why do systems vary so widely with respect to these factors?

2. Michael Flynn defined the terms SISD, SIMD, MISD, and MIMD to represent certain classes of computer architectures that have been built or at least considered. Tell what each of these abbreviations stands for, describe the general characteristics of each of these architectures, and explain how they are similar to and different from one another. If possible, give an example of a specific computer system fitting each of Flynn's classifications.

3. What is the main difference between a vector computer and the scalar architectures that we studied in Chapters 3 and 4? Do vector machines tend to have a high or low degree of generality as defined in Section 1.4? What types of applications take best advantage of the properties of vector machines?

4. How are array processors similar to vector processors and how are they different? Explain the difference between fine-grained and coarse-grained array processors. Which type of array parallelism is more widely used in today's computer systems? Why?

5. How are graphics processing units (GPUs) similar to, yet different from, SIMD array processors? What types of computational problems are well suited to run on GPU-accelerated systems?

6. Explain the difference between multiprocessor and multicomputer systems. Which of these architectures is more prevalent among massively parallel MIMD systems? Why? Which architecture is easier to understand (for programmers familiar with the uniprocessor model)? Why?

7. Explain the similarities and differences between UMA, NUMA, and COMA multiprocessors.

8. What does "cache coherence" mean? In what type of computer system would cache coherence be an issue? Is a write-through strategy sufficient to maintain cache coherence in such a system? If so, explain why. If not, explain why not and name and describe an approach that could be used to ensure coherence.

9. What are the relative advantages and disadvantages of write-update and write-invalidate snoopy protocols?

10. What are directory-based protocols, and why are they often used in CC-NUMA systems?

11. Explain why synchronization primitives based on mutual exclusion are important in multiprocessors. What is a read-modify-write cycle, and why is it significant?

12. Describe the construction of a cluster system (once commonly referred to as a "Beowulf cluster" after the NASA project that spawned the idea). Architecturally speaking, how would you classify such a system? Explain.

13. Describe the similarities and differences between circuit-switched networks and packet-switched communications networks. Which of these network types is considered static, and which is dynamic? Which type is more likely to be centrally controlled, and which is more likely to use distributed control? Which is more likely to use asynchronous timing, and which is more likely to be synchronous?

14. What type of interconnection structure is used most often in small systems? Describe it and discuss its advantages and disadvantages.

15. Describe the operation of a static network with a star topology. What connection degree do its nodes have? What is its communication diameter? Discuss the advantages and disadvantages of this topology.

16. How are torus and Illiac networks similar to a two-dimensional nearest-neighbor mesh? How are they different?

17. Consider a message-passing multicomputer system with 16 computing nodes.
    a. Draw the node connections for the following connection topologies: linear array, ring, two-dimensional rectangular nearest-neighbor mesh, binary n-cube.
    b. What is the connection degree for the nodes in each of the above interconnection networks?
    c. What is the communication diameter for each of the above networks?
    d. How do these four networks compare in terms of cost, fault tolerance, and speed of communications? (For each of these criteria, rank them in order from most desirable to least desirable.)

18. Describe, compare, and contrast store-and-forward routing with wormhole routing. Which of these approaches is better suited to implementing communications over a static network with a large number of nodes? Why?

19. In what type of system would one most likely encounter a full crossbar switch interconnection? Why is this type of network not usually found in larger (measured by number of nodes) systems?

20. Consider the different types of dynamic networks discussed in this chapter. Explain the difference between a blocking network and a nonblocking network. Explain how a rearrangeable network compares to these other two dynamic network types. Give an example of each.

21. Choose the best answer to each of the following questions:
    a.  Which of the following is *not* a method for ensuring cache coherence in a multiprocessor system? (1) Write-update snoopy cache, (2) write-through cache, (3) write-invalidate snoopy cache, (4) full-map directory protocol.
    b.  In a 16-node system, which of these networks would have the smallest communication diameter? (1) n-cube, (2) two-dimensional nearest-neighbor mesh, (3) ring, (4) torus.
    c.  Which of the following is a rearrangeable network? (1) Illiac network, (2) multistage cube network, (3) crossbar switch, (4) Benes network, (5) none of the above.
    d.  In a 64-node system, which of the following would have the smallest node connection degree? (1) ring, (2) two-dimensional nearest-neighbor mesh, (3) Illiac network, (4) n-cube.
22. Fill in the blanks below with the most appropriate term or concept discussed in this chapter:

    _____ A parallel computer architecture in which there are several processing nodes, each of which has its own local or private memory modules.

    _____ A parallel computer architecture in which there are several processing nodes, all of which have access to shared memory modules.

    _____ Another name for an array processor.

    _____ A batch of threads that run together on one of a GPU's streaming multiprocessors.

    _____ A widely-used cross-platform language for writing code to do nongraphical tasks on GPUs.

    _____ A proprietary language for writing code to do nongraphical tasks on the most popular brand of GPUs.

    _____ A relatively small MIMD system in which the uniform memory access property holds.

    _____ A situation in which messages on a network cannot proceed to their destinations because of mutual or cyclic blocking.

    _____ An interconnection network in which any node can be connected to any node, but some sets of connections are not simultaneously possible.

    _____ The maximum number of hops required to communicate across a network.

    _____ Multicomputers with many nodes would be interconnected by this.

    _____ The classic example of a nonblocking, circuit-switched interconnection network for multiprocessor systems.

_____ A method of message passing in which flits do not continue toward the destination node until the rest of the packet is assembled.

_____ A method used for ensuring coherence of data between caches in a multiprocessor system in which a write hit by one CPU causes other processors' caches to receive a copy of the written value.

_____ The basic unit of information transfer through the network in a multicomputer system using wormhole routing.

# chapter seven

# Special-purpose
# and future architectures

In the previous six chapters, we discussed the characteristics common to the vast majority of computer architectures in use today, as well as most historical architectures since the dawn of modern computing in the 1940s. We could stop at this point, and the typical reader would be prepared to understand, compare, and evaluate most, if not all machines he or she might encounter in at least the first several years (if not more) of a career in the computing field. It has been said, however, that there is an exception to every rule, and so it is in computer architecture. For all the rules, or at least standard design practices, that we have discussed, there are at least a few exceptions: unique types of computer architectures designed in special ways for special purposes—past, present, and possibly future.

As we learned in Chapter 1, most single-processor systems use either a Princeton or modified Harvard (split cache, unified main memory) architecture. Chapters 3 and 4 revealed that these architectures may be designed using a hardwired or microprogrammed control unit, using a nonpipelined, pipelined, or superscalar implementation; yet their similarities (as seen by the user) outweigh their differences. Machines based on the Princeton and Harvard architectures are similar in design and programming, differing only in the aspect of having single versus separate paths between the CPU and (at least the first level of) memory for accessing instructions and data. In Chapter 6, we studied conventional parallel architectures based on the message-passing and shared memory models and built with a wide variety of interconnection networks; again, in each case the individual processors that made up the parallel system were based on the traditional, von Neumann, sequential execution model. In this final chapter, we consider several types of computing systems that are not based on the typical single or parallel processing architectures that perform conventional operations on normal binary data types, but instead have their own characteristics that may render them better suited to certain applications.

## 7.1 Dataflow machines

One type of computer architecture that eschews the traditional, sequential execution model of the von Neumann architecture is a *dataflow* machine. Such a machine does not rely on a sequential, step-by-step algorithm as exemplified by the machine, assembly, and high-level languages used to program conventional computers. Execution is not instruction-driven, but rather data-driven; in other words, it is governed by the availability of operands and hardware to execute operations. When a hardware execution unit finishes a previous computation and is ready to perform another, the system looks for another operation of the same type that has all operands available and schedules it into that unit.

Dataflow computers are not controlled by a program in the sense that term is understood by most computer programmers. No program counter is required to sequence operations. Instead, execution is controlled according to a *dataflow graph* that represents the dependencies between computations. Figure 7.1 depicts a few of the more common symbols that can be used to construct such a graph.

Each *node*, or *actor*, in the dataflow graph represents an operation to be performed (or a decision to be made). Each *arc* (arrow) connecting nodes represents a result from one operation that is needed to perform the other (with the dependency indicated by the direction of the arrow). *Tokens* representing values are placed on the arcs as operations are performed; each node requires certain tokens to be available as inputs before it can *fire* (be executed). In other words, any time all the previous operations leading to a given node in the dataflow graph have been completed, the operation represented by that node may be dispatched to the processing element (or, preferably, to one of a number of parallel processing elements) for execution. After the operation executes, tokens representing its results are placed on its output arcs and are then available for other operations that

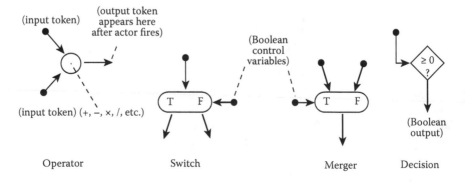

**Figure 7.1** Examples of dataflow graph symbols.

were waiting on them. Figure 7.2 shows a dataflow graph corresponding to the following high-level programming construct:

```
if (m < 5)
{
n = (m * 5) - 2;
}
else
{
n = (m * 3) - 2;
}
```

Memory accesses can be represented by load and store nodes (actors) that have *predicate* token inputs that restrain them from firing until the proper time for the read or write to occur. As long as the predicate is false, the memory operation does not proceed.

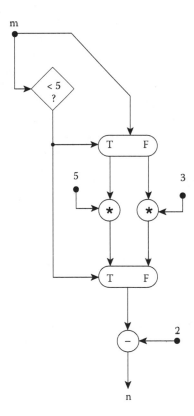

*Figure 7.2* Example dataflow graph.

An alternative representation of a program for a dataflow machine uses *activity templates* to represent each operation. (See Figure 7.3 for a template corresponding to the graph shown in Figure 7.2.) Each template has a code representing the operation to be performed, fields to store input tokens, and an address field indicating where the output tokens should be sent. Whether activity templates or a dataflow graph are used to indicate the precedence of operations, the important concept is that execution is not inherently sequential (as in the von Neumann model) but is controlled by the availability of operands and functional hardware.

Although the concept of dataflow processing may be an intuitively appealing model of computation, and the dataflow graph and activity template models are interesting ways to represent the methods for carrying out computational tasks, the reader is no doubt wondering what, if any, advantage such an architecture might have in comparison to a machine using conventional von Neumann–style programming. It turns out that the chief potential advantage of a dataflow architecture is an increased ability to exploit parallel execution hardware without a lot of overhead. If only one hardware unit is available to execute operations, we might as well write programs using the time-honored sequential model, as going to a dataflow model can do nothing to improve performance. However, if we have the ability to construct multiple functional units, then it is possible that a dataflow approach to scheduling them may relieve some of the bottleneck that we usually (and artificially) impose by adopting the von Neumann programming paradigm.

Dataflow machines have the potential to outperform traditional architectures because, in many computational tasks, a number of operations logically could be performed concurrently, but the von Neumann–style programmer is given no way to explicitly specify that concurrency. Unless the latent instruction-level parallelism inherent to the task can be detected by the control unit hardware (as in a superscalar architecture) or by the compiler (as in a very long instruction word [VLIW] architecture), the system may not make maximum use of parallel resources. In particular, dataflow machines take advantage of fine-grained parallelism (many simple processing elements) and do so in a more general way than single instruction stream, multiple data stream (SIMD) machines, which only

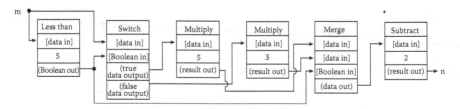

*Figure 7.3* Example activity template.

take advantage of the explicit parallelism inherent to array computations, or superscalar/VLIW machines, which only take advantage of the (usually coarse-grained) instruction-level parallelism that can be found in and extracted from a sequential-instruction program.

With this in mind, consider an example dataflow machine with the general structure shown in Figure 7.4. Note the availability of a number of functional units (processing elements) similar to those that might be found in an SIMD array processor but, in this case, not constrained only to be used in lockstep for array computations. Available functional units are given operation packets corresponding to actors that are ready to fire and, upon performing the required operations, produce result packets whose contents, of course, may become operands for subsequent computations.

Although it has many processing elements, a dataflow machine is clearly not an SIMD architecture. In fact, although dataflow machines are invariably parallel in their construction, they have no instruction stream in the conventional sense. Therefore, strictly speaking, they are neither

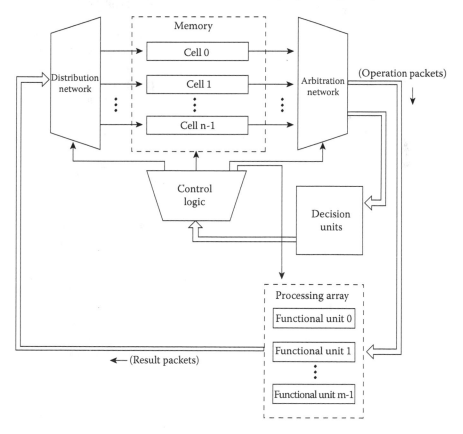

*Figure 7.4* Example dataflow machine.

SIMD nor MIMD (as some authors classify them) but instead can be said to form their own unique class of systems outside Flynn's taxonomy. (If one had to try to pigeonhole them into one of Flynn's classifications, a reasonable argument could be made for considering dataflow computers as MISD machines because the flow of data, in lieu of an instruction stream, passes through multiple actors that correspond to instructions.) This ambiguity in classifying dataflow machines explains the dashed line and question marks in the previous chapter's Figure 6.1.

Dataflow architectures can be classified as either *static* or *dynamic*. In a static dataflow machine (the type illustrated in Figures 7.1 through 7.4), only one token at a time can exist on any arc in the dataflow graph. An actor (node) is scheduled to fire when all of its required input tokens are present but no result token exists. The static dataflow approach has the benefit of (relatively) simpler implementation but tends to limit the amount of parallelism that can be exploited. Iterative computations (the equivalent of loops) can only be executed in pipelined or staged, rather than truly parallel, fashion. Static dataflow also does not allow for the equivalents of procedure calls and recursion, which are useful and important programming constructs.

To get around these problems and take better advantage of inherent parallelism, Arvind, Kim Gostelow, and Wil Plouffe (of the University of California at Irvine) developed the dynamic dataflow model. In a dynamic dataflow machine, more than one token can exist on any arc in the dataflow graph. In order for this to work, values are identified by tags to distinguish them from other, dynamically created, instances of the same variable. (Because of this, dynamic dataflow machines are also called *tagged-token* dataflow machines.) An actor only fires when all of its input arcs contain tokens with the same tag. The benefits of dynamic dataflow include better exploitation of concurrency as well as the ability to implement recursive/reentrant constructs. However, these advantages are counterbalanced somewhat by the added overhead of generating and matching the tags for the data tokens and the need to provide storage for multiple tagged tokens at the inputs of each node.

The idea of dataflow graphs and dataflow machines to execute them dates back to the late 1960s. Richard Karp, Ray Miller, and Shmuel Winograd developed the dataflow idea as a graphical method for representing computations, although Duane Adams of Stanford University is credited with coining the term *dataflow*. The first (1971) design for a computer based on dataflow principles was the MIT Static Dataflow Architecture, which was described by Jack Dennis and David Misunas. The first operational dataflow machine was the Data Driven Machine-1 (DDM-1), a static dataflow machine designed by Al Davis of the University of Utah and built in cooperation with Burroughs Corporation. This project was started in 1972, but the machine was not completed until 1976—a testimony, perhaps, to the difficulty of building dataflow computers.

These early dataflow machines and their successors have generally been one-of-a-kind, specially designed, prototype or research machines. A number were built during the 1980s and early 1990s (the boom period, to the extent there was one, for dataflow computers), mainly in Europe and Japan. Some of the notable examples include the (static) Data Driven Processor (or DDP) built by Texas Instruments, as well as the University of Manchester Dataflow Computer, the MIT Tagged-Token Dataflow Machine, the Japan Electro-Technical Laboratory SIGMA-1, and the Motorola/MIT Monsoon (four dynamic dataflow computers). Although these were all significant research projects, none of them was ever brought successfully to the general computing market.

What are some of the problems that have kept dataflow machines from widespread use? One can point to several stumbling blocks to their acceptance. First, dataflow machines have generally required special programming languages. (Dataflow graphs, activity templates, and similar data-driven constructs are represented in arcane languages such as Id, VAL, SISAL, and Lucid that are not familiar to most programmers.) Conversely, programs written in traditional, standard, high-level languages like Fortran, C, C++, and Java do not map easily to dataflow graphs because control dependencies do not easily map to data dependencies. Given that the cost of software development is usually a dominant—if not the largest—cost component of most computing solutions, few commercial developers want to bother to create dataflow implementations of their applications.

It is also true that although dataflow machines are in theory well suited to exploit the parallelism inherent to many tasks, they only do a really good job on applications with a level of data parallelism that happens to be a good match for the parallelism of the underlying machine hardware. A dataflow approach can be good for speeding up scalar computations with a moderate to high degree of unstructured parallelism, but it provides no significant performance advantage for the vector and array computations that are typically of interest in the high-performance, scientific computing market. (Vectors and arrays are not the only problems; multi-element data structures in general, especially large ones, are not handled efficiently by dataflow architectures, in which resource scheduling is done on the level of individual operations.) Even with regard to more general scalar computing applications, some studies have found that dataflow machines may perform worse than typical superscalar processors on algorithms that are control-intensive (depend on a lot of looping and branching). In addition—and this is not a trivial problem—the lack of locality of reference inherent in a dataflow graph makes it difficult to take advantage of hierarchical storage systems (featuring fast cache memory at the highest levels) to improve system performance.

Finally, high-performance dataflow computers are just not easy to build. It has historically been difficult to keep sufficient quantities of data in fast memory near the dataflow machine's many processing elements to keep them busy and take full advantage of the parallelism inherent in the application. Instead, it has thus far proven to be easier to improve performance by building superscalar processors with large cache memories. However, with the many advances made in microelectronics in the past few years, the ability to integrate large numbers of processing elements (and large amounts of fast memory) in a small space has alleviated some of the previous implementation problems. Thus, despite the several drawbacks that have restricted their use in the past, it is possible that dataflow machines (like SIMD machines before them) may yet experience a surge in popularity.

Even if they are never adopted widely in their own right, however, dataflow architectures have had—and may continue to have—influence with regard to other types of machine architectures. As we noted in Chapter 4, computer architects have discovered that the dataflow approach has certain benefits that can be applied to traditional processor design to try to maximize the utilization of resources and thus processing performance. One example was seen in Section 4.3.7 where we described the control strategy devised by Robert Tomasulo to schedule the multiple functional units of the IBM 360/91. This approach was later adapted for use in a number of internally parallel (a.k.a. superscalar) microprocessors. Tomasulo's method is essentially a dataflow approach to scheduling hardware with register renaming and data forwarding used to optimize operations originally specified in a sequential program. Thus, although they outwardly perform as von Neumann machines and are programmed using the conventional, sequential programming paradigm, processors with Tomasulo scheduling operate internally as dataflow machines. Another example of dataflow's influence was the development of multithreaded processors, which we discussed in Section 4.5.4. These improvements to conventional CPU design may prove to be the most lasting legacies of the dataflow approach.

## 7.2   *Artificial neural networks*

*Artificial neural networks* (ANNs) are another special class of computer architecture outside Flynn's taxonomy, at least as far as most computer scientists and engineers are concerned. (ANNs may, in some sense of the terms, be considered MIMD or even MISD machines as described in Section 6.1, but the comparisons are very loose, and it is probably better to think of them as exemplifying a unique type of architecture with its own attributes.) Like dataflow machines, ANNs usually employ a large number of simple hardware processing elements and are data-driven rather than relying on a sequential, algorithmic programming paradigm.

Artificial neural networks are not based on the von Neumann execution model, but rather on a biological model: the organization of the human nervous system. The fundamental processing unit in the human brain is a *neuron* or nerve cell. Processing in an artificial neural network is based on the functionality of neurons in a real (biological) neural network, but these functions are implemented digitally using interconnected processing elements.

Neurons in the human brain are composed of a cell body, or *soma*, along with fibers called *dendrites* that receive electrical impulses from other neurons, and other, long fibers known as *axons* that conduct impulses away from the cell (see Figure 7.5). Phrased in computer terminology, dendrites act as input devices for a neuron, and output to other neurons occurs via axons. The interface between an axon and another neuron to which it transmits information occurs across a tiny gap called a *synapse*. An electrical impulse sent down an axon causes the release of certain chemicals, called *neurotransmitters*, into the synapse. Depending on the nature and amount of these chemicals that are released, the receiving neuron is either *excited* (made more likely) to "fire" (transmit its own electrical impulse) or *inhibited* (made less likely to fire).

The many neurons in the human nervous system are connected in complex, three-dimensional patterns. Each neuron has a large number (possibly many thousands) of synapses that conduct impulses to it from other neurons; it may also send its output via axons to many other neurons. Each individual dendrite may receive excitatory or inhibitory stimuli, and the overall effect of these stimuli on the neuron is algebraically additive. If the net effect of the excitatory neurotransmitters minus the net effect of the inhibitory ones exceeds a certain electrical threshold called the *action potential*, then the neuron will fire (and thus affect other neurons to which its output is connected); otherwise, it will remain dormant. Although the functionality of an individual neuron is simple, the connections between neurons are very complex and organized into hierarchical

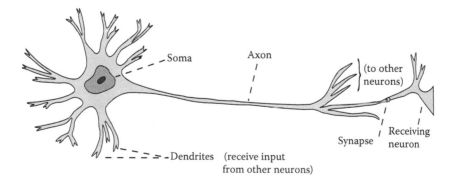

*Figure 7.5* Basic structure of a neuron.

layers. It is these connections that define the functionality of the nervous system.

In an artificial neural network, the neurons of a biological neural network are typically simulated with a large number of simple processing elements that compute the weighted sum of several inputs as illustrated in Figure 7.6. The programming of the ANN lies in the interconnections between the elements and the assignment of the weights to the inputs of each element. The connections are usually numerous and dense; for this reason, ANNs are also sometimes known as *connectionist systems*. The set of inputs, multiplied by their respective weights, are summed together and passed through an *activation* or transfer function that is used to determine the neuron's output. In some cases, the activation function may be a simple step function, but more often it is a logistic (or sigmoid) curve, exponential (Gaussian) function, etc., depending on the behavior desired of the network.

Generally the "programming," if one can call it that, of an ANN is done in an iterative fashion. The network is presented with a variety of inputs and their corresponding desired output values; when in learning mode, it "explores" the domain and range of inputs and outputs. As we alter weights and connections, reiterate and get the network closer and closer to the desired output behavior, it adapts its function to suit the task (thus, yet another name for an artificial neural network is an *adaptive system*). In effect, we are teaching or training it to perform the task, just as we might train a human (or other) biological system, controlled by real neurons, to throw a ball, fetch a stick, walk, count, or perform some other activity.

After some number of iterations of training (note that training performance can be significantly affected by the particular activation function used in the simulated neurons), the network (we hope) eventually achieves *convergence*, which means it develops the ability to represent the complete gamut of inputs and corresponding outputs and does not forget its

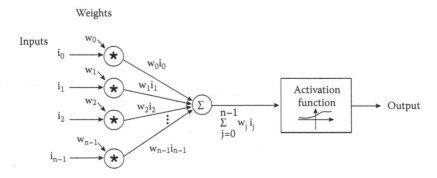

*Figure 7.6* Simulated neuron in an artificial neural network.

previous training. Effectively, the network learns to give a correct output in response to any given set of inputs without needing to algorithmically model the desired behavior. Then, upon switching from its learning mode to retrieving, or operating, mode, the network can accept new inputs and provide outputs corresponding to its training even though the inputs presented in operating mode may not be exactly the same as any of the inputs on which the network was trained (to take a human example, a player needing to make a throw from shortstop to first base, instead of from second base or third base to first base as he or she had previously learned to do). It is this ability to handle different inputs that makes artificial neural networks so useful. After all, we already know the proper outputs for the training inputs. Although there are many potential problems, including "overfitting" the training data, in many applications ANNs can provide good, or at least acceptable, outputs across a wide spectrum of inputs.

Various models can be used to make the network adapt; the simplest, dating back to the 1960s, is called the perceptron learning rule. The network is trained by changing weights by amounts proportional to the difference between the desired output and the actual output. The earliest type of neural network, invented by Frank Rosenblatt based on this model, is the *Single-Layer Perceptron* (SLP). As its name implies, it is constructed with a single set of simulated neurons between its inputs and its outputs as shown in Figure 7.7.

The SLP is simple to construct and train, but not versatile enough to solve many problems. In particular, Marvin Minsky and Seymour Papert showed that SLPs are not able to solve tasks that are "linearly inseparable" in a mathematical sense. Perhaps the most popular refinement of the basic

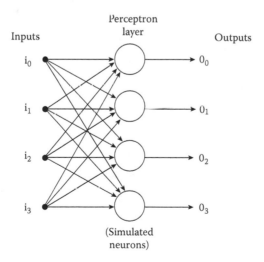

**Figure 7.7** Single-layer perceptron example.

method is the *Multi-Layer Perceptron* (MLP). Figure 7.8 shows an example of such a network. Note that the number of inputs, outputs, and layers, as well as the number of artificial neurons making up each layer, may vary considerably depending on the characteristics of the problem to be solved.

Historically, the problem with MLPs was difficulty in training. The output layer (like the single layer in an SLP) presented little difficulty because the network's computed outputs could be compared to known good values; however, there was no good way to adjust the weights on the previous layers. This problem was eventually (in the mid-1980s) solved through the use of *back-propagation*, a procedure in which for each forward pass (training iteration), the output errors are used in a reverse pass through the layers of the network with weights adjusted to minimize errors at each step until the input layer is reached. Then another forward and backward pass are made with the process iterated until convergence is achieved. On each reverse pass, as one works backward through the layers, the weights $w_{ij}$ of each neuron $i$ in the current layer with respect to neuron $j$ in the following layer are adjusted according to the generalized Delta Rule:

$$w_{ij} = r\delta_j o_i$$

where $r$ is the learning rate (varying this can affect whether and how fast the network achieves convergence), $\delta_j$ is the error for neuron $j$ in the following layer, and $o_i$ is the output of neuron $i$. The error function $\delta_j$ is defined as $(o_j)$ times $(1 - o_j)$ times the summation of the errors in the following layer multiplied by their weights (or simply the output errors, if the following layer is the output layer). Because they typically use the

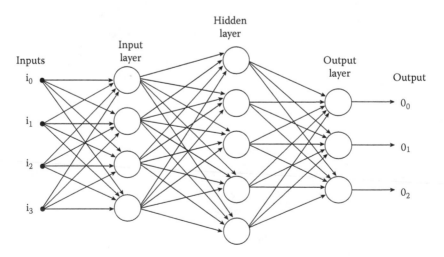

***Figure 7.8*** Multilayer perceptron example.

training method just described, MLPs are also known as *back-propagated Delta Rule networks*.

SLPs, MLPs, and many other types of artificial neural networks, including *radial basis function* (RBF) networks, are *supervised* neural networks. This means that before being put into operation, they must be trained (as described above) by a human user to perform the desired function. This is a fairly straightforward process as long as there exist sufficient examples of desired input and output combinations that can be used to train the system.

In some situations, however, the desired behavior of the system may not be well defined. We may be able to classify an output as "good" or "bad" but not know much more about it; therefore, the supervised learning model may not be feasible. In such cases, artificial neural networks may be designed to use a modified form of supervision known as the *reinforcement* model of learning. Reinforcement learning uses criticism instead of teaching. In other words, after each learning trial, the only feedback provided to guide the system is a single reinforcement variable often expressed as a value in the range from +1 (perfect) to –1 (completely wrong). As the reader might expect, systems using reinforcement learning generally tend to converge on a solution more slowly than those that use supervised learning.

For situations in which no training feedback—not even mere "up or down" criticism—is available, yet another class of ANNs known as *unsupervised* neural networks (sometimes called *competitive learning* networks) have been developed. As the name implies, these are systems that learn on their own without user intervention. Examples of such networks include Kunihiko Fukushima's *Neocognitron* and the *Self-Organizing Map* (SOM) invented by Teuvo Kohonen of the Helsinki (Finland) University of Technology. Although the details of these advanced ANN techniques are beyond the scope of this book, the curious reader may find them worthy of further inquiry.

The strength of all types of artificial neural networks lies in the same areas in which human beings are strong: being able to "see" that a relationship or pattern exists in data and make correlations between inputs and desired outputs without necessarily being able to express those correlations in an explicit description. In particular, ANNs are well suited to those applications with parameters that are difficult to measure and structures that are too complex to model (and thus that do not lend themselves well to an algorithmic solution), but in which the "programmers" have access to examples of the desired behavior (or at least some idea of what that behavior should look like). As we mentioned previously, the machine is not so much programmed as trained (either by a user or by itself) to perform the correct functions to arrive at a desired result. Take the examples of throwing and catching a ball or recognizing an image of a certain animal, such as a cat. Algorithmically programming a robot arm to throw a ball is a complex exercise involving precise timing and the articulation of multiple joints. Programming a robot to catch a thrown ball (which may approach at different speeds, with different spins, from different directions, etc.) may be even more difficult, and writing an algorithm to recognize a photo of an animal more difficult still. Yet almost any healthy six-year-old child can be taught (using his or her biological neural network) to throw and catch a ball fairly reliably with enough iterations of practice and can recognize a cat, having seen several examples of cats. Artificial neural networks offer great advantages for robotic control and pattern recognition as well as many other applications, including process monitoring and control, image processing, market forecasting, etc., that are difficult to implement with conventional computers using traditional programming methods.

Artificial neural networks can theoretically be implemented in software running on any computing platform, but few general-purpose processors or embedded control processors (another typical application of ANNs) have the computational horsepower to achieve the performance levels typically desired of neural network applications. Thus, ANNs are best implemented on specialized machines specifically intended for such use—ideally, a highly parallel machine with many simple processing elements and a high degree of interconnection between the processing elements. There have been a number of neural network–based systems built over the years. Many have been university research machines, often in the form of custom integrated circuits that implement neural network processors or building blocks, such as the FIRE1 chip being developed at the University of Manchester. Increasingly in recent years, however, neural network processors have found niches in the commercial market. One example of a neural network chip suitable for multiple applications was the VindAX processor, made by Axeon Ltd. (Scotland). This processor was intended for use in embedded control systems in automotive applications as well as image and audio processing. VindAX implemented artificial neurons using an array of 256 identical parallel processing elements implemented in CMOS logic, although the manufacturer claimed that the chip hid the details of its neural architecture from the user. Axeon did not market the device as an artificial neural network processor at all; instead, the manufacturer downplayed that angle and referred to its operation as being based on an advanced statistical modeling technique. The effect, however, is the classical ANN learning pattern: by training on actual data, the machine determined the correlations between inputs and outputs in complex, difficult-to-model systems without attempting to describe them algorithmically, and once trained, it could efficiently and effectively classify data or control a system.

Other commercial neural network examples include the Neural Network Processor (NNP) built by Accurate Automation Corporation; Irvine Sensors Corporation's 3D Artificial Neural Network (3DANN) chip, which served as a building block for their Silicon Brain project; and the IBM ZISC (Zero Instruction Set Computer) and its successor the General Vision NeuroMem CM1K chip. Qualcomm is currently developing its "brain-inspired" family of Zeroth processors, and in 2014 IBM announced its TrueNorth chip that contains one million simulated neurons and 256 million synapses.

In Section 6.1.1, we noted that although expensive, massively parallel SIMD array processing supercomputers have fallen out of favor, low-cost, coarse-grained SIMD technology has gained acceptance in a variety of application areas, including multimedia and graphics processing in desktop PCs and workstations. It appears that, in much the same way, artificial neural network technology (which is also based on arrays of simple processing elements)—although it is unlikely to take over the world of computing—will probably continue to be used, and perhaps even increase its market share, in a variety of niche applications, such as those discussed here. In that case, the reader may find this brief introduction to ANNs to have been well worth the time spent reading it.

## 7.3   Fuzzy logic architectures

All the architectures we considered in the first six chapters of this book and even (for the most part) those discussed in the preceding two sections share the common feature of performing operations on digital operands in binary form. Ultimately, every datum in the system resolves to a true or false value, represented by binary 1 or 0. Multiple-bit integers and even real numbers in floating-point format are usually supported, but all *logical* variables are either true or false, on or off. Although this maps well to the preferred physical implementations of computers in hardware, it does not always exemplify the way knowledge exists or decisions are made by human beings, in the real world. Propositions are not always demonstrably 100% true or 100% false, and inputs may be subject to noise or not known precisely. (For example, when driving, how fast is that other vehicle approaching ours from behind? "Pretty fast" may be all we can determine.) Instead of an absolute certainty, we may only know the truth of a logical proposition with some relative degree of confidence. It may be true with probability 0.3, or 0.7, or 0.99, etc. Architectures based on the principles of *fuzzy logic* address the uncertainty inherent in all real-world situations by making it a design feature. In a fuzzy logic system, instead of just the discrete values 0 and 1, logical variables can take on a continuum of values between 0 and 1, inclusive.

The concept of fuzzy logic was first described in the 1960s by Lotfi Zadeh of the University of California at Berkeley. Zadeh's research deals with principles of *approximate reasoning*, rather than *precise reasoning* as used in systems based on Boolean logic. Fuzzy logic allows the designer of a system to express not only on/off, true/false, or other binary notions, but also "high versus medium versus low" or "very much versus considerable versus somewhat versus not much" and so on. When applied to quantities of interest, such as speed, color, and height, these imprecise modifiers give rise to *linguistic variables* that can take on values such as "very fast," "moderately heavy" and "slightly dirty." Although they are vague compared to the binary values assigned to Boolean variables, these concepts are readily understandable to human beings and are frequently used in everyday life.

The idea of fuzzy logic is to create a digital system that functions more like the way humans make decisions, especially in cases in which their knowledge is incomplete. Like the artificial neural networks described in the previous section, fuzzy logic was an outgrowth of research into artificial intelligence. Both fuzzy systems and neural networks are considered *soft computing* approaches because they not only tolerate but embrace the uncertainty and lack of precision inherent in real-world scenarios. Unlike ANNs, however, fuzzy logic systems do not attempt to model the actual, physical function of the human nervous system. They try to achieve similar results without using the connectionist architecture of a neural network.

To understand how fuzzy logic works, we start with the idea of a *universe of discourse*. This is simply the set of all things under consideration; for example, all men, all automobiles, or all rocks. The universe of discourse is made up of any number of *fuzzy subsets*. A fuzzy subset is any group of elements of the universe whose membership cannot be defined precisely. For example, the universe of men is made up of subsets of tall men, thin men, bald men, young men, handsome men, and so on. How many men belong to the subset of tall men? It depends on what you consider "tall." The perceived likelihood that an element of the universe belongs to a given fuzzy subset is embodied in the idea of a *membership function*, which produces a *truth value* indicating the degree of membership in the fuzzy subset. The truth value produced by the membership function for a given member of the universe is not restricted only to the binary values 0 and 1 but can be any value in between.

Consider the concept of "heaviness" as applied to the set (or universe) of all human beings. Is a given person a member of the subset of heavy people or not? There is most likely some chance that some given observer would classify almost anyone as heavy. (For example, a child might consider any adult heavy by comparison to himself or herself.) The membership function in this case is intended to provide a reasonable approximation, over the continuous interval from 0 to 1, of the degree to which a given person is a member of the subset of heavy people. (Note that a membership function is not the same thing, mathematically speaking, as a probability distribution; it is more accurately described as a *possibility* distribution.) A simple example membership function *heavy()* that assigns a given person $x$ a likelihood of being considered heavy could be specified as follows:

heavy(x) = {0, if weight(x) < 40 kg;
(weight(x) – 40 kg)/100 kg,
if 40 kg ≤ weight(x) ≤ 140 kg;
1, if weight(x) > 140 kg.}

A graph of this function is shown in Figure 7.9. Applying this function, we could say that a 65-kg person has a degree of membership in the set of heavy people (or a truth value of being heavy) of 0.25, and a 110-kg person is heavy with a truth value of 0.7.

This is an extremely simple membership function (a simple linear function of one variable with lower and upper limits at 0 and 1, respectively). Other commonly used piecewise linear membership functions include the triangular and trapezoidal functions shown in Figure 7.10. Nonlinear relationships, such as the logistic and exponential functions (mentioned in Section 7.2 in the context of activation functions for simulated neurons), can also be used although they may be more expensive to implement in hardware or software. In many cases, it might be desirable for a fuzzy subset's membership function to depend on more than one variable (although, again, this tends to increase computational cost). For example, a better definition of *heavy()* might consider a person's height, age, and gender in addition to his or her weight in determining how

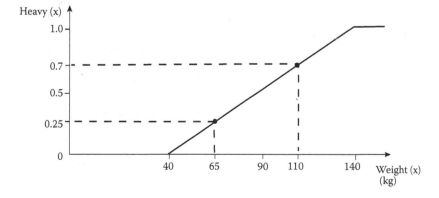

*Figure 7.9* Simple membership function.

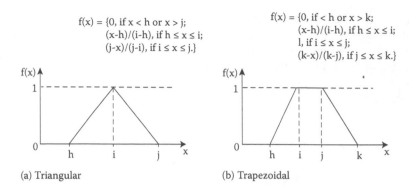

$f(x) = \{0, \text{ if } x < h \text{ or } x > j;$
$(x-h)/(i-h), \text{ if } h \le x \le i;$
$(j-x)/(j-i), \text{ if } i \le x \le j.\}$

$f(x) = \{0, \text{ if } < h \text{ or } x > k;$
$(x-h)/(i-h), \text{ if } h \le x \le i;$
$1, \text{ if } i \le x \le j;$
$(k-x)/(k-j), \text{ if } j \le x \le k.\}$

(a) Triangular                                    (b) Trapezoidal

*Figure 7.10* Commonly used membership functions.

heavy he or she is. To take another example, the degree to which the distance between two cars on a highway is close or far away might depend not only on their separation in feet or meters, but also on the speed they are traveling.

Conventional computer hardware (and software) can be very complex but ultimately bases its operation on the individual Boolean operations AND, OR, and NOT as applied to binary operands. Given that fuzzy variables, unlike Boolean variables, can take on a continuum of values from 0 to 1, can we apply these same logical operations, and if so, how? It turns out that we can indeed perform logic operations on fuzzy subsets if we define those operations appropriately. What are suitable definitions? NOT is fairly straightforward; we can simply say that if *truth(x)* has a value in the range of 0 to 1, then we can say *truth(NOT x)* is equal to 1.0 – *truth(x)*. The AND and OR functions have been subject to various interpretations, but the most common definitions are *truth(x* AND *y)* = *min(truth(x), truth(y))* and *truth(x* OR *y)* = *max(truth(x), truth(y))*. Notice that if one were to restrict the fuzzy variables to only the discrete values 0 and 1 instead of a continuum of values, these definitions would yield the same results as the traditional Boolean algebra definitions of NOT, AND, and OR. Thus, one can say that Boolean logic is a subset of fuzzy logic, or alternatively, that fuzzy logic is an extension of Boolean (or "traditional" or "crisp") logic from the discrete set {0, 1} to cover the range of real numbers between 0 and 1, inclusive.

Where is the use of fuzzy logic valuable? Perhaps the most common application is in *expert systems*, which try to replicate the decisions made by a human who is knowledgeable about the chosen application area. A human subject matter expert formulates fuzzy *rules* that reflect his or her understanding of the workings of the system. The fuzzy expert system then makes *inferences* about data and ultimately chooses a course of action, using those fuzzy logic rules rather than Boolean (crisp) logic.

Fuzzy expert systems (like artificial neural networks) may be used in business decision support systems, financial investment trading programs, weather prediction, and many other areas in which precise, algorithmic modeling is difficult.

Another typical application of fuzzy logic is in control systems (which some authors consider to be a subset of expert systems). Figure 7.11 shows a block diagram of a control system based on fuzzy logic. The fuzzy controller (the operation of which will be discussed below) sits within a closed feedback loop, just as would a typical analog or digital servo control unit. Although traditional analog and digital control design approaches are often best for control of systems that are linear (or can be adequately approximated as linear systems), fuzzy logic control—again like neural network control—is particularly useful for controlling systems that are nonlinear, complex, or have poorly specified characteristics. (Due to their similar strengths, some control systems combine aspects of fuzzy logic and artificial neural networks.)

Whether it is part of an expert system, a control system, or something else, the process of computation in a fuzzy system generally proceeds in the same sequence of steps as illustrated in Figure 7.11. The first step is called *fuzzification*. During fuzzification, the membership functions that are defined as part of the system's knowledge base are applied to the values of the input variables. In other words, fuzzification maps the raw, numeric input data into linguistic values that represent membership in

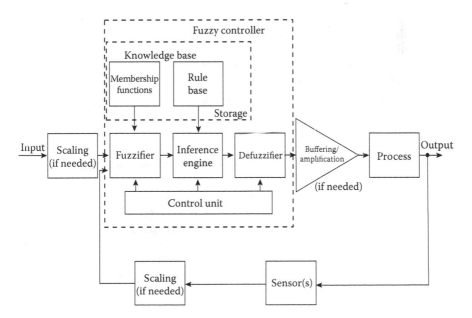

*Figure 7.11* Fuzzy logic control.

fuzzy subsets. This allows the system to determine the truth value for each *premise* or precondition for each rule that has been defined to govern the operation of the system.

The second step is generally referred to as *inferencing* (also sometimes referred to as *reasoning* or fuzzy processing). Inferencing is the process of rule evaluation and composition. The set of rules that were mentioned earlier, which reflect expert knowledge of system behavior, are normally expressed in terms of if–then relationships between fuzzy output subsets and linguistic variables derived from the inputs in such forms as "if *speed* is low and *distance* is high, then make *throttle* = medium." Any given application may have a number of rules (collectively known as the *rule base*) of a form similar to this, describing the relationship between input and output parameters. Thus, rule evaluation means that the truth values for each premise for every rule are combined and used in computing the conclusions of the rules. This results in the creation of a number of fuzzy subsets (one for each rule that is used in the determination of each output). Composition takes the results of rule evaluation for all rules that can possibly affect each output and combines them to form a single fuzzy subset for each output variable. (It is interesting that the steps of rule evaluation and composition are analogous to the AND and OR operations in Boolean logic, so the end result of the inferencing process is the evaluation of a sort of sum of products expression in fuzzy logic for each output.) Inferencing evaluates the *fuzzy relation* (effectively, the model of system behavior) that exists between an output and the fuzzified input variables, determining the fuzzy output that corresponds to the current inputs.

Finally, in cases in which the end result of the fuzzy processing must be a single, definite, numeric value (or a binary either/or decision) rather than a fuzzy or linguistic value, we have the process of *defuzzification*. The need for defuzzification in a control system is fairly obvious; an electric motor, for example, does not understand the fuzzy command "slow down a little," but it will respond to a change in its input voltage. In this case, defuzzification involves determining the specific voltage value to send to the motor based on the fuzzy output obtained from rule evaluation and composition. In addition to control systems, many other applications also require defuzzification. A number of techniques can be used for defuzzification, each with its own advantages and disadvantages; most amount to taking some sort of a weighted average of the fuzzy output subset. Two of the most common are the *mean of maxima* method and the *centroid* method. The mean of maxima method selects the crisp output value for which the fuzzy output takes on its maximum truth value. If there is more than one such value, as its name implies, this method calculates the average of the values for which maxima occur. The centroid method uses the "center of gravity" of the fuzzy output (the center of the area under its membership function) to determine the crisp output value.

Like artificial neural networks, fuzzy systems do not absolutely require a special hardware architecture for implementation. The fuzzy logic operations of fuzzification, inferencing, and defuzzification can be implemented in software on top of a conventional SISD, SIMD, or MIMD computer architecture. This is good enough for many applications, but not very efficient. If the desired application makes extensive use of fuzzy logic techniques, it may make sense to try to maximize performance by optimizing the hardware to implement them.

When a fuzzy logic algorithm is run on a conventional, von Neumann CPU, the membership functions and the rules in the rule base are evaluated sequentially. A machine optimized for fuzzy control would have a dedicated *fuzzifier* (or membership function unit) that would likely be implemented with parallel hardware to speed up the process. The *inference engine* could also use replicated hardware to enable it to evaluate multiple rules simultaneously, and the *defuzzifier* would probably have several methods built in for the system designer to choose from. The resulting machine, although it would have a control unit and use internal storage to hold the membership functions and rules in its knowledge base, would function differently from most machines based on the von Neumann architecture. Such machines do exist; although no conventional, commercial microprocessors (let alone supercomputers) have yet been designed around a fuzzy logic core, it is possible to add a fuzzy logic accelerator to a general-purpose CPU if needed. One example of this approach is the Siemens 81C99, a coprocessor that was developed in the mid-1990s to offload calculations for fuzzy logic systems from the main system processor. It was not manufacturer-specific and could be used with a variety of general-purpose microprocessors to implement high-performance, fuzzy logic–based applications.

Although general-purpose machines rarely make use of special fuzzy logic features even in the form of coprocessors, that is not to say that such features do not play a significant role in some systems. It turns out that because fuzzy logic techniques are used most frequently in control systems, one is most likely to find hardware support for fuzzy logic in *microcontrollers* (single-chip computers used in embedded control applications). The first standard, commercial MCU (microcontroller unit) to incorporate built-in hardware support for fuzzy systems was the Motorola 68HC12. This 16-bit chip, first produced in 1996, was the successor to the 68HC11 (one of the most widely used 8-bit microcontrollers dating back to the mid-1980s).

In addition to its general-purpose instruction set, the 68HC12 implemented four operations specifically intended for use in developing fuzzy logic applications. The MEM instruction computed a membership function to fuzzify an input, while REV performed unweighted rule evaluation on a rule list. The REVW instruction did the same thing as REV and also allowed the programmer to specify a per-rule weighting function. Finally, WAV computed the sum of products and sum of weights needed to calculate a weighted average for defuzzification purposes. (WAV must be followed by a "normal" EDIV instruction to compute the average and thus complete the defuzzification.) Although the 68HC12 instruction set was optimized for the implementation of fuzzy systems, it is worth noting that these instructions were not carried out by physically separate hardware. Instead, the chip used the main CPU core logic to perform them. Still, the manufacturer claimed that a fuzzy inference kernel could be implemented on the 68HC12 in only 20% of the code space required by, and executing 10 times faster than, its predecessor, the 68HC11, which did not have these special instructions.

Some other microcontrollers have gone the extra mile for even better performance and incorporated hardware fuzzy inference engine modules. For example, members of the FIVE family of 8-bit ICUs (intelligent controller units) manufactured by STMicroelectronics contained both a general-purpose, 8-bit CPU core and a hardware "decision processor." The decision processor included both a hardware fuzzy inference unit that could be used to accelerate rule evaluation and composition, and an optimized arithmetic/logic unit used to perform fast multiply, divide, accumulate, minimum, and maximum calculations. (Like many microcontrollers, the chip also incorporated analog and digital peripheral interfaces so that in many applications it could function as a single-chip solution.) Having both a generic CPU and a decision processor on the chip allowed the designer to combine fuzzy and traditional control approaches if desired. The ST FIVE's intended applications included motor control, appliance control, and thermal regulation. Finally, it is worth noting that some embedded control processors have combined aspects of fuzzy logic and artificial neural networks to support "neuro-fuzzy" control. An example of such a device was the NeuraLogix (later Adaptive Logic, Inc.) NLX230 Fuzzy Microcontroller. This chip (developed around 1994) featured a general-purpose fuzzy logic engine with an added neural network processor to enhance performance.

The outlook for continued, and even more widespread, use of fuzzy logic is promising—perhaps even more so than for artificial neural networks, because fuzzy logic systems are typically simpler and cheaper to implement than ANNs. This is especially important for high-volume production systems that require only moderate performance but need to be implemented with low cost. In recent years, fuzzy technology has been successfully used (in preference to more conventional approaches) in a variety of applications ranging from household appliances to consumer electronics to navigation systems to automobiles. Japanese companies have been particularly aggressive in making use of fuzzy logic systems; Nissan, Mitsubishi, and others have built fuzzy logic controllers for antilock brake systems, engines, transmissions, and other vehicle systems. Although the average reader should not expect to encounter many fuzzy logic–based systems in the business or scientific computing environments, a familiarity with their characteristics could come in handy, especially for those who work in the growing field of embedded systems.

## 7.4    Quantum computing

Perhaps the most difficult type of alternative computer architecture to understand, yet the one that holds the most promise to revolutionize computing in the professional lifetime of today's student, is the *quantum* computer. A quantum computer is a device based not on Boolean algebra, but on the principles of quantum physics. These principles, in effect, allow the same physical device to simultaneously compute a vast number of possibilities as though it were massively parallel hardware. Although they are still mostly in the experimental stage and no one knows whether they will ultimately achieve commercial success, if they can be reliably constructed, quantum computers may prove to be orders of magnitude more powerful than today's fastest supercomputers, performing some types of calculations millions or even billions of times faster. They may one day even render most of the material in the first six chapters of this book obsolete.

Conventional computers have increased by orders of magnitude in speed over their 70-plus years of existence. Much of this increase in speed and computing power has been made possible by continuing reductions in the sizes of the individual components used in their construction. Vacuum tubes gave way to discrete semiconductor devices, which, in turn, were replaced by integrated circuits. Over the years, the sizes of the individual transistors that make up an integrated circuit have shrunk to the point at which each individual device is a tiny fraction of a *micron* (more properly known as a micrometer or one millionth of a meter) across. In fact, Intel's latest microprocessors based on the Broadwell (2014) and Skylake (2015) microarchitectures use 14-nanometer transistors (a nanometer is one billionth, or $10^{-9}$, of a meter). It has been estimated that integrated circuit transistor sizes will shrink to single-digit nanometers by the year 2018 or 2019.

The performance achievable with computers based on integrated circuits has historically followed a pattern known as Moore's Law (named after Gordon Moore, one of the cofounders of Intel, who first expressed it). Moore's Law says that the continually shrinking sizes of semiconductor devices results in an exponential growth in the number of transistors that can feasibly be integrated on a single chip. According to Moore's 1965 prediction (which has turned out to be amazingly accurate over a 50-year period), the number of transistors on a single integrated circuit would continue to double on approximately a yearly basis, with a corresponding doubling of computational power approximately every 18 to 24 months. However, the limits achievable under Moore's Law may be reached in just a few years; Moore himself, in 2015, estimated that "an insurmountable barrier" based on fundamental limits would be reached within 5 to 10 years.

What does it mean to run up against "fundamental limits" in the design of traditional, Boolean algebra–based computers? As performance continues to increase mainly because components continue to shrink, what happens when the transistor sizes are reduced to the size of just a few atoms each? (A 2-nanometer transistor would only be 10 silicon atoms wide.) Ultimately, the semiconductor industry will reach a stopping point at which the transistors that make up the basic logic gates can be made no smaller while continuing to work as binary switches and thus, effectively, classical computer architectures can be made no faster. It is at this point, which by all estimates is not too many years away, that we will have to turn away from semiconductor-based computing as we know it today to some new technology or even a whole new paradigm—perhaps quantum computing—to achieve further increases in performance.

The idea of building computers based on the principles of quantum mechanics goes back to the late 1970s and early 1980s. Several scientists were already considering the fundamental limits of semiconductor-based computation 30 to 40 years ago. They saw that if implementation technology continued to advance according to Moore's Law, then the ever-shrinking size of silicon transistors must eventually result in devices no larger than a few atoms. At this point, problems would arise because on an atomic scale, the laws that govern device properties and behavior are those of quantum mechanics, not classical (Newtonian) physics. This observation led these researchers to wonder whether a new, radically different type of computer could be devised based on quantum principles.

Paul Benioff, a scientist at the Argonne National Laboratory, is generally credited with being the first to apply the ideas of quantum physics to computers. Other scientists who did early work on the idea included Richard Feynman of the California Institute of Technology, who conceived the idea of a quantum computer as a simulator for experiments in quantum physics, and David Deutsch of the University of Oxford, who expanded Feynman's idea and showed that any physical process could, in theory, be modeled by a quantum computer. Deutsch's work was very important: His findings showed that not only was a general-purpose quantum computer possible, but that such computers could have capabilities greatly surpassing those of conventional machines and could solve classes of problems that are impractical or impossible to solve with even the fastest supercomputers of today.

How do quantum computers work? One important feature of all conventional computers, whether they are uniprocessor (SISD) or parallel

(SIMD or MIMD) machines, is that the basic unit of information they store or process is a binary digit, or *bit*. These bits may be grouped together to form bytes or words, but all the basic combinational and sequential logic devices (gates, latches, flip-flops, etc.) operate on or store individual bits. Quantum computers are different from all the other machines we have studied in that their basic unit of information is a *quantum bit* (*qubit*, for short). Unlike a bit, which can take on only one or the other of two states (1 or 0, true or false, on or off, etc.) at a given time, a qubit is an entity that can assume not only the logical states corresponding to one or zero, but also other states corresponding to both 1 and 0 at the same time, or a *superposition* or blend of those states with a certain probability of being either.

All conventional computers, even though their components are very tiny, obey the laws of classical physics, just like all the other larger, visible objects with which we interact on a daily basis. Quantum computers operate at the level of molecules, or even individual atoms or ions, and their component particles, which obey the laws of quantum physics. In many cases, these laws give rise to effects that are extremely counterintuitive to those of us more familiar with classical physics. For example, experiments with photon beam splitting have demonstrated the phenomenon of *quantum interference*, which results from the superposition of the multiple possible quantum states. Without delving too deeply into theoretical physics, which is beyond the scope of this book, one can say that subatomic particles do not have a definite existence in the sense that macroscopic objects do (e.g., a chair is either in the room or not). Rather, such a particle can be said to exist at a certain place and time with a given statistical probability. It does not have a definite existence or nonexistence (or, alternatively, it both exists and does not exist) until someone observes it, at which time the probability resolves to 1 (it exists) or 0 (it does not).

In a quantum computer, the atoms (and the subatomic particles that comprise them) are essentially used as processors and memories at the same time. Even though there is only one set of some number of qubits in the machine, the qubits are not restricted to be in only one state at a time as the bits in a binary register would be. Although a 3-bit binary register can only take on one of the eight states 000 through 111 at any given time, a 3-qubit quantum register can be in all eight states at once in *coherent superposition*. Using *quantum parallelism*, a set of $n$ qubits can store $2^n$ numbers at once, and once the quantum register is initialized to the superposition of states, operations can be performed on all the states simultaneously. Thus, adding more qubits makes a quantum computer exponentially more powerful as shown in Figure 7.12.

The process of computation using a quantum computer is very different than in a von Neumann machine with sequential programming. In today's research machines, a computation is typically initiated by providing a tuned pulse of energy (for example, laser energy or radio-frequency

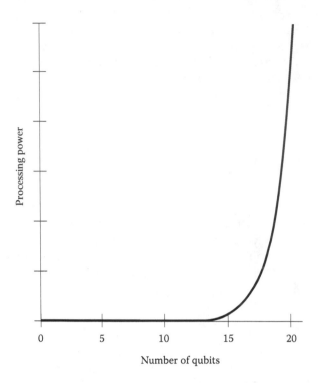

*Figure 7.12* Potential processing power versus number of qubits.

energy applied via *nuclear magnetic resonance* [NMR]) to change the energy state of an atom. The energized atom then engages in controlled interactions with several other atoms, going through sequences of *quantum gates* (or unitary transformations on qubits) and thus changing superpositions of quantum states into other superpositions. In doing so, the atoms acting as the computer establish a resulting pattern of quantum states that corresponds to results that might be generated by conventional computers. The resulting superposition of qubits represents a massively parallel set of computations all done by a single piece of (quantum) hardware. In effect, in one time step an $n$-qubit quantum computer can perform the same calculation on $2^n$ values, something that would take $2^n$ time steps on a uniprocessor machine, or require an SIMD machine with $2^n$ processors to accomplish in one time step.

If all this were easy to do, we would already see quantum computers in widespread use. Not only are quantum computers difficult and expensive to build, at least for now (it is hard to separate one or a small number of atoms from others to use for computation using techniques such as nuclear magnetic resonance, ion traps, optical lattices, or cavity quantum electrodynamics), but it is also very difficult to keep an individual atom in a steady state while its energy levels are manipulated and its spin directions or other properties are observed. One of the most significant problems in making practical application of quantum computers is retrieving the results of the computations. Measuring the results of a quantum computation is problematic because interfering in any way with the atoms being used for computation can change their value, collapsing them back into the single state 1 or 0 instead of their coherent superposition of many states, which is what is so valuable for parallel computation. Therefore, measurements on qubits must be made indirectly. Most approaches that have been tried so far involve cooling the atoms in the computer to very low temperatures, near absolute zero. This is obviously not an inexpensive technique, nor one that would work well for production machines.

One method for determining the states of the qubits uses the phenomenon of *quantum entanglement*. The quantum principle of entanglement allows one to apply an external force to a pair of atoms; once they are entangled, the second atom takes on the properties of the first atom but with opposite spin. This linkage allows the qubit value stored by a given atom to be read indirectly via techniques (such as NMR) used to measure the spin of its entangled cohort. Thus, the results of a quantum computation can be read without interfering with the atom doing the computation. Another, more recently devised approach involves using the interference between quantum states, rather than entanglement, to retrieve information from the qubits. None of the approaches developed so far are ideal, and research is ongoing.

Another, related problem that must be solved to make quantum computing practical is error correction. In a quantum computer, errors can be introduced due to *decoherence*, or the tendency for unavoidable interactions with the surrounding environment to affect the qubits, destroying the constructive interference patterns that are used in the computation and decaying the quantum state to an incoherent mixed state. This tends to break down the information stored in the quantum computer and induce errors in computations. As in the case of information retrieval, research into effective methods of error correction is under way. Substantial progress in fighting decoherence and other potential error sources will need to be made before robust quantum computers, capable of solving computationally difficult problems, can be built.

As the reader has no doubt concluded, research into quantum computers is still very much at the theoretical, not practical, stage. Some of the leading investigators are at IBM, Los Alamos National Laboratories, Caltech, and the University of Oxford. The most advanced quantum computing research machines of today contain only a few qubits. Researchers at the IBM Almaden Research Center constructed a 5-qubit computer in 2000 and used it to compute a single-step solution to an "order-finding" equation that would have required repeated iterations to solve on a conventional machine. Since then, the same researchers have produced a 7-qubit machine that has run Shor's factoring algorithm (the first algorithm designed specifically for quantum computers; see below). A 7-qubit computer has also been built (in 2000) at the Los Alamos National Laboratory. Other researchers subsequently created coherent quantum registers of 12 and 14 qubits, although it doesn't appear that these have been used for practical computation. These examples illustrate that the development of quantum computers, while the subject of intense research, is still in its infancy. It has been estimated by physicist David Deutsch, a pioneer of quantum computing, that a quantum computer would need to have 30 qubits to possess the equivalent computing power of a 10-teraflop conventional supercomputer. Given the cost and difficulty of constructing quantum computers today, it would take even more—dozens—of qubits to make quantum computers worthwhile compared to conventional, integrated circuit–based computers. Although practical, marketable quantum computers suitable for tackling large-scale scientific problems are probably still years—maybe decades—away, they may (if we are lucky) become available just as they are really needed; in other words, just as Moore's Law finally runs its course.

It must be mentioned that one commercial manufacturer—D-Wave Systems—has designed and sold machines that it claims are quantum computers with far more qubits than have ever been implemented in research machines. In 2007, D-Wave unveiled a 16-qubit prototype machine based on "quantum annealing," a specialized architecture suitable for solving certain classes of optimization problems; later that year, the company demonstrated what it claimed to be a 28-qubit computer. In 2011, D-Wave introduced what it called "the world's first commercially available quantum computer," the D-Wave One, featuring 128 qubits; it was followed in 2013 by the D-Wave Two (512 qubits) and in 2015 by the D-Wave 2X (1000 qubits). These are impressive—one could even say amazing—achievements if true, and D-Wave's customers include Google, NASA, and Lockheed Martin Corporation. However, scientists are skeptical that the machines being built and sold by D-Wave are actually quantum computers in the same sense as the other machines discussed above. For now, it appears that although D-Wave's computers may indeed be using some quantum effects, they are not achieving the exponential speedup that has been the main goal of research into quantum computing; nor are they able to perform the most challenging and potentially game-changing tasks, such as large number factorization (see below).

Assuming the problems can be worked out and practical quantum computers can be built, how will they be used? Probably not as general-purpose machines. By their nature, quantum computers are not likely to be useful or practical for mundane tasks, such as e-mail and word processing, which are quite handily accomplished using low-cost, semiconductor-based computers. It appears that the most appropriate application of quantum computers will be extremely numerically intensive computations, such as the factoring of large integers. This was their first application. After David Deutsch published his paper on the capabilities of quantum computers, scientists began to search for problems that would be a good fit for such machines and for algorithms that could be used to solve those problems. For a while, it appeared that quantum computers might remain an academic curiosity without much in the way of practical uses.

Then, in 1994, Peter Shor (a research scientist at AT&T Bell Laboratories) wrote a paper in which he described the first algorithm specifically intended for quantum computation. Shor's algorithm takes advantage of the power of quantum superposition to implement a set of mathematical operations that enable a machine to rapidly factor very large integers. Given a quantum computer of sufficient power, the factorization could be accomplished orders of magnitude faster than is possible with even the most advanced silicon-based computers. Shor's algorithm has already been run on the 7-qubit computer developed by IBM, factoring the number 15 into the numbers 3 and 5. (Later, other researchers factored 143 into 11 and 13 using an "adiabatic factoring" algorithm and four qubits.) Although these computations seem trivial, they represent proofs of the concept of factorization, with staggering implications if the same algorithms can be run on quantum computers with many more qubits.

Although at first blush, number factorization sounds like a rather academic exercise in itself, it has important applications, especially in the field of cryptography. The difficulty of using conventional computers to factor large integers is what makes present crypto systems, such as RSA, extremely hard to crack. (The RSA algorithm, named for its inventors Rivest, Shamir, and Adelman, is the most widely used public key encryption method, and the principal technology behind secure transactions on the Internet.) A person, organization, or government with access to a powerful quantum computer could use Shor's (or some other) factoring algorithm to break into any encrypted message and recover the supposedly secure information; thus, today's encryption methods could be made obsolete. Conversely, quantum computers could potentially be used to implement new codes that would encode information with far stronger encryption than is possible with even the fastest supercomputers of today and thus be essentially impossible to crack.

Given the ability of quantum computers to factor very large numbers, cryptography could very well turn out to be the "killer application" that drives the continued development of quantum computers. However, making and breaking codes is not the only area in which these revolutionary machines may make an impact. Other potential applications of quantum computers include signal processing, solving differential equations, and searching large databases. (The potential for this last application should be easy to see if the reader recalls the advantages of an associative search in hardware as described in Section 2.3.3. Because quantum computers inherently look at all the data at the same time, they could possibly perform the same type of associative search without the cost of the parallel search hardware.) Although the potential of quantum computers has barely begun to be tapped, their possible benefits are almost limitless. As with so many other aspects of computer architecture, the future awaits our discovery.

## 7.5   *Chapter wrap-up*

Only time will tell whether the venerable von Neumann and Harvard architectures and their parallel descendants will persist or go the way of horse-drawn carriages. It appears that conventional computer architectures will remain dominant for at least the next 10 years; but already, alternative architectures, such as dataflow machines, artificial neural networks, and fuzzy logic systems have found market niches in which, for particular applications, they offer advantages that make them more suitable than traditional SISD, SIMD, or MIMD systems. It seems likely that at least some of these alternative architectures will continue to be used, perhaps even increasing in popularity over the next few years.

Looking farther down the road, the looming brick wall at the end of the Moore's Law road implies that new architectures, radically different from those of today, may be needed if further performance improvements are desired (and they always will be). Quantum computers, with their unique promise, may prove to be the architecture (or one of the architectures) of the future if their many practical difficulties can be overcome. Other possibilities on the horizon include optical, molecular, and biological computers (which may include genetic or even DNA-based machines). Although we have not explored their properties or uses here, developments in those fields will bear watching. Considering the ongoing research in these fields that represent the frontiers of computing, it is entirely possible that a brand-new paradigm of architecture or implementation may render today's supercomputers as obsolete as the ENIAC. Any way you slice it, the next 20 to 30 years will certainly be an interesting time to be a computer scientist or computer engineer.

### REVIEW QUESTIONS

1. Explain how a dataflow machine avoids the von Neumann bottleneck.
2. Draw a dataflow graph and an activity template for the following programming construct:
   if (x > = 0)
   {

```
z = (x + y) * 4;
}
else
{
z = (y - x) * 4;
}
```

3. If you had a scientific application that involved a large number of matrix manipulations, would you rather run it on a dataflow computer or an SIMD computer? Explain.

4. What do you think is the main reason dataflow computers have so far not been widely adopted?

5. Give an example of how dataflow techniques have influenced and/or been used in conventional computer design.

6. Would you classify an ANN as an SISD, SIMD, MISD, or MIMD system or something else? Make a case to support your choice.

7. Explain how the processing elements and interconnections in an ANN relate to the structure of the human nervous system.

8. How is a supervised ANN programmed to carry out a particular task? What is the difference between a supervised versus unsupervised ANN?

9. Why are ANNs well suited to applications such as robotic control? Give an example of an application for which you do not think an ANN would be a good choice.

10. What is different about logical variables in a fuzzy system compared to a conventional computer system?

11. Both ANNs and fuzzy logic systems attempt to mimic the way human beings make decisions. What is the main difference between the two approaches?

12. What is a fuzzy subset and how does the idea of a membership function relate to it? Propose a simple membership function *rich()* that deals with the concept of a fuzzy subset of wealthy people.

13. Can the Boolean, or crisp, logic operations AND, OR, and NOT be defined in regard to fuzzy logic? If so, explain how; if not, explain why not.

14. Explain, in the context of a fuzzy expert system, what rules are and how they are used.

15. For what type(s) of physical system is fuzzy control particularly well suited?

16. What is Moore's Law, and how has it related to advances in computing over the past 50 years? Is Moore's Law expected to remain true forever or lose its validity in the future? Explain your answer and discuss the implications for the design of future high-performance computer systems.

17. How does a quantum computer fundamentally differ from all the other computer architectures discussed in this book? What allows a quantum computer to achieve the effect of a massively parallel computation using a single piece of hardware?

18. What are some of the problems scientists must solve in order to make supercomputers based on the principles of quantum mechanics practical?

19. What application(s) are expected to be a good match for the unique capabilities of quantum computers? Explain.

20. Fill in the blanks below with the most appropriate term or concept discussed in this chapter:

_____ A type of computer architecture in which execution depends on the availability of operands and execution units rather than a sequential instruction program model.

_____ An element in a dataflow graph that represents an operation to be performed on data.

_____ These are used to represent data values (operands and results) in algorithms for a dataflow architecture.

_____ This (outwardly) von Neumann machine made use of dataflow techniques for internal scheduling of operations.

_____ A type of computer architecture with a structure based on that of the human nervous system.

_____ The fundamental units that make up a biological neural network.

_____ These are fibers that act as input devices for neurons in human beings.

_____ When an artificial neural network achieves this, it is trained and ready to be put into operating mode.

_____ The earliest and simplest type of artificial neural network.

_____ A type of artificial neural network that does not require user intervention for training.

_____ A type of computer architecture in which logical values are not restricted to purely "true" or "false" (1 or 0).

_____ A type of variable that expresses a fuzzy concept; for example, "slightly dirty" or "very fast."

_____ The set of all objects under consideration in the design of a fuzzy system.

_____ The numerical degree (between 0 and 1, inclusive) of membership that an object has in a fuzzy subset.

_____ The first step performed in doing fuzzy computations for an expert system, control system, etc.

_____ This is necessary if a fuzzy result must be converted to a crisp output.

_____ A type of computer architecture in which the same physical hardware can be used to simultaneously compute many results as though it were parallel hardware; its operation is not based on Boolean algebra, but on the physics of subatomic particles.

_____ A prophetic observation of the fact that conventional computers would tend to grow exponentially more powerful over time as integrated circuit features got smaller and smaller.

_____ The basic unit of information in a quantum computer.

_____ This phenomenon results from the superposition of multiple possible quantum states.

_____ A state in which an atom's properties are identically assumed by another atom but with opposite spin.

_____ The tendency for interactions with the surrounding environment to disturb the state of qubits, possibly resulting in computational errors.

_____ A quantum computer with this many qubits has been estimated to have 10 TFLOPS of computational power.

_____ So far, this appears to be the most likely application for supercomputers based on quantum principles.

# Appendix: reference and further reading materials with web links

## Chapter 1: Introduction to computer architecture

A Brief History of Computing Technology, Part 1—Smith, Derek J. [University of Wales Institute, Cardiff], http://www.smithsrisca.co.uk/STMsubtypes-pt1.html

A Brief History of the Abacus—Fernandes, Luis [Ryerson University], http://www.ee.ryerson.ca:8080/~elf/abacus/history.html

Accelerating High Performance Linpack (HPL) With GPUs—Iqbal, Saeed and Gao, Shawn [Dell Inc.], http://en.community.dell.com/cfs-file/__key/telligent-evolution-components-attachments/13-4491-00-00-20-36-16-06/Accelerating-High-Performance-Linpack-_2800_HPL_2900_-with-GPUs.pdf

Active TPC Benchmarks—[Transaction Processing Performance Council], http://www.tpc.org/information/benchmarks.asp

All the Best Big Data Tools and How to Use Them—[Import.io], https://www.import.io/post/all-the-best-big-data-tools-and-how-to-use-them/

An Overview of Computational Science—Computational Science Education Project [Oak Ridge National Laboratory], http://www.phy.ornl.gov/csep/ov/ov.html

Archives: Mainframes: Basic Information Sources—[IBM Corporation], http://www.03.ibm.com/ibm/history/exhibits/mainframe/mainframe_basinfo.html

Benchmark Programs and Reports—[Netlib.org], http://www.netlib.org/benchmark/

Big Data: What It Is and Why It Matters—[SAS Institute], http://www.sas.com/sk_sk/insights/big-data/what-is-big-data.html

Classic Benchmarks—Longbottom, Roy [Roybot Ltd.], http://freespace.virgin.net/roy.longbottom/classic.htm

Cleve's Corner: The LINPACK Benchmark—Moler, Cleve [MathWorks], http://blogs.mathworks.com/cleve/2013/06/24/the-linpack-benchmark/

Computer Systems Performance Analysis and Benchmarking—Kurmann, Christian [Computer Systems Institute, Switzerland], http://www.cs.inf.ethz.ch/37-235/vorl/vorl03-02.pdf

CPU Performance—Madison, David Ljung [DaveFAQ.com], http://davefaq.com/Opinions/CPU_Performance.html

DEC VAX History—Bader, William, http://williambader.com/museum/vax/vax history.html

Development of Computers—[The Columbia Electronic Encyclopedia, 6th Edition], http://www.infoplease.com/encyclopedia/science/computer-development -computers.html

Frequently Asked Questions on the Linpack Benchmark—Dongarra, Jack [University of Tennessee, Knoxville], http://www.netlib.org/utk/people /JackDongarra/faq-linpack.html

Hadoop: What Is It and Why Does it Matter?—[SAS Institute], http://www.sas .com/en_us/insights/big-data/hadoop.html

Highlights, November 2015—[top500.org], http://www.top500.org/lists/2015/11 /highlights/

History of Java Technology—[Oracle Corporation], http://www.oracle.com/tech network/java/javase/overview/javahistory-index-198355.html

History of the Microcomputer Revolution—Delaney, Frank [MTA Micro Technology Associates], http://www.mtamicro.com/microhis.html

HPL: A Portable Implementation of the High-Performance Linpack Benchmark for Distributed-Memory Computers—Petitet, A.; Whaley, R. C.; Dongarra, J.; and Cleary, A. [University of Tennessee, Knoxville], http://www.netlib.org /benchmark/hpl/

IBM Mainframes: 45+ Years of Evolution—Elliott, Jim [IBM Corporation], http:// www.vm.ibm.com/devpages/jelliott/pdfs/zhistory.pdf

IBM PC DOS—[Wikipedia, the Free Encyclopedia], https://en.wikipedia.org /wiki/IBM_PC_DOS

Information Technology Industry TimeLine—Bellec, Jean [Fédération des Equipes Bull], http://jeanbellec.pagesperso-orange.fr/information_technology_1.htm

Intel Enters Billion-Transistor Processor Era—[CRN], http://www.crn.com/news /channel-programs/172301096/intel-enters-billion-transistor-processor -era.htm

Intel Itanium 9300 Processor Raises Bar for Scalable, Resilient Mission-Critical Computing—Intel Corporation], http://www.intel.com/pressroom/archive /releases/2010/20100208comp.htm

Introduction—Smotherman, Mark [Clemson University], http://people.cs.clem son.edu/~mark/464/intro.html

Introduction to Storage Area Networks—Tate, Jon; Beck, Pall; Ibarra, Hector Hugo; Kumaravel, Shanmuganathan; and Miklas, Libor [IBM Redbooks], https:// www.redbooks.ibm.com/redbooks/pdfs/sg245470.pdf

John von Neumann—O'Connor, John J. and Robertson, Edmund F. [University of St. Andrews, Scotland], http://www.groups.dcs.st-and.ac.uk/~history /Biographies/Von_Neumann.html

John von Neumann—[Wikipedia, the Free Encyclopedia], http://en.wikipedia .org/wiki/John_von_Neumann

Livermore Loops—Dongarra, Jack and Luszczek, Piotr [Springer Link], http:// link.springer.com/referenceworkentry/10.1007%2F978-0-387-09766-4_161

Microprocessor Types and Specifications: CPU Operating Voltages—Mueller, Scott and Soper, Mark Edward [InformIT], http://www.informit.com/articles /article.aspx?p=130978&seqNum=18

Microsoft MS-DOS Early Source Code—Shustek, Len [Computer History Museum], http://www.computerhistory.org/atchm/microsoft-ms-dos-early-source -code/

Oracle Cranks Up The Cores To 32 With Sparc M7 Chip—Morgan, Timothy P. [EnterpriseTech], http://www.enterprisetech.com/2014/08/13/oracle-cranks -cores-32-sparc-m7-chip/

ORNL Debuts Titan Supercomputer—[Oak Ridge National Laboratory], https:// www.olcf.ornl.gov/wp-content/themes/olcf/titan/Titan_Debuts.pdf

Overclocking for Beginners—Walton, Mark [GameSpot], http://www.gamespot .com/articles/overclocking-for-beginners/1100-6421190/

Personal Computers: History and Development—Archee, Raymond [Western Sydney University, Australia], http://stc.uws.edu.au/ERM/hist_pc.htm

Selected Historical Computer Designs—Smotherman, Mark [Clemson University], http://people.cs.clemson.edu/~mark/hist.html

Smartphone Users Worldwide Will Total 1.75 Billion in 2014—[eMarketer], http:// www.emarketer.com/Article/Smartphone-Users-Worldwide-Will-Total -175-Billion-2014/1010536

SPEC Benchmarks—[Standard Performance Evaluation Corporation], https:// www.spec.org/benchmarks.html

The Analytical Engine Table of Contents—Walker, John [Fourmilab Switzerland], http://www.fourmilab.ch/babbage/contents.html

The CDC 6600—Scientific Computing Division [National Center for Atmospheric Research], http://www.cisl.ucar.edu/computers/gallery/cdc/6600.jsp

The Green500 List—[Green500.org], http://www.green500.org/

The LINPACK Benchmark: Past, Present, and Future—Dongarra, Jack J.; Luszczek, Piotr; and Petitet, Antoine [University of Tennessee, Knoxville], http:// www.netlib.org/utk/people/JackDongarra/PAPERS/hpl.pdf

The SPEC Organization—[Standard Performance Evaluation Corporation], http:// www.spec.org/spec/spec.html

The Supercomputer Arrives with the CDC 6600 in 1964—Howe, Tom [CEDMagic .com], http://www.cedmagic.com/history/cdc-6600.html

The Supercomputer Company—[Cray, Inc.], http://www.cray.com/

Titan Cray XK7—[Oak Ridge National Laboratory], https://www.olcf.ornl.gov /computing-resources/titan-cray-xk7/

Top 500 Supercomputer Sites—[Top500.org], http://www.top500.org/

Types of Computers—Gandon, Fabien [Carnegie Mellon University], https:// www.cs.cmu.edu/~fgandon/lecture/uk1999/computers_types/

UNIVAC I—Thelen, Ed [ed-thelen.org], http://ed-thelen.org/comp-hist/UNIVAC -I.html

What Cloud Computing Really Means—Knorr, Eric [InfoWorld], http://www .infoworld.com/article/2683784/cloud-computing/what-cloud-computing -really-means.html

What Is a Storage Area Network—[Storage Networking Industry Association], http://www.snia.org/education/storage_networking_primer/san/what_san

What Is Cloud Computing?—[IBM Corporation], http://www.ibm.com/cloud -computing/what-is-cloud-computing.html

## *Chapter 2: Computer memory systems*

Blu-ray Information—[CD-info.com], http://www.cd-info.com/blu-ray/

Bubble Memory—[Wikipedia, the Free Encyclopedia], https://en.wikipedia.org /wiki/Bubble_memory

Cache Memory—Anthes, Gary H. [*Computerworld*, May 29, 2000], http://www
.computerworld.com.au/article/54693/cache_memory/

Cache Memory Systems—Storr, Phil [Phil Storr's PC Hardware Book], http://phil
ipstorr.id.au/pcbook/book2/cache.htm

Caches: What Every OS Designer Must Know—Heiser, Gernot [University of
New South Wales], http://www.cse.unsw.edu.au/~cs9242/08/lectures/04
-cachex6.pdf

Concerning Caches—Dugan, Ben [University of Washington], https://courses.cs
.washington.edu/courses/cse378/02sp/sections/section9-2.html

Data Organization on Disks—Rafiei, Davood [University of Alberta], https://web
docs.cs.ualberta.ca/~drafiei/291/notes/6-disks.pdf

Definition of SDRAM—[PC Magazine Encyclopedia], http://www.pcmag.com
/encyclopedia/term/50982/sdram

Disk Array Performance—Burkhard, Walt [University of California, San Diego],
http://cseweb.ucsd.edu/classes/wi01/cse102/sol2.pdf

DRAM—[The PC Technology Guide], https://www.pctechguide.com/computer
-memory/dram-dynamic-random-access-memory

DVD Information—[CD-info.com], http://www.cd-info.com/dvd/

Hardware Components: Semiconductor Digital Memory Manufacturers—Davis,
Leroy [InterfaceBus.com], http://www.interfacebus.com/memory.html

Instant Access Memory—Voss, David [*Wired Magazine* Issue 8.04], http://www
.wired.com/2000/04/mram/?pg=1&topic=&topic_set=

Introduction: What is MRAM (Magnetic RAM)?—[MRAM-Info], http://www
.mram-info.com/introduction

Memory—Sorin, Daniel J. [Duke University], http://people.ee.duke.edu/~sorin
/prior-courses/ece152-spring2008/lectures/6.9-memory.pdf

Memory 1996: Complete Coverage of DRAM, SRAM, EPROM, and Flash Memory
ICs—Griffin, Jim; Matas, Brian; and de Suberbasaux, Christian [Smithsonian
Institution, National Museum of American History: Chip Collection], http://
smithsonianchips.si.edu/ice/cd/MEM96/title.pdf

Memory Classification—Puchala, Dariusz [Instytut Informatyki, Poland], http://
www.ics.p.lodz.pl/~dpuchala/CompArch/Lecture_6.pdf

Memory Interleaving—Matloff, Norman [University of California at Davis], http://
heather.cs.ucdavis.edu/~matloff/154A/PLN/Interleaving.pdf

Memory Interleaving—Wang, Ruye [Harvey Mudd College], http://fourier.eng
.hmc.edu/e85_old/lectures/memory/node2.html

MRAM—[Webopedia.com], http://www.webopedia.com/TERM/M/MRAM.html

MRAM Replaces FRAM (FeRAM)—[Everspin Technologies], https://www.ever
spin.com/mram-replaces-fram-feram

Researchers Overcome Show-Stopping Problems with Ferroelectric RAM—Hruska,
Joel [ExtremeTech], http://www.extremetech.com/computing/158590-research
ers-overcome-show-stopping-problems-with-ferroelectric-ram

Segmentation and Paging—Pasquale, Joe [University of California, San Diego],
http://cseweb.ucsd.edu/classes/fa03/cse120/Lec08.pdf

Segmentation and Paging—Yang, Junfeng [Columbia University], http://www.cs
.columbia.edu/~junfeng/13fa-w4118/lectures/l05-mem.pdf

Solaris 64-bit Developer's Guide—[Oracle Corporation], http://docs.oracle.com
/cd/E18752_01/html/816-5138/toc.html

Solid State Memory Development in IBM—Pugh, E. W; Critchlow, D. L.; Henle, R. A.; and Russell, L. A. [IBM Corporation], http://www.research.ibm.com /journal/50th/hardware/pugh.html

Static RAM (SRAM)—[The PC Guide], http://www.pcguide.com/ref/ram/typesS RAM-c.html

Systems and Components Reference Guide: Hard Disk Drives—Kozierok, Charles M. [The PC Guide], http://www.pcguide.com/ref/hdd/

Taking Disk Drive Speeds for a Spin—Barrall, Geoff [Blue Arc Corporation], http:// searchstorage.techtarget.com/tip/Taking-disk-drive-speeds-for-a-spin

Types of Cache Misses: The Three C's—Shaaban, Muhammad [Rochester Institute of Technology], http://meseec.ce.rit.edu/eecc551-winter2001/551-1-30-2002. pdf

What Is the Difference Between SDRAM, DDR1, DDR2, DDR3 and DDR4?— [Transcend Information, Inc.], http://www.transcend-info.com/Support /FAQ-296

# Chapter 3: Basics of the central processing unit

A Brief History of Microprogramming—Smotherman, Mark [Clemson University], http://people.cs.clemson.edu/~mark/uprog.html

A Generalized Carry-Save Adder Array for Digital Signal Processing—Karlsson, Magnus [OKG AB, Sweden], http://citeseerx.ist.psu.edu/viewdoc/download ?doi=10.1.1.571.5825&rep=rep1&type=pdf

Applets: The Wallace Tree Simulator—Carpinelli, John D. [New Jersey Institute of Technology], https://web.njit.edu/~carpinel/Applets.html

Arithmetic Circuits & Multipliers—Hom, Gim [Massachusetts Institute of Technology], http://web.mit.edu/6.111/www/f2012/handouts/L08.pdf

Booth's Algorithm for Binary Multiplication Example—Misurda, Jonathan [University of Pittsburgh], https://people.cs.pitt.edu/~jmisurda/teaching /cs447/examples/Booth%20Example.pdf

Carry Look Ahead Adder—Balsara, Poras T. [University of Texas at Dallas], http:// www.utdallas.edu/~poras/courses/ee3320/xilinx/upenn/lab4-CarryLoo kAheadAdder.htm

Carry Save Arithmetic—Knagge, Geoff [GeoffKnagge.com], http://www.geoff knagge.com/fyp/carrysave.shtml

Division algorithm—[Wikipedia, the Free Encyclopedia], https://en.wikipedia .org/wiki/Division_algorithm#SRT_division

Fast Addition: Carry Lookahead Adders—Lin, Charles [University of Maryland], https://www.cs.umd.edu/class/sum2003/cmsc311/Notes/Comb/looka head.html

IEEE 754 Revision—[Wikipedia, the Free Encyclopedia], https://en.wikipedia .org/wiki/IEEE_754_revision

IEEE 754: Standard for Binary Floating-Point Arithmetic—[Institute of Electrical and Electronics Engineers], http://grouper.ieee.org/groups/754/

IEEE Standard for Floating-Point Arithmetic—[*IEEE Computer Society*], http:// www.csee.umbc.edu/~tsimo1/CMSC455/IEEE-754-2008.pdf

Instruction Set Architecture (ISA)—Citron, Daniel [Jerusalem College of Technology], https://homedir.jct.ac.il/~citron/ca/isa.html

Instruction Set Design—Weems, Charles [University of Massachusetts], https://peo
ple.cs.umass.edu/~weems/homepage/CmpSci535_635_files/Instruction%20
Set%20Design.pdf

Integer and Floating Point Arithmetic—Smotherman, Mark [Clemson University],
http://people.cs.clemson.edu/~mark/464/fp.html

Integer Division—Kaplan, Ian [Bearcave.com], http://www.bearcave.com/soft
ware/divide.htm

Lecture Notes for CSC 252—Scott, Michael L. [University of Rochester], http://
www.cs.rochester.edu/courses/252/spring2014/notes/04_architecture

Maurice Wilkes—[Wikipedia, the Free Encyclopedia], https://en.wikipedia.org
/wiki/Maurice_Wilkes

Multiplication—Johnson, Martin [Massey University], http://cs-alb-pc3.massey
.ac.nz/notes/59304/l5.html

Multiplication and Division—Wiseman, Yair and Lohev, Rafi [Coherence, Ltd.],
http://u.cs.biu.ac.il/~wiseman/co/co9.pdf

Multiplication in Binary—[Xilinx, Inc.], http://www.xilinx.com/univ/teaching
_materials/dsp_primer/sample/lecture_notes/FPGAArithmetic_mult.pdf

Other CPU Architectures: 0, 1, 2 and 3 Address Machines—Bjork, Russell C. [Gordon
College], http://www.math-cs.gordon.edu/courses/cs222/lectures/other
_architectures.html

RISC Maker—Perry, Tekla S. [*IEEE Spectrum*], http://spectrum.ieee.org/geek-life
/profiles/risc-maker

Sequential Multiplication and Division—Koren, Israel [University of Massachusetts],
http://www.ecs.umass.edu/ece/koren/arith/slides/Part3-seq.ppt

The MIPS Processor—Pfieffer, Joe [New Mexico State University], http://www.cs
.nmsu.edu/~pfeiffer/classes/473/notes/mips.html

The Pentium SRT Flaw—Edelman, Alan [Massachusetts Institute of Technology],
http://www-math.mit.edu/~edelman/homepage/talks/pentium.ppt

## *Chapter 4: Enhancing CPU performance*

A Study of Hyper-Threading in High-Performance Computing Clusters—Leng, Tau;
Ali, Rizwan; Hsieh, Jenwei; and Stanton, Christopher [*Dell PowerSolutions*,
November 2002], http://ftp.dell.com/app/4q02-Len.pdf

Access Ordering Hazards—Glew, Andy [Paul A. Clayton Place], https://
sites.google.com/site/paulclaytonplace/andy-glew-s-comparch-wiki
/access-ordering-hazards

Alpha 21064—[Wikipedia, the Free Encyclopedia], https://en.wikipedia.org
/wiki/Alpha_21064

An Efficient Algorithm for Exploiting Multiple Arithmetic Units—Tomasulo, R. M.
[*IBM Journal*, January 1967], http://courses.cs.washington.edu/courses
/cse548/11au/Tomasulo-An-Efficient-Algorithm-for-Exploiting-Multiple
-Arithmetic-Units.pdf

An Eight-Issue Tree-VLIW Processor for Dynamic Binary Translation—Ebcioglu,
Kemal; Fritts, Jason; Kosonocky, Stephen; Gschwind, Michael; Altman, Erik;
Kailas, Krishnan; and Bright, Terry [IBM T. J. Watson Research Center],
http://ieeexplore.ieee.org/xpls/abs_all.jsp?arnumber=727094

Architectural and Organizational Tradeoffs in the Design of the MultiTitan CPU—
Jouppi, Norman P. [Digital Equipment Corporation, WRL Research Report
89/9], http://www.hpl.hp.com/techreports/Compaq-DEC/WRL-89-9.pdf

Branch Prediction Strategies and Branch Target Buffer Design—Lee, Johnny K. F. and Smith, Alan J. [*IEEE Computer*, January 1984], http://web.ece.ucda vis.edu/~vojin/CLASSES/EEC272/S2005/Papers/Lee-Smith_Branch -Prediction.pdf

Cache Missing for Fun and Profit—Percival, Colin [*Proceedings of BSDCan 2005*], http://www.daemonology.net/papers/cachemissing.pdf

CDC 6600—Pfieffer, Joe [New Mexico State University], http://www.cs.nmsu .edu/~pfeiffer/classes/473/notes/cdc.html

Chapter 1: Introduction: CISC, RISC, VLIW, and EPIC Architectures—[Intel Press], http://www.csit-sun.pub.ro/~cpop/Documentatie_SMP/Intel_Micropro cessor_Systems/IA64%20Intel%20Itanium%20Processor/instruction-level -parallelism-and-the-itaniumr-architecture.pdf

Chapter 4–Instruction Level Parallelism and Dynamic Execution—Kelly, Paul H. J. [Imperial College of Science, Technology and Medicine], http://www.doc .ic.ac.uk/~phjk/AdvancedCompArchitecture/2003-04/Lectures/Ch04 /ACA-Ch04-CurrentVersion_files/v3_document.htm

Computer Architecture Tutorial: Forwarding—Prabhu, Gurpur M. [Iowa State University], http://web.cs.iastate.edu/~prabhu/Tutorial/PIPELINE/forward .html

DAISY: Dynamic Compilation for 100% Architectural Compatibility—Ebcioglu, Kemal and Altman, Erik R. [IBM Research Report RC 20538], http://cite seerx.ist.psu.edu/viewdoc/download?doi=10.1.1.153.4968&rep=rep1&type =pdf

DAISY: Dynamically Architected Instruction Set from Yorktown—[IBM Corporation], http://researcher.watson.ibm.com/researcher/view_group.php ?id=2909

Data Hazard Classification—Prabhu, Gurput [Iowa State University], http:// www.cs.iastate.edu/~prabhu/Tutorial/PIPELINE/dataHazClass.html

DEC Alpha—[Wikipedia, the Free Encyclopedia], https://en.wikipedia.org/wiki /DEC_Alpha

Decoding Instructions: Intel Core versus AMD's K8 architecture—De Gelas, Johan [AnandTech], http://www.anandtech.com/show/1998/3

Design of a Computer: The Control Data 6600—Thornton, J. E. [Control Data Corporation], http://www.cs.nmsu.edu/~pfeiffer/classes/473/notes/Design OfAComputer_CDC6600.pdf

Earth Simulator—SX-9/E/1280M160—[top500.org], http://www.top500.org/sys tem/176210

Floating Point Adder Unit—Joseph, Vincy, http://www.vincyjoseph.files.word press.com/2013/03/lecture25a.ppt

Historical Background for EPIC—Smotherman, Mark [Clemson University], http://people.cs.clemson.edu/~mark/epic.html

Hyper-Threading Technology—[*Intel Technology Journal*, Vol. 6, Issue 1], http:// www.ece.cmu.edu/~ece742/f12/lib/exe/fetch.php?media=marr_hyper thread02.pdf

IA-32 Execution Layer: A Two-Phase Dynamic Translator Designed to Support IA-32 Applications on Itanium-Based Systems—Baraz, Leonid; Devor, Tevi; Etzion, Orna; Goldenberg, Shalom; Skaletsky, Alex; Wang, Yun; and Zemach, Yigal [*Proceedings of the 36th Annual Symposium on Microarchitecture*, 2003], http://www.microarch.org/micro36/html/pdf/goldenberg-IA32Execu tionLayer.pdf

IA-64 Assembler User's Guide—[Intel Corporation], http://www.intel.com /design/itanium/downloads/asmusrgd.pdf

IBM Archives: System/360 Model 91—[IBM Corporation], http://www-03.ibm .com/ibm/history/exhibits/mainframe/mainframe_PP2091.html

IBM Research: VLIW Architecture—[IBM Corporation], http://researcher.watson .ibm.com/researcher/view_group.php?id=2831

IBM Stretch (7030)—Aggressive Uniprocessor Parallelism—Smotherman, Mark [Clemson University], http://people.cs.clemson.edu/~mark/stretch.html

Improving Branch Prediction Performance with a Generalized Design for Dynamic Branch Predictors—Lin, Wei-Ming; Madhavaram, Ramu; and Yang, An-Yi [Informatica 29], http://www.informatica.si/index.php/informatica/arti cle/viewFile/52/45

Information System Hardware—[Indira Gandhi National Open University, India], https://webservices.ignou.ac.in/virtualcampus/adit/course/cst101 /block1/unit4/cst101-bl1-u4-08.htm

Inside the IBM PowerPC 970—Part I: Design Philosophy and Front End—Stokes, Jon [Ars Technica], http://arstechnica.com/features/2002/10/ppc970/

Instruction-Level Parallelism and Its Exploitation—Panda, Dhabaleswar K. [Ohio State University], http://web.cse.ohio-state.edu/~panda/775/slides/Ch2 _6.pdf

Intel Hyper-Threading Technology: Your Questions Answered—Cepeda, Shannon [Intel Communities: The Data Stack], https://communities.intel .com/community/itpeernetwork/datastack/blog/2009/06/02/intel -hyper-threading-technology-your-questions-answered

Introduction to Multithreading, Superthreading and Hyperthreading—Stokes, Jon [ArsTechnica.com], http://arstechnica.com/features/2002/10/hyper threading/

Itanium—[Wikipedia, the Free Encyclopedia], https://en.wikipedia.org/wiki /Itanium

Itanium Architecture Software Developer's Manual—[Intel Corporation], http:// www.intel.com/design/itanium/manuals/iiasdmanual.htm

Lecture 3: Pipelining Basics—Balasubramonian, Rajeev [University of Utah], http://www.cs.utah.edu/~rajeev/cs6810/pres/07-6810-03.pdf

Microprocessor Design/Hazards—[Wikibooks.org], https://en.wikibooks.org /wiki/Microprocessor_Design/Hazards

Oracle's SPARC T4-1, SPARC T4-2, SPARC T4-4, and SPARC T4-1B Server Architecture—[Oracle Corporation], http://www.oracle.com/technetwork /server-storage/sun-sparc-enterprise/documentation/o11-090-sparc-t4 -arch-496245.pdf

Organization of Computer Systems: Pipelining—Schmalz, M. S. [University of Florida], https://www.cise.ufl.edu/~mssz/CompOrg/CDA-pipe.html

Out-of-Order Architectures: Understanding the Cell Microprocessor—Shimpi, Anand Lal [AnandTech.com], http://www.anandtech.com/show/1647/7

Parallel Operation in the Control Data 6600—Thornton, James E. [*Proceedings of the Spring Joint Computer Conference*, 1964], http://dl.acm.org/citation .cfm?doid=1464039.1464045

Pipeline Architecture—Ramamoorthy, C. V. and Li, H. F. [*Computing Surveys*, Vol. 9, No.1], http://web.ece.ucdavis.edu/~vojin/CLASSES/EEC272/S2005 /Papers/Ramamoorthy_mar77.pdf

Pipeline Hazards—Pfieffer, Joe [New Mexico State University], http://www.cs
.nmsu.edu/~pfeiffer/classes/473/notes/hazards.html

Pipelining: Basic Concepts—Wilsey, Philip A. [University of Cincinnati], http://
www.ece.uc.edu/~paw/classes/eecs3026/lectureNotes/pipelining/pipe
lining.pdf

POWER7 Performance Guide—[EECatalog, Extension Media], http://eecatalog
.com/power/2010/08/04/power7-performance-guide/

Reduced Instruction Set Computers (RISC): Academic/Industrial Interplay Drives
Computer Performance Forward—Joy, William N. [Sun Microsystems, Inc.],
http://homes.cs.washington.edu/~lazowska/cra/risc.html

Reduced Instruction Set Computing—[Wikipedia, the Free Encyclopedia], https://
en.wikipedia.org/wiki/Reduced_instruction_set_computing

Reducing the Branch Penalty in Pipelined Processors—Lilja, David J. [*IEEE Computer*,
July 1988], https://courses.cs.washington.edu/courses/csep548/00sp/hand
outs/lilja.pdf

Register Renaming—[Wikipedia, the Free Encyclopedia], https://en.wikipedia
.org/wiki/Register_renaming

RISC I: A Reduced Instruction Set VLSI Computer—Patterson, David A. and Sequin,
Carlo H. [*Proceedings of the 8th Annual Symposium on Computer Architecture*],
http://web.cecs.pdx.edu/~alaa/courses/ece587/spring2011/papers/patter
son_isca_1981.pdf

RISC vs. CISC: The Post-RISC Era—Hannibal [ArsTechnica.com], http://archive
.arstechnica.com/cpu/4q99/risc-cisc/rvc-1.html

Russian Microprocessor Firms to Challenge Intel and AMD on Domestic
Market—Pototsky, Dan [Russia Beyond the Headlines], http://rbth.com
/science_and_tech/2014/07/10/russian_microprocessor_firms_to_chal
lenge_intel_and_amd_on_d_38095.html

Shadows of Itanium: Russian Firm Debuts VLIW Elbrus 4 CPU with Onboard x86
Emulation—Hruska, Joel [ExtremeTech], http://www.extremetech.com
/computing/205463-shadows-of-itanium-russian-firm-debuts-vliw-elbrus
-4-cpu-with-onboard-x86-emulation

Simultaneous Multithreading: A Platform for Next-Generation Processors—
Eggers, Emer, Levy, Lo, Stamm, and Tullsen [University of Washington],
http://www.cs.washington.edu/research/smt/papers/ieee_micro.pdf

The Design of the Microarchitecture of UltraSPARC-I—Tremblay, Marc; Greenley,
Dale; and Normoyle, Kevin [*Proceedings of the IEEE*, Vol. 83, No. 12], http://
cseweb.ucsd.edu/classes/wi13/cse240a/pdf/03/UltraSparc_I.pdf

The IBM RT Information Page—Brashear, Derrick [Carnegie Mellon University],
http://www.contrib.andrew.cmu.edu/~shadow/ibmrt.html

The Microarchitecture of Intel, AMD and VIA CPUs—Fog, Agner [Technical
University of Denmark], http://www.agner.org/optimize/microarchitec
ture.pdf

The MIPS R10000 Superscalar Microprocessor—Yeager, Kenneth C. [*IEEE Micro*, April
1996], http://people.cs.pitt.edu/~cho/cs2410/papers/yeager-micromag96.pdf

The MIPS R4000 Processor—Mirapuri, Sunil; Woodacre, Michael; and Vasseghi,
Nader [*IEEE Micro*, April 1992], http://home.deib.polimi.it/silvano
/FilePDF/ARC-MULTIMEDIA/mipsr4000_00127580.pdf

The Pentium: An Architectural History of the World's Most Famous Desktop
Processor (Part I)—Stokes, Jon [Ars Technica], http://arstechnica.com
/features/2004/07/pentium-1/

The RISC Concept: A Survey of Implementations—Esponda, Margarita; and Rojas, Raul [Freie Universitat Berlin], http://www.inf.fu-berlin.de/lehre/WS94/RA/RISC-9.html

The SPARC Architecture Manual, Version 9—Weaver, David L. and Germond, Tom [SPARC International], http://pages.cs.wisc.edu/~fischer/cs701.f08/sparc.v9.pdf

TMS320C67x/C67x+ DSP CPU and Instruction Set Reference Guide—[Texas Instruments], http://www.ti.com/lit/ug/spru733a/spru733a.pdf

Understanding EPIC Architectures and Implementations—Smotherman, Mark [Clemson University], http://people.cs.clemson.edu/~mark/464/acmse_epic.pdf

Understanding Stacks and Registers in the Sparc Architecture(s)—Magnusson, Peter [Swedish Institute of Computer Science], http://icps.u-strasbg.fr/people/loechner/public_html/enseignement/SPARC/sparcstack.html

VLIW Processors—[University of Washington], http://courses.cs.washington.edu/courses/csep548/06au/lectures/vLIW.pdf

VLIW Processors and Trace Scheduling—Mathew, Binu K. [University of Utah], http://www.siliconintelligence.com/people/binu/coursework/686_vliw/

VLIW Processors: From Blue Sky to Best Buy—Fisher, Faraboschi, and Young [*IEEE Solid-State Circuits Magazine*, Spring 2009], http://ieeexplore.ieee.org/stamp/stamp.jsp?arnumber=5116831

VLIW: The Unlikeliest Computer Architecture—Colwell, Robert P. [*IEEE Solid-State Circuits Magazine*, Spring 2009], http://ieeexplore.ieee.org/stamp/stamp.jsp?tp=&arnumber=5116832

What Is Pipelining?—Definition—Rouse, Margaret [WhatIs.com], http://whatis.techtarget.com/definition/pipelining

Which Machines Do Computer Architects Admire?—Smotherman, Mark [Clemson University], http://people.cs.clemson.edu/~mark/admired_designs.html

# Chapter 5: Exceptions, interrupts, and input/output systems

8259A Interrupt Controller on the PC—Frank, Cornelis [The Nondotted Group], http://himmele.googlecode.com/svn/trunk/Operating%20Systems/Build%20Your%20Own%20OS/8259A%20PIC.pdf

Assembly Language Trap Generating Instructions—Milo [OSdata.com], http://www.osdata.com/topic/language/asm/trapgen.htm

DMA and Interrupt Handling—[EventHelix.com], http://www.eventhelix.com/RealtimeMantra/FaultHandling/dma_interrupt_handling.htm

DMA Controller—[Eagle Planet], http://members.tripod.com/~Eagle_Planet/dma_controller.html

Exceptions and Interrupts 1—Walker, Hank [Texas A&M University], http://courses.cs.tamu.edu/cpsc462/walker/Slides/Exceptions_Interrupts_1.pdf

FreeBSD Handbook: DMA: What It Is and How It Works—Durda, Frank IV [Nemesis.Lonestar.org], http://docs.freebsd.org/doc/3.3-RELEASE/usr/share/doc/handbook/dma.html

How USB Ports Work—Brain, Marshall [HowStuffWorks], http://computer.howstuffworks.com/usb.htm

Interrupt Controllers—Kozierok, Charles M. [The PC Guide], http://www
.pcguide.com/ref/mbsys/res/irq/funcController-c.html

Introduction to Watchdog Timers—Barr, Michael [Embedded], http://
www.embedded.com/electronics-blogs/beginner-s-corner/4023849
/Introduction-to-Watchdog-Timers

Operating Systems and Kernels: Concepts, Direct Memory Access—[CyberiaPC
.com], http://www.cyberiapc.com/os/concepts-dma.htm

Programmed I/O, Interrupt & Direct Memory Access (DMA)—Louie Wong,
http://www.louiewong.com/archives/137

UF 68HC12 Development Kit Manual—Schwartz, Eric M. [University of Florida],
http://www.mil.ufl.edu/4744/labs/UF6812BoardV30_ManualV34.pdf

Undocumented Windows NT: Chapter 10: Adding New Software Interrupts—
Dabak, Prasad; Borate, Milind; and Phadke, Sandeep, http://wordbook.xyz
/books/Programming/Undocumented%20Windows%20NT/part_II_a.pdf

Universal Serial Bus—[OSDev Wiki], http://wiki.osdev.org/Universal_Serial
_Bus

USB in a NutShell—Peacock, Craig [Beyond Logic], http://www.beyondlogic.org
/usbnutshell/usb1.shtml

USB Made Simple—[MQP Electronics Ltd.], http://www.usbmadesimple.co.uk
/ums_1.htm

x86 Architecture Exceptions—[Sandpile.org], http://www.sandpile.org/x86/except
.htm

# Chapter 6: Parallel and high-performance systems

A Hybrid Interconnection Network for Integrated Communication Services—
Chen, Yi-long and Lyu, Jyh-Charn [Northern Telecom, Inc./Texas A&M
University], https://pdfs.semanticscholar.org/13fc/7e0c6fb91977bfc8c7a92
ca14a57f5b41250.pdf

A Shading Language on Graphics Hardware: The PixelFlow Shading System—
Olano, Marc; and Lastra, Anselmo [University of North Carolina], http://
www.cs.unc.edu/~pxfl/papers/pxflshading.pdf

A Survey of Routing Techniques in Store-and-Forward and Wormhole
Interconnects—Holman, David M. and Lee, David S. [Sandia National
Laboratories], http://prod.sandia.gov/techlib/access-control.cgi/2008/080068
.pdf

Cache Coherence in Shared—Memory Architectures—Watson, Ian [University of
Manchester], https://www.cs.utexas.edu/~pingali/CS378/2015sp/lectures
/mesi.pdf

Cache-Coherent Distributed Shared Memory: Perspectives on its Development
and Future Challenges—Hennessy, John; Heinrich, Mark; and Gupta, Anoop
[*Proceedings of the IEEE*, Vol. 87, No. 3], http://ieeexplore.ieee.org/stamp
/stamp.jsp?arnumber=747863

Cache Only Memory Architecture (COMA)—Pimentel, Andy [University of
Amsterdam, Netherlands], https://staff.fnwi.uva.nl/a.d.pimentel/apr/node74
.html

Cache-Only Memory Architecture (COMA)—Torrellas, Josep [Encyclopedia of
Parallel Computing], http://link.springer.com/referenceworkentry/10.100
7%2F978-0-387-09766-4_166

Classification of Parallel Computers—Vainikko, Eero [University of Tartu, Estonia], http://kodu.ut.ee/~eero/PC/Lecture4.pdf

Clear Speed Revises Graphics Engine to Process Packets—Edwards, Chris [EE Times/CommsDesign.com], http://www.eetimes.com/document.asp?doc_id=1180991

Coherent Threading—Glew, Andy [University of California, Berkeley], http://parlab.eecs.berkeley.edu/sites/all/parlab/files/20090827-glew-vector.pdf

Comp.compilers: History of Supercomputing—Wilson, Greg [Australian National University], http://compilers.iecc.com/comparch/article/93-08-095

Computer Architecture: SIMD/Vector/GPU—Mutlu, Onur [Carnegie Mellon University], https://www.ece.cmu.edu/~ece740/f13/lib/exe/fetch.php?media=seth-740-fall13-module5.1-simd-vector-gpu.pdf

Cray Gemini Interconnect—Sobchyshak, Denys [Technical University of Munich], http://richardcaseyhpc.com/wp-content/uploads/2015/04/denyssobchyshak-gemini-interconnect.pdf

CUDA C Programming Guide—[NVIDIA Developer Zone], http://docs.nvidia.com/cuda/cuda-c-programming-guide/#axzz40eIoAeb0

Current Trends in Parallel Computing—Khan, Rafiqul Zaman and Ali, Md Firoj [*International Journal of Computer Applications*, Vol. 59, No. 2], http://research.ijcaonline.org/volume59/number2/pxc3883923.pdf

Data Diffusion Machine Home Page—[University of Bristol], http://www.cs.bris.ac.uk/Research/DDM/index.html

Deadlock-Free Message Routing in Multiprocessor Interconnection Networks—Dally, William J. and Seitz, Charles L. [California Institute of Technology], http://authors.library.caltech.edu/26930/1/5231-TR-86.pdf

DSP & SIMD—[ARM Ltd.], https://www.arm.com/products/processors/technologies/dsp-simd.php

GPGPU and Stream Computing—Fietkau, Julian [University of Hamburg], https://wr.informatik.uni-hamburg.de/_media/teaching/sommersemester_2011/paps11-fietkau-gpgpu_and_stream_computing-presentation.pdf

GPU Computing—Owens, John D.; Houston, Mike; Luebke, David; Green, Simon; Stone, John E.; and Phillips, James C. [*Proceedings of the IEEE*, Vol. 96, No. 5], http://cs.utsa.edu/~qitian/seminar/Spring11/03_04_11/GPU.pdf

Hardware: G4 Executive Summary: What Is AltiVec?—[Apple Computer, Inc.], http://mirror.informatimago.com/next/developer.apple.com/hardware/ve/summary.html

History of GPU Computing—Zahran, Mohamed [New York University], http://cs.nyu.edu/courses/spring12/CSCI-GA.3033-012/lecture2.pdf

History of Supercomputing—Farber, David [University of Pennsylvania], http://seclists.org/interesting-people/1993/Aug/116

ILLIAC IV—Thelen, Ed [Ed-Thelen.org], http://ed-thelen.org/comp-hist/vs-illiac-iv.html

ILLIAC IV CFD—Carpenter, Bryan [Indiana University], http://grids.ucs.indiana.edu/ptliupages/projects/HPJava/talks/beijing/hpf/introduction/node4.html

Interfacing the ADSP-21161 SIMD SHARC DSP to the AD1836 (24-bit/96kHz) Multichannel Codec—Tomarakos, John [Analog Devices], http://www.analog.com/media/en/technical-documentation/application-notes/TN-AD1836_21161.pdf

Is the AMD vs. Nvidia conflict coming to an end?—Smith, Matt [Digital Trends], http://www.digitaltrends.com/computing/is-the-eternal-pc-graphics-war -coming-to-an-end/

Lecture 19: Multiprocessors—Balasubramonian, Rajeev [University of Utah], http://www.cs.utah.edu/classes/cs6810/pres/6810-19.pdf

Lecture 30: Multiprocessors-Flynn Categories, Large vs. Small Scale, Cache Coherency—Katz, Randy H. [University of California, Berkeley], http:// www.cs.berkeley.edu/~randy/Courses/CS252.S96/Lecture30.pdf

Lists: November 2015: Highlights—[top500.org], http://www.top500.org/lists /2015/11/highlights/

Maspar Machines—Whaley, Tom [Washington and Lee University], http://home .wlu.edu/~whaleyt/classes/parallel/topics/maspar/maspar.html

Memory Part 4: NUMA Support—Drepper, Ulrich [LWN.net], https://lwn.net /Articles/254445/

Message Passing Architectures—Bestavros, Azer [Boston University], http://cs -www.bu.edu/faculty/best/crs/cs551/lectures/lecture-15.html

MMX Technology: SSE And 3DNow!—Bestofmedia Team [Tom's Hardware], http://www.tomshardware.com/reviews/processors-cpu-apu-features -upgrade,3569-3.html

More Than 100 Accelerated Systems Now on TOP500 List—[NVIDIA Corporation], https://news.developer.nvidia.com/more-than-100-accelerated-systems -now-on-top500-list/

Multi-Processor Architectures—Narahari, Bhagirath [George Washington University], http://www.seas.gwu.edu/%7Enarahari/cs211/materials/lectures/multipro cessor.pdf

Multiprocessors and Multiprocessing—Thornley, John [California Institute of Technology], http://users.cms.caltech.edu/~cs284/lectures/7oct97.ppt

Multiprocessors and Multithreading—Lebeck, Alvin R. [Duke University], http:// www.cs.duke.edu/courses/cps220/fall04/lectures/6-mt.pdf

Networks and Topologies—Pfieffer, Joe [New Mexico State University], http:// www.cs.nmsu.edu/~pfeiffer/classes/573/notes/topology.html

Next Generation Supercomputers: Cluster of Multi-processor Systems Project Launched—Kahn, Jeffery [Lawrence Berkeley National Laboratory], http:// www.lbl.gov/Science-Articles/Archive/COMPS-collaboration.html

Notes on Concurrency—Chase, Jeffrey S. [Duke University], https://users.cs.duke .edu/~chase/systems/concurrency.html

NUMA for Dell PowerEdge 12G Servers—Beckett, John [Dell, Inc.], http:// en.community.dell.com/.cfs-file/__key/telligent-evolution-components -attachments/13-4491-00-00-20-26-69-46/NUMA-for-Dell-PowerEdge-12G -Servers.pdf

NUMA Frequently Asked Questions—[SourceForge.net], http://lse.sourceforge .net/numa/faq/

Numascale, Supermicro, and AMD Announce the World's Largest Shared Memory System to Date—Carlsen, Espen [Numascale AS], http://www.numascale .com/numascale-supermicro-amd-announce-worlds-largest-shared-memory -system-date-2/

On MMX, 3DNow!, and Katmai—Karbo, Michael B. [KarbosGuide.com], http:// www.karbosguide.com/hardware/module3e09.htm

Parallel Architectures—Senning, Jonathan [Gordon College], http://www.math -cs.gordon.edu/courses/cps343/presentations/Parallel_Arch.pdf

QuickPath Interconnect vs. HyperTransport—Lloyd, Chris [TechRadar], http://www.techradar.com/us/news/computing-components/quick path-interconnect-vs-hypertransport-909072

SIMD Processing (Vector and Array Processors)—Mutlu, Onur [Carnegie Mellon University], http://www.ece.cmu.edu/~ece447/s14/lib/exe/fetch .php?media=onur-447-spring14-lecture16-simd-afterlecture.pdf

SMP and NUMA Multiprocessor Systems—Hwang, Kai [University of Southern California], http://gridsec.usc.edu/files/EE657/Hwang-EE657-Lectture5 -SMPccNUMA-Sept14-2007.pdf

STiNG Revisited: Performance of Commercial Database Benchmarks on a CC-NUMA Computer System—Clapp, Russell M. [IBM Corporation], http://iacoma.cs .uiuc.edu/caecw01/sting.pdf

Supercomputers: The Amazing Race—Bell, Gordon [Microsoft Corporation], http://research.microsoft.com/en-us/um/people/gbell/MSR-TR-2015-2 _Supercomputers-The_Amazing_Race_Bell.pdf

The Cray XT4 and Seastar 3-D Torus Interconnect—Abts, Dennis [Google, Inc.], http://static.googleusercontent.com/media/research.google.com/en //pubs/archive/36896.pdf

The Design of the MasPar MP-1: A Cost Effective Massively Parallel Computer— Nickolls, John R. [MasPar Computer Corporation], https://courses.engr.illi nois.edu/cs533/reading_list/14a.pdf

The Effect of NUMA Tunings on CPU Performance—Hollowell, Christopher [Brookhaven National Laboratory], https://indico.cern.ch/event/304944 /session/8/contribution/3/attachments/578723/796898/numa.pdf

The Gemini Network—[Cray, Inc.], https://wiki.alcf.anl.gov/parts/images/2/2c /Gemini-whitepaper.pdf

The History of the Development of Parallel Computing—Wilson, Gregory [University of Toronto], http://parallel.ru/history/wilson_history.html

The Performance and Scalability of Distributed Shared Memory Cache Coherence Protocols—Heinrich, Mark [Stanford University], http://csl.cs.ucf.edu /~heinrich/papers/Dissertation.pdf

The Pixel-Planes Family of Graphics Architectures—Cohen, Jonathan [Johns Hopkins University], http://www.cs.jhu.edu/~cohen/VW2000/Lectures /Pixel-Planes.color.pdf

Understanding the CUDA Data Parallel Threading Model—Wolfe, Michael [The Portland Group], https://www.pgroup.com/lit/articles/insider/v2n1a5 .htm

Vector and SIMD Processors—Welch, Eric and Evans, James [Rochester Institute of Technology], http://meseec.ce.rit.edu/756-projects/spring2013/2-2.pdf

Vector Processor—[Wikipedia, the Free Encyclopedia], http://en.wikipedia.org /wiki/Vector_processor

Virtual-Channel Flow Control—Dally, William J. [Massachusetts Institute of Technology], http://www.eecs.berkeley.edu/~kubitron/courses/cs258-S08 /handouts/papers/dally-virtual.pdf

Visual Instruction Set—[Wikipedia, the Free Encyclopedia], http://en.wikipedia .org/wiki/Visual_instruction_set

VMware vSphere 4: What is NUMA?—[VMware, Inc.], https://pubs.vmware.com /vsphere-4-esx-vcenter/index.jsp?topic=/com.vmware.vsphere.resource management.doc_41/using_numa_systems_with_esx_esxi/c_what_is _numa.html

WildFire: A Scalable Path for SMPs—Hagersten, Erik and Koster, Michael [Sun Microsystems, Inc.], http://user.it.uu.se/~eh/papers/wildfire.ps

Wormhole Routing in Parallel Computers—Seydim, Ayse Yasemin [Southern Methodist University], https://www.cs.hmc.edu/~avani/wormhole/1999 -wormhole-routing-in-parallel.pdf

You Can Do Any Kind of Atomic Read-Modify-Write Operation—Preshing, Jeff [Preshing on Programming], http://preshing.com/20150402/you-can-do -any-kind-of-atomic-read-modify-write-operation/

## Chapter 7: Special-purpose and future architectures

12-Qubits Reached in Quantum Information Quest—[ScienceDaily], https:// www.sciencedaily.com/releases/2006/05/060508164700.htm

14-Qubit Entanglement: Creation and Coherence—Monz, Schindler, Barreiro, Chwalla, Nigg, Coish, Harlander, Hansel, Hennrich, and Blatt [*Physical Review*], http://journals.aps.org/prl/abstract/10.1103/PhysRevLett.106.130506

3DANN-R Vector Image Processing System—[Irvine Sensors Corporation], http:// www.ax-09.com/gruppa/materials/biblioteka/Shemotehnika/Sensors /Vector%20Image%20Processing%20System.pdf

50 Years of Moore's Law—[Intel Corporation], http://www.intel.com/content /www/us/en/silicon-innovations/moores-law-technology.html

A Practical Architecture for Reliable Quantum Computers—Oskin, Mark; Chong, Frederic T.; and Chuang, Isaac L. [*IEEE Computer*, January 2002], http:// homes.cs.washington.edu/~oskin/Oskin-A-Practical-Architecture-for -Reliable-Quantum-Computers.pdf

A Turning Point for Quantum Computing—Hamm, Steve [IBM THINK], http:// www.ibm.com/blogs/think/2015/12/07/a-turning-point-for-quantum -computing/

Advanced Topics in Dataflow Computing and Multithreading—Gao, Guang R.; Bic; Lubomir; and Gaudiot, Jean-Luc [*IEEE Computer Society*/John Wiley & Sons], http://www.wiley.com/WileyCDA/WileyTitle/productCd-0818665424, miniSiteCd-IEEE_CS2.html

AI–CS364 Fuzzy Logic Fuzzy Logic 3—Vrusias, Bogdan L. [University of Surrey], http://www.docfoc.com/ai-cs364-fuzzy-logic-fuzzy-logic-3-03-rd-october -2006-dr-bogdan-l-vrusias

An Idiot's Guide to Neural Networks—Bowles, Richard, http://bowles.byethost3 .com/neural/neural.htm

An Introduction to Neural Networks—Smith, Leslie [University of Stirling], http://www.cs.stir.ac.uk/~lss/NNIntro/InvSlides.html

An Introduction to Quantum Computing—West, Jacob [California Institute of Technology], http://www.cs.rice.edu/~taha/teaching/05F/210/news/2005 _09_16.htm

Artificial Neural Network—[Wikipedia, the Free Encyclopedia], https:// en.wikipedia.org/wiki/Artificial_neural_network

Axeon and Infineon Unveil Embedded Machine Learning System—[Design & Reuse S.A.], http://www.design-reuse.com/news/10024/axeon-infineon -unveil-embedded-machine-learning-system.html

Biological Computing Fundamentals and Futures—Akula, Balaji and Cusick, James [Wolters Kluwer Corporate Legal Services], http://arxiv.org/ftp/arxiv /papers/0911/0911.1672.pdf

Biological Transistor Enables Computing Within Living Cells, Study Says—
Myers, Andrew [Stanford Medicine], https://med.stanford.edu/news
/all-news/2013/03/biological-transistor-enables-computing-within-living
-cells-study-says.html

Dataflow: A Complement to Superscalar—Budiu, Mihai; Artigas, Pedro V.; and
Goldstein, Seth Copen [Microsoft Research/Carnegie Mellon University],
http://www.cs.cmu.edu/~mihaib/research/ispass05.pdf

Dataflow Architectures—Silc, Jurij [Jozef Stefan Institute], http://csd.ijs.si
/courses/dataflow/

Dataflow Computers: Their History and Future—Hurson, Ali R. and Kavi, Krishna
M. [Wiley Encyclopedia of Computer Science and Engineering], http://csrl
.unt.edu/~kavi/Research/encyclopedia-dataflow.pdf

Dataflow Machine Architecture—Veen, Arthur H. [*Center for Mathematics and
Computer Science—ACM Surveys*, Vol. 18, No. 4], http://web.cecs.pdx.edu
/~akkary/ece588/p365-veen.pdf

Demonstration Model of fuzzyTECH Implementation on Motorola 68HC12 MCU—
Drake, Philip; Sibigtroth, Jim; von Altrock, Constantin; and Konigbauer, Ralph
[Motorola, Inc./Inform Software Corporation], http://www.fuzzytech.com
/e/e_a_mot.html

Editorial: Fuzzy Models-What Are They, and Why?—Bezdek, J. C. [*IEEE
Transactions on Fuzzy Systems*, Vol. 1, No. 1], https://www.researchgate.net
/publication/224259085_Fuzzy_models-What_are_they_and_why_Editorial

Executing a Program on the MIT Tagged-Token Dataflow Architecture—Arvind
and Nikhil, Rishiyur S. [*IEEE Transactions on Computers*, Vol. 39, No. 3],
http://pages.cs.wisc.edu/~isca2005/ttda.pdf

FAQ: Fuzzy Logic and Fuzzy Expert Systems—Kantrowitz, Mark; Horstkotte,
Erik; and Joslyn, Cliff [Carnegie Mellon University], http://www.faqs.org
/faqs/fuzzy-logic/part1/

Fuzzy Estimation Tutorial—[*Electronics Now*, May, 1996], http://www.fuzzysys
.com/fuzzyestimationtutorial.htm

Fuzzy Expert System for Navigation Control—Hoe, Koay Kah [University of
Technology, Malaysia], http://www.geocities.ws/kh_koay/khkoay_project
.html

Fuzzy Logic and Neural Nets: Still Viable After All These Years?—Prophet, Graham
[EDN.com], http://www.edn.com/design/systems-design/4331487/Fuzzy
-logic-and-neural-nets-still-viable-after-all-these-years-

Fuzzy Logic Design: Methodology, Standards, and Tools—von Altrock, Constantin
[EE Times], http://www.fuzzytech.com/e/e_a_eet.html

Fuzzy Logic for "Just Plain Folks"—Sowell, Thomas, https://web.fe.up
.pt/~asousa/sbld/recursos/FuzzyLogicJumpStart_free%20.pdf

Fuzzy Logic in Knowledge Builder—[XpertRule Software Ltd.], http://www.xper
trule.com/pages/fuzzy.htm

Fuzzy Logic Microcontroller Implementation for DC Motor Speed Control—
Tipsuwan, Yodyium and Chow, Mo-Yuen [North Carolina State University],
http://www4.ncsu.edu/~chow/Publication_folder/Conference_paper
_folder/1999_IECon_FZ_Impl_Tipsuwan.pdf

Fuzzy Logic Resources—[Byte Craft Limited], http://www.bytecraft.com
/Fuzzy_Logic_Resources

Fuzzy Logic Toolbox: Dinner for Two, Reprise—[The MathWorks, Inc.], http://
www-rohan.sdsu.edu/doc/matlab/toolbox/fuzzy/fuzzytu7.html

Gordon Moore Predicts 10 More Years for Moore's Law—Poeter, Damon [*PC Magazine*], http://www.pcmag.com/article2/0,2817,2484098,00.asp

Home—David Deutsch [University of Oxford], http://www.daviddeutsch.org.uk/

How Quantum Computers Work—Bonsor, Kevin and Strickland, Jonathan [Howstuffworks.com], http://computer.howstuffworks.com/quantum-computer.htm

IBM Unveils a 'Brain-Like' Chip With 4,000 Processor Cores—[*WIRED Magazine*], http://www.wired.com/2014/08/ibm-unveils-a-brain-like-chip-with-4000-processor-cores/

IBM's Test-Tube Quantum Computer Makes History—[IBM Corporation], https://www-03.ibm.com/press/us/en/pressrelease/965.wss

If the End of Moore's Law Is Near, What's Next?—Wladawsky-Berger, Irving [*The CIO Report–Wall Street Journal*], http://blogs.wsj.com/cio/2015/08/07/if-the-end-of-moores-law-is-near-whats-next/

Intel Fuzzy Logic Tool Simplifies ABS Design—[Intel Corporation], http://notes-application.abcelectronique.com/027/27-46432.pdf

Introducing Qualcomm Zeroth Processors: Brain-Inspired Computing—[Qualcomm Technologies Inc.], https://www.qualcomm.com/news/onq/2013/10/10/introducing-qualcomm-zeroth-processors-brain-inspired-computing

Introduction to Artificial Neural Networks, Part 1—Jacobson, Lee [The Project Spot], http://www.theprojectspot.com/tutorial-post/introduction-to-artificial-neural-networks-part-1/7

Life Beyond Moore's Law—Feldman, Michael [Intersect360 Research/top500.org], http://www.top500.org/blog/life-beyond-moores-law/

Light-speed Computing Now Only Months Away—[Optalysys], http://optalysys.com/light-speed-computing-now-only-months-away/

Model Extremely Complex Functions, Neural Networks—[Dell Software], http://www.statsoft.com/Textbook/Neural-Networks

Moore's Law 40th Anniversary—[Intel Corporation], http://www.intel.com/pressroom/kits/events/moores_law_40th/

Moore's Law Is 40—Oates, John [*The Register*], http://www.theregister.co.uk/2005/04/13/moores_law_forty/

Moore's Law Keeps Going, Defying Expectations—Sneed, Annie [*Scientific American*], http://www.scientificamerican.com/article/moore-s-law-keeps-going-defying-expectations/

Moore's Law Really Is Dead This Time—Bright, Peter [Ars Technica], http://arstechnica.com/information-technology/2016/02/moores-law-really-is-dead-this-time/

Moore's Law: The Rule That Really Matters in Tech—Shankland, Stephen [CNET], http://www.cnet.com/news/moores-law-the-rule-that-really-matters-in-tech/

Neural Network Chip: Patented Parallel Architecture—[General Vision Inc.], http://general-vision.com/cm1k/

Neural Networks in Hardware: A Survey—Liao, Yihua [University of California, Davis], http://apt.cs.manchester.ac.uk/intranet/csonly/nn/NNHSurvey.pdf

Neuron Basics—AnimatLab [NeuroRobotic Technologies LLC], http://animatlab.com/Help/Documentation/Neural-Network-Editor/Neural-Simulation-Plug-ins/Firing-Rate-Neural-Plug-in/Neuron-Basics

NNW in HEP: Hardware—Lindsey, Clark S.; Denby, Bruce; and Lindblad, Thomas [Royal Institute of Technology, Sweden], http://neuralnets.web.cern.ch/NeuralNets/nnwInHepHard.html

Oxford Quantum [Oxford University], http://oxfordquantum.org/

Parallel Processing Architectures—Manzke, Michael [Trinity College Dublin], https://www.cs.tcd.ie/Michael.Manzke/3ba5/3BA5_third_lecture.pdf

Processor Allocation in a Multi-ring Dataflow Machine—Barahona, Pedro and Gurd, John R. [University of Manchester], http://www.sciencedirect.com/science/article/pii/0743731586900183

Programmable Cells: Engineer Turns Bacteria Into Living Computers—Quinones, Eric [Princeton University], http://www.eurekalert.org/pub_releases/2005-04/pu-pce042505.php

Quantum Computing—[Wikipedia, the Free Encyclopedia], https://en.wikipedia.org/wiki/Quantum_computing

Quantum Leap: IBM Scientists Lay the Foundations for a Practical, Scalable Quantum Computer–Borghino, Dario [gizmag.com], http://www.gizmag.com/quantum-computer-error-correction/37249/

RSA Encryption: Keeping the Internet Secure—Blanda, Stephanie [AMS Graduate Student Blog], http://blogs.ams.org/mathgradblog/2014/03/30/rsa/#sthash.eq9kK4i0.dpbs

Scalable Quantum Computing Using Solid-State Devices—Kane, Bruce [*National Academy of Engineering: The Bridge*, Vol. 32, No. 4], http://www.nae.edu/Publications/Bridge/ExpandingFrontiersofEngineering7308/ScalableQuantumComputingUsingSolid-StateDevices.aspx

Scheduling Dynamic Dataflow Graphs With Bounded Memory Using The Token Flow Model—Buck, Joseph Tobin [University of California at Berkeley], http://ptolemy.eecs.berkeley.edu/publications/papers/93/jbuckThesis/thesis.pdf

STMicroelectronics Introduces a New Family of Microcontrollers—[Anglia Components Ltd.], http://www.anglia.com/newsarchive/pdfs/434.pdf

Synapse—Jones, Paul [Multiple Sclerosis Information Trust], http://www.mult-sclerosis.org/synapse.html

The D-Wave 2X System—[D-Wave Systems], http://www.dwavesys.com/d-wave-two-system

The Economist Explains: The End of Moore's Law—[*The Economist*], http://www.economist.com/blogs/economist-explains/2015/04/economist-explains-17

The End of Moore's Law—BottleRocket [Kuro5hin.org], http://www.kuro5hin.org/story/2005/4/19/202244/053

The End of Moore's Law?—Mann, Charles C. [*MIT Technology Review*], http://www.technologyreview.com/featuredstory/400710/the-end-of-moores-law/

The FIRE Neural Network Project—Furber, Steve [University of Manchester], http://apt.cs.manchester.ac.uk/projects/fire/

The Future of CPUs in Brief—Essex, David [TechnologyReview.com], http://www.technologyreview.com/news/401342/the-future-of-cpus-in-brief/?p=0

The Manchester Prototype Dataflow Computer—Gurd, J. R.; Kirkham, C. C.; and Watson, I. [*Communications of the ACM*, Vol. 28, No. 1], https://courses.cs.washington.edu/courses/csep548/05sp/gurd-cacm85-prototype.pdf

The WaveScalar Instruction Set—[University of Washington], http://wavescalar.cs.washington.edu/wavescalar.shtml

Thinking Like a Human: Computer Technology—[National Aeronautics and Space Administration], https://spinoff.nasa.gov/spinoff1999/ct1.htm

Using Quantum Effects for Computer Security—Hartgroves, Arran; Harvey, James; Parmar, Kiran; Prosser, Thomas; and Tucker, Michael [University of Birmingham], http://www.cs.bham.ac.uk/~mdr/teaching/modules04/secu rity/students/SS1-quantum.pdf

Why Nobody Can Tell Whether the World's Biggest Quantum Computer is a Quantum Computer—Mirani, Leo and Lichfield, Gideon [Quartz.com], http://qz.com/194738/why-nobody-can-tell-whether-the-worlds-biggest -quantum-computer-is-a-quantum-computer/

Zero Instruction Set Computer—[General Vision Inc.], http://general-vision.com /zisc/

## Computer architecture (general)

Computer Architecture—[Wikipedia, the Free Encyclopedia], https://en.wikipedia .org/wiki/Computer_architecture

Computer Architecture and Organization Notes—Kumar, Vasantha, http://www .pdf-archive.com/2015/11/07/85478771-coa-141403-and-cs2253-questions -and-answers-1/85478771-coa-141403-and-cs2253-questions-and-answers .pdf

Computer Architecture Educational Tools—Koren, Israel [University of Massachusetts, Amherst], http://www.ecs.umass.edu/ece/koren/architecture/

Computer Architecture: Single and Parallel Systems—Zargham, Mehdi R. [Prentice Hall], http://prenhall.com/books/esm_0130106615.html

Computer Architecture: The Anatomy of Modern Processors—Morris, John [University of Western Australia], https://www.cs.auckland.ac.nz/~jmor159/363 /html/ca_preface.html

Course: CS301: Computer Architecture—[Saylor Academy], https://learn.saylor .org/course/cs301

Great Microprocessors of the Past and Present—Bayko, John [The CPU Shack], http://www.cpushack.com/CPU/cpu.html

IBM Research: Computer Architecture—[IBM Corporation], http://researcher .watson.ibm.com/researcher/view_group.php?id=138

Lecture Notes: Computer System Architecture—Arvind and Emer [MIT Open CourseWare], http://ocw.mit.edu/courses/electrical-engineering-and-com puter-science/6-823-computer-system-architecture-fall-2005/lecture-notes/

WWW Computer Architecture Page—Hower, Derek; Yen, Luke; Xu, Min; Martin, Milo; Burger, Doug; and Hill, Mark [University of Wisconsin-Madison /University of Pennsylvania/University of Texas at Austin], http://www .cs.wisc.edu/~arch/www/

# *Index*

Page numbers followed by f and t indicate figures and tables, respectively.